Gas Treating

Gas Treating

Gas Treating

Absorption Theory and Practice

DAG A. EIMER

Tel-Tek and Telemark University College, Norway

WILEY

This edition first published 2014
©2014 John Wiley & Sons, Ltd

Registered office
John Wiley & Sons Ltd, The Atrium, Southern Gate, Chichester, West Sussex, PO19 8SQ, United Kingdom

For details of our global editorial offices, for customer services and for information about how to apply for permission to reuse the copyright material in this book please see our website at www.wiley.com.

Library of Congress Cataloging-in-Publication Data

Eimer, D. (Dag)
  Gas treating : absorption theory and practice / Professor D. Eimer.
      pages cm
  Includes index.
  ISBN 978-1-118-87773-9 (cloth)
  1. Gases–Absorption and adsorption.   2. Gases–Purification.   3. Gases.   I. Title.
  TP242.E34 2014
  536′.412–dc23
                                                                                2014014243

A catalogue record for this book is available from the British Library.

ISBN: 9781118877739

Set in 10/12pt TimesLTStd-Roman by Laserwords Private Limited, Chennai, India.

1   2014

To the people of Glasgow and Scotland for making me a chemical engineer, and for providing my life-long wife Barbara, and in turn two daughters and grandchildren.

Contents

Preface

This book came about because of the lack of a suitable text from which to lecture post-graduate students the topic of absorption/desorption mass transfer in combination with chemical reaction. This book would, however, also be suitable to teach mass transfer in a broader way to undergraduates who have a wish to enlarge upon this topic relative to a typical unit operations course. From my industrial experience, I would also say that those engineers who specialise, or who are heavily involved in this field, would benefit from this text.

The book starts by setting gas treating into perspective in Chapter 1 by discussing a few drivers, providing a feel for the size of the problem and the challenges in the form of gas specifications to achieve. In Chapter 2 the absorption process is put into context by discussing alternatives for treating natural gas, synthesis gas and exhaust gas. The latter is a subject that has gained importance in recent years and means that many more engineers work with this process. Chapter 2 also explains a number of other processes that an engineer is likely to come across in this field, and also one or two that are rare but are useful to know. Sulfur is a theme in its own right in the hydrocarbon industry and a quick overview is provided. It should be possible for the interested reader to quickly reel in more information by starting with the references provided.

The text introduces the alternative treatments of rate based or equilibrium based mass transfer analysis as a basis and then goes on to discuss the chemistry of acid gas absorption into alkaline solvents and physical chemistry topics that are important for fundamental understanding and that are somewhat special to this field.

Diffusion is given its own chapter to underline its importance. This is followed by a discussion of the traditional concepts for estimating the relative height requirements of separations in columns in Chapter 7, while the actual mass transfer coefficient estimates are delayed until later in the text. Hardware types are then discussed in Chapters 8 and 9 before correlations of mass transfer coefficients related to these are discussed in Chapter 11. The alternative ideas behind mass transfer coefficients are discussed in Chapter 10.

Chapter 12 develops the concepts and equations for limiting behaviour of mass transfer and the influence of chemical reaction. It is aimed at doing this in more detail than in previous texts to meet the needs of the non-specialists and students who want to gain a proper understanding. This is followed up in Chapter 13 which discusses the particular situation when both CO_2 and H_2S are being absorbed. The absorption process discussion is rounded off in Chapter 14 with absorption of water in glycol which is a very important process, and is also very different due to the extremely high solubility of water in glycol that renders this process essentially gas side limited.

At this stage it is clear that there is a great need for data to compute solutions. Techniques for measuring these are discussed in Chapter 15. Even when literature data can be found and there is no need for making our own measurements, it is useful to gain insight in these techniques and be able to form an opinion of uncertainties involved, and how data models become integrated when data are interpreted. Chapter 16 discusses absorption equilibria and models to represent them to provide an introduction to this extensive field.

Having discussed the absorption process to a great extent, it is appropriate to discuss the desorption process specifically, and this is dealt with in Chapter 17. There is more to this process than being the reverse of absorption.

Chapters 18–22 discuss various topics that are important in the absorption-desorption process, but are often neglected or overviews missing in related literature. These include heat exchange, solution management, flow sheet variations, degradation of solvent and choice of materials in view of corrosion issues.

The text is rounded off by discussing technological fronts in general and issues related to flue gas treating and treating of natural and synthesis gas in Chapters 23–26.

A chance happening in my life provided me with the time to write this book. It has been lectured to postgraduate students, and has since been strengthened. I am indebted to my students and colleagues for invaluable discussions on these topics, not the least Professor Klaus Jens, who has elevated my insight to the chemistry aspects and John Arild Svendsen, with whom I have shared joy and despair while modelling. Thanks are also due to Zulkifli Idris who has helped with chemistry figures and proofreading.

Online Supplementary Material

Concept checklists, review questions and PowerPoint slides of all figures from this book can be found online at http://booksupport.wiley.com.

List of Abbreviations

ACS	American Chemical Society
AEE	2-(2-AminoEthylamino)Ethanol
AEEA	2-(2-AminoEthyl)EthanolAmine
AEPD	2-Amino-2-Ethyl-1,3-PropanDiol
AEPDNH2	2-Amino-2-Ethyl-1,3-PropaneDiamine
AHPD	2-Amino-2-Hydro-xymethyl-1,3-PropanDiol
AIChE	American Institute of Chemical Engineers
AMP	Amino-Methyl-Propanol
AMPD	2-Amino-2-Methyl-1,3-PropanDiol
ASU	Air Separation Unit
BFW	Boiler Feed Water
BTEX	VOC emissions from glycol plants
C2+	Implying ethane (C_2H_6) and heavier alkanes
C3-MR	Propane (C3) Mixed Refrigerant process
CAPEX	CAPital EXpenditure
CCGT	Combined Cycle Gas Turbine
CCS	Carbon Capture and Storage
CFZ	Controlled Freeze Zone process
COP	Conference Of Parties
CS	Carbon Steel
CSIRO	Commonwealth Scientific and Industrial Research Organisation. (An Australian research organisation).
CWHE	Coil-Wound Heat Exchanger
DEA	DiEthanolAmine
DEG	DiEthylene Glycol
DEMEA	DiEthylMonoEthanolAmine
DETA	DiEthyleneTriAmine
DGA	DiGlycolAmine
DIPA	DiIsoPropanolAmine
EDA	EthylDiAmine
EEA	Ethyl EthanolAmine
EMEA	EthylMonoEthanolAmine
EOR	Enhanced Oil Recovery

GPA	Gas Processors Association
GPSA	Gas Processors Suppliers Association
GTI	Gas Technology Institute
HETP	Height Equivalent of a Theoretical Plate
HTU	Height of Transfer Unit
IEA	International Energy Agency
IEAGHG	International Energy Agency GreenHouse Gas program
IGU	International Gas Union
IPCC	Intergoverntal Panel Climate Change
IUPAC	International Union of Pure and Applied Chemistry
JT valve	Joule-Thompson valve
LNG	Liquefied Natural Gas
LPG	Lliquefied Petroleum Gas
MDEA	MethylDiEthanolAmine
MEA	MonoEthanolAmine
MEG	MonoEthylene Glycol
MMSCFD	Million Standard Cubic Feet per Day (i.e. 24 hours)
MPA	MonoPropanolAmine (3-amino-1-propanol)
NETP	Number of Equivalents of a Theoretical Plate
NG	Natural Gas
NGL	Natural Gas Liquids
NGO	Non-Governmental Organisation
NGSA	Natural Gas Supply Association
NMR	Nuclear Magnetic Resonance
NPV	Net Present Value
NRU	Nitrogen Rejection Unit
NTP	Normal Temperature and Pressure (0 °C and 1.013 bar)
NTU	Number of Transfer Units
OPEC	Organization of the Petroleum Exporting Countries. There are presently 12 member countries.
OPEX	OPerational EXpenditure
PCHE	Printed Circuit Heat Exchanger
PE	PiperidineEthanol
PFHE	Plate & Frame Heat Exchanger
PFHE	Plate-Fin Heat Exchanger
PSA	Pressure Swing Adsorption
PZ	PiperaZine
RPB	Rotating Packed Bed
SAFT	Statistical Associating Fluid Theory
SCF	Standard Cubic Feet
SCOT	Shell Claus Off-gas Treatment
SS	Stainless Steel
STP	Standard Temperature and Pressure (1.013 bar and 15 °C **or** 60 °F)
SWHE	See CWHE

TEA	TriEthanolAmine
TEG	TriEthylene Glycol
UNEP	UN Environental Program
VOC	Volatile Organic Compounds
VSA	Vacuum pressure Swing Adsorption
W.C.	Water Column
WMO	World Meteorological Organization

Nomenclature List

(Local variables used in specific equation are not included)

a	nominal specific contact area of mass transfer equipment, m^2/m^3
a_c	acceleration in an RPB, m/s^2
a_e	effective interfacial area for packing, m^2/m^3
a'_p	nominal specific surface area of 2 mm diameter beads, m^2/m^3
A	cross-sectional area perpendicular to flow or flux, m^2
A	constant in Debye–Hückel equation
A	heat exchanger area, m^2
A_i	regression constant(s) in equation 5.72
b_k	constant in Debye–Hückel equation
B	used to represent a generic base
C	concentration of species, $kmol/m^3$. (See Figure 3.1)
	Superscripts: L for liquid, G for gas, * if an equilibrium value
	Subscript: denotes component. i is usually the volatile, j the absorbent, tot for total (molar density), otherwise as given on a case to case basis
	0 or i as subscripts also relates to $z = 0$ or the gas-liquid interface
C_S	a factor used in relation to capacity of columns
C_{SB}	Souders–Brown constants, see equation 8.8
d_p	packing size, or dimension for a structured packing, m
d_{eq}	equivalent diameter of flow channel, m
D	diffusion coefficient, m^2/s
E	enhancement factor. Subscript refers to component
E_∞	enhancement factor for infinitely fast reaction
F	packing factor. Specific to packing
Fr	Froude number. Defined for each application
	Subscript indicates component, a second subscript indicates solvent or gas
g	gravitational acceleration, m/s^2
G	gas flow, kmol/s if not otherwise indicated by subscript
	Subscript V for volumetric m^3/s, w for mass kg/s, m for molar kmol/s
	Second subscript f for flux indicating the above per m^2.
ΔG	Gibbs free energy change
	Superscript: f for formation
h	enthalpy, kJ/kmol
h	liquid hold-up fraction in a packed column, dimensionless
H	Henry's coefficient, bar
	Subscript indicates component, a second subscript indicates solvent

H_c Henry's coefficient on concentration basis, $bar.m^3/kmol$

        Additional subscript indicates component

ΔH heat of what is specified by subscript, kJ/kmol

        Subscript ABS for absorption, REAC for reaction

I ionic strength, $kmol/m^3$

k reaction kinetics constant

        a number is either associated with an equation number or reaction order

        a minus indicates a reversing reaction

        ps1 for pseudo first order reaction

K equilibrium constant

        Subscripts I and II are used for first and second dissociation of diprotic acids

        Subscript HYD for hydrolysis

K constant. See equation 8.10

K_a acid protonation equilibrium constant

K_{ap} autoprotolysis equilibrium constant including $[H_2O]$

        Superscript ' indicates $K_{ap}/[H_2O]$

K_b basicity equilibrium constant

        Superscript ' indicates $K_b[H_2O]$

k_G gas side mass transfer coefficient, m/s (see equation 7.1)

K_G overall mass transfer coefficient referred to the gas side, m/s (see equation 7.4)

k_L^0 liquid side mass transfer coefficient, m/s (see equation 7.2)

K_L overall mass transfer coefficient referred to the liquid side, m/s (see equation 7.5)

K_w K_{ap} for water

l lower case l used for thickness of mass transfer film, *m*

L a length to be specified, *m*

L liquid flow, kmol/s if not otherwise indicated by subscript

        Subscript *V* for volumetric m^3/s, *w* for mass kg/s, *m* for molar kmol/s

        Second subscript *f* for flux indicating the above per m^2.

m dimensionless partition coefficient between gas and liquid, defined by equation 5.58

        Subscript indicates component

mj molality, mol/1000 g. See equation 16.34

M molecular weight, kg/kmol

        Subscript indicate component and average values, see equation 6.21

n number of moles of component indicated by subscript

N_i molar flux of component indicated by subscript, $kmol/m^2 \cdot s$

p partial pressure of component indicated in subscript, bar or kPa

p^0 vapour pressure of pure component indicated in subscript, bar or kPa

P total pressure, bar or kPa

P_A Parachor, see equation 6.23

ΔP pressure drop.

        Subscript *Fl* for flooding

pH   acidity scale. Defined in chapter 5.3.3

pK $-log_{10}(K)$, used for K_a, K_b, K_w, K_{ap} as indicated by subscript

r rate of reaction, $kmol/m^3 \cdot s$

	Subscript indicates component
	Subscript 'obs' indicates observed value
	Subscripts: o to situation with no net reaction,
	D for ratio defined by equation 12.53
R	gas constant, see Table 1.5
Re	Reynolds number. Defined for each application
RMT	rate of mass transfer
s	surface renewal rate, 1/s. (surface renewal theory)
s	corrugation side length, m (see equations 11.7 to 11.10)
t	time, s
T	absolute temperature, K
	Superscript G for gas, L for liquid, Ref for reference temperature
ΔT	temperature difference, K
T*	parameter defined by equation 6.18
u	velocity m/s.
	subscript L for liquid, G for gas,
	second subscript e for effective, see equation 11.5 and 11.6
U	overall heat transfer coefficient, kW/m^2.K
v	velocity of convective flux, see equation 6.13, m/s
V	volume of liquid, m^3
V_i	partial molar volume of component i
V_i, V_o, V_t	see equation 9.3
V	molar volume, m^3/kmol. (subscript G for gas and L for liquid)
	Subscript indicate component,
x	mole fraction in a liquid phase
	subscript indicates component
X_i	molar ratio, see equation 7.16 or 7.17
X	partial molar property of component subscripted, see equation 5.70
	Superscript E indicates excess property of mixture (see subscript)
y	mole fraction in a gas phase
	subscript indicates component, solute, solvent
z	distance, m
Z	used to represent the zwitterion

Greek letters:

α	loading of acid gas, mol acid gas per mol absorbent
β	defined by equation 12.36
β_{kj}	interaction parameter in equation 16.34
γ_k	activity coefficient, see equation 16.34 etc.
Γ	perimeter flow of liquid, kg/s.m
Δ	difference operator. See ensuing symbol.
ε	parameter, see equation 6.19
ε	fractional free volume in a packed bed
ε	eddy diffusivity, m^2/s, see equation 10.12
ε*	parameter defined by equation 6.26
$\in$	ratio defined by equation 12.82

ζ	reaction order with respect to component, equation 4.2
ζ	substitute variable defined by equation 6.11
ζ	variable defined by equation 12.66
θ	exposure time between gas and liquid of element, s (penetration theory)
μ	viscosity, kg/m.s (unless otherwise specified)
	subscript indicates component, solvent, water
μ	coefficient defined by equation 12.47
v	stoichiometric coefficient
ρ	density, kg/m^3
	subscript: L for liquid, G for gas
$\Delta\rho$	the difference between liquid and gas densities, kg/m^3
σ	surface tension, dyn/cm or N/m
σ_C	critical surface tension between liquid and material, dyn/cm or N/m
σ	characteristic length, see equation 6.20
Σadv	parameter, see equation 6.16
ϕ	association factor, see equation 6.22
ϕ	factor defined by equation 9.10 with inputs defined in text
ψ	sphericity of packing, see equation 9.9
ψ	ratio defined by equation 12.70
Ω	collision integral, see equation 6.17

1

Introduction

Gas treating is featured in many process plants in many contexts. There are almost always unwanted components that need to be removed from a gas stream. These components may need to be removed for a number of reasons like:

- Contamination of product
- Catalyst poison
- Reaction by-product
- Corrosion
- Dew point, unwanted condensation downstream
- Environmental considerations.

The challenges are many, and they occur when dealing with natural gas, synthesis gas, air and latterly, the challenge associated with CO_2 abatement. Different settings, seemingly different challenges, but for the chemical engineer there is a common denominator as shall become clear by the end of this book.

No matter what the application is, and no matter what the treatment needs are, cost effective solutions are always targeted. Having said that, it must be remembered that operational costs and any lost production are also factors included in this equation. There is always competition and the operator with the best profit margin will be better off in the longer term.

1.1 Definitions

Natural gas is the gas produced from hydrocarbon reservoirs. Some fields are gas fields producing nothing but natural gas, but natural gas is also produced as so-called associated gas where the gas comes from the reservoir along with the oil. The composition of natural gas varies, but is dominated by the presence of methane. It may be contaminated by CO_2 and H_2S, and there may be more or less of ethane and heavier hydrocarbons.

Gas Treating: Absorption Theory and Practice, First Edition. Dag Eimer.
© 2014 John Wiley & Sons, Ltd. Published 2014 by John Wiley & Sons, Ltd.

Most natural gas is transported to its point of use by pipeline, but there are markets that are too far away from the natural gas source. Japan is a case, and is served by liquefied natural gas, LNG, that is shipped in on gas tankers. LNG is mostly methane, it is made to be liquid at atmospheric pressure and requires a temperature down towards 111 K. The low temperature requires that higher boiling components must be removed in order not to precipitate, and water, CO_2 and H_2S must naturally be removed in order not to freeze out in the condensation process and thus block the flow channels.

Natural gas liquids, or NGLs, is a term that is used to describe the hydrocarbon condensate separated from natural gas on cooling. It is essentially ethane and heavier. In the case of NGL there is no particular refinement of the product such that there can be a tail of heavier hydrocarbons. This is different from liquefied petroleum gas, LPG, which is a tailored product that is mainly ethane and propane and may also contain a little butane, but nothing heavier.

Natural gas may also be referred to as lean or rich. A rich gas implies that there are significant amounts of ethane and heavier components that may be recovered for extra value. If a gas is lean, no such condensate would be economical to recover and the gas is sold for fuel.

Next there is synthesised gas, often referred to as syngas. This is gas that has been synthetically manufactured. Often natural gas has been the raw material, but it could also be produced as part of the activity in an oil refinery although this is more likely referred to as refinery gas. Ammonia production involves the making of syngas. Here natural gas is heated in the presence of steam and methane is converted to hydrogen, carbon monoxide and carbon dioxide. This gas is further processed with steam to convert monoxide to hydrogen and CO_2 and so on.

Flue gas or exhaust gas is the waste stream coming off a power plant. Offgases from syngas plants are usually referred to as bleeds or waste stream. The flue is usually the chimney, or at least the exhaust channel.

In the natural gas industry the Wobbe number is sometimes used. (Geoffredo Wobbe was an Italian Physicist who experimented with combustion of gases.) It is a way of judging if two fuel gases may be interchanged without affecting the performance of the burner. This number is defined as

$$\text{Wobbe no} = (\text{Upper heating value of the gas}) / \sqrt{(\text{specific gravity of the gas})} \quad (1.1)$$

This sounds simple enough, but specific gravity (note: not 'specific weight') is a ratio. For a gas it is the ratio of the density of the gas and that of air. The gas is usually at atmospheric conditions as is the case for the reference air. The related temperatures and pressures must be defined, and in the case of air its water content is also important for its density. Specific gravity is dimensionless. The upper heating value is used but the lower may also be specified. The units should in any case be given, but it has been practised not to do this in order not to get it confused with the gas' volumetric heating value. It is practised to quote the Wobbe number in Btu/ft^3, but in Europe it is more common to use MJ/Nm^3. Common values of the Wobbe number are $39-45$ MJ/Nm^3. It is heavily influenced by the gas' content of nitrogen and $C_2{}^+$. If it is specified in a gas sales contract, it is important to understand the implications. Further discussions on this subject may be found in a couple of documents issued by the American Gas Association (Ennis, Botros and Engler, 2009; Halchuk-Harrington and Wilson, 2007).

1.2   Gas Markets, Gas Applications and Feedstock

The natural gas market world-wide is huge. Although there is a need to provide a standardised gas such that all the end users' gas burners will function as intended, there are regional differences in specifications. The US market has this challenge that makes the interchangeability of gases difficult, and the cost and feasibility of standardising has been considered but discarded. In the UK, however, a similar conversion was done area by area in the 1960s and 1970s as the market was converted from 'town gas' to 'North Sea gas'. (Town gas was synthesised by gasification of coal.) Town gas was common in Europe until the advent of gas finds in the North Sea. Pipelines from these and Russian fields serves this market today. North America has had a change of fortune in recent years by technology enabling the production of so-called shale gas. There have also been LNG projects developed, with more coming on stream in the next few years. Gas is challenged by other forms of energy. Although existing users are to an extent 'sitting ducks' due to investments made, provision costs of gas must be kept in check to keep its market share. Electricity is the immediate competitor in the retail market, and that in turn could be provided through the combustion of gas, coal or oil, and other sources are nuclear power plants and hydroelectricity. The more alternatives that are available in any one market, the more the focus on provision cost of energy in the market. Deeper discussions of these issues may be found elsewhere (BP, 2011; IGU, 2013a,b; Natural Gas Supply Association, 2005).

Specifications of natural gas as a product is a very interesting topic in many ways and the specifications really determine what treatment a gas eventually needs. There are two dimensions to this. One is the transport system that supplies a market and what treatment the gas needs to uphold flow assurance in the supply chain. The other is the end market with its appliances where gas burners have been fitted with certain gas properties in mind. Interchangeability of gas cannot be taken for granted. There are many stumbling blocks to this (IGU, 2011).

Methane, or natural gas, is less reactive than their heavier analogues like ethane, propane and so on. As feedstock for making hydrogen as in the ammonia process it is the preferred starting point as the ratio of hydrogen to carbon is highest in methane. For this reason, and because of the pricing, natural gas is the feedstock of choice for this purpose.

The C_2+ fraction of the natural gas has in the main a higher market value as feedstock than as fuel. Hence the opportunity to separate these components from the gas is often taken. The economics of this has varied over time though.

1.3   Sizes

For various assessments it is valuable to have a feel for sizes of plants and associated variables. The question being, what is big, what is small, what is a challenge and what is trivial. Plant sizes and complexities will vary widely. Perhaps the simplest gas treating plant to be encountered in this context will the end-of-pipe solution scrubber where some contaminant is to be removed from an effluent gas stream before being released. Maybe this scrubber has a packing height of 3 m and a diameter of 2 m, and furthermore when the absorbent has done its job, it may be returned to the process without further ado. A 400 MW CCGT

(Combined Cycle Gas Turbine) power plant that needs CO_2 abatement will have a gas stream in the order of 1.8 million m^3/h, and the absorber would have a diameter around 17 m if there is one train only.

A large synthesis gas train may have a gas flow in the order of 10 000 kmol/h. This would be 224 000 Nm^3/h. However, the pressure could be around 25 bar if this was an ammonia plant, and this would imply a real gas stream in the order of 10 000 m^3/h.

In natural gas treating there is a wide range of plants. A fairly small one might be 10 MMSCFD. This is a typical way of specifying plant size in North America. MM stands for 'mille-mille', which is Latin inspired, meaning 1000×1000 (or a million). SCF is Standard Cubic Feet, and D implies per 24 hours (a Day). In North America 'Standard' means the gas volume is at 60°F and an absolute pressure of 14.696 psi (psi = pounds per square inch). Wikipedia points out that the 'standard' pressure may also be 14.73 psi, which is based on a pressure of 30 in. of a mercury column. Beware; if you are buying gas the difference in what you get is 0.23%, which is not to be given away easily in negotiations.

A large gas plant could be in the region of 2 million Sm^3/day. This is typical of a gas field in the North Sea. This is in metric units, and the 'standard' now implies 15°C and 1.013 bar. If this was indeed the gas's temperature and pressure it would be at its 'standard conditions.' Note that 15°C and 60°F are not identical. European and American standard conditions are not equal: something to be kept in mind when selling and buying.

An often used specification for H_2S allowed in natural gas is 0.25 grain per 100 SCF. This is a US term. One 'grain' is 1/7000th of a pound (lb).

LNG plants are usually referred to in million tonnes of LNG per year. A plant of 3 million tonnes per year was considered big less than 10 years ago, but one-train capacities have been stretched to 5–7 and there is a new generation of plants with a third refrigeration loop that could take the capacity to 10 million or more.

A large ammonia plant today would typically be 2000 tonnes per day. This is almost the double of what was usual around 1970. Cryogenic air separation units (ASU) could be as big as 3500 tonne of oxygen per day, but this size of plant is rare. Traditionally they have been built to provide oxygen for steel works. However, they figure in present day studies on oxy-fuel plants. That is, power plants where hydrocarbons, or coal more likely, is combusted with oxygen to make the CO_2 resulting more easily accessible for capture and storage.

It is good to develop an intuitive sense for plant sizes and put them into perspective. The ability to distinguish between the various 'standard' units of gas quantity is a must. To help in this direction and to summarise the earlier discussion of plant sizes, Table 1.6 is provided at the end of the example problems.

1.4   Units

There are a number of units being used in the industry that are not intuitive and will be unfamiliar to newcomers. To fill in the void, this section will go through a number of such units. The reader will undoubtedly come across further units before finishing and these will need to be deciphered using reference works.

Table 1.1   *Imperial volumetric relations. (from 1824 onwards in the brewery business.)*

	Pint	Gall (imp)	Firkin	Kilderkin	Barrel	Hogshead
Pint	1	8	72	144	288	432
Gall (imp)		1	9	18	36	54
Firkin			1	2	4	6
Kilderkin				1	2	3
Barrel					1	1.5
Hogshead						1

Let us start with the measurement of liquid. For most purposes a chemical engineer could use m^3 for volume and be done. However, the oil and gas business has a few special quirks when it comes to volumetric units, and oil, in North America in particular, is reported in barrels. Barrels are part of an old system of volumetric units where the sizes have changed over the centuries, and they have also differed between businesses. Today, a barrel as used in the oil industry is 158.9873 l. This is supposed to represent exactly 42 US gal. The reader will no doubt have come across various non-metric units of volume in non-professional context, and Table 1.1 is included to put these volumes into perspective. Oil density varies significantly and there will typically be 6–8 barrels per tonne.

Gas volumes are straightforward in the sense that either metric or well-defined Anglo–American measures are used. The important part here is to be able to distinguish between the various 'standard' or 'normal' conditions used. These must be defined in any gas sales contract to avoid legal disputes later. This is discussed previously to the necessary extent. However, the reader may well meet further definitions in the future since IUPAC changed their recommendation for standard pressure to 1 bar (100 kPa) in 1982. In other fields of gas processing so-called 'normal' conditions are also in use. These are defined as 0°C and 1 atm = 1.01325 bar.

It is prudent to mention that absolute temperatures in the 'Fahrenheit spirit' is known as °R (degrees Rankine) and

$$°R = °F + 459.67$$

A final topic worth mentioning is pressure units. They are mostly self-explanatory, but there is a unit called 'atmosphere.' If this is spelt 'atm', it is 'one' when the pressure is 760 mm Hg or 1.013 bar. However, if it is merely spelt 'at', we are talking about a 'technical atmosphere' (which is an old European tradition). This equals 'one' when the pressure is 1 kp/cm². (kp, or kilopond, is the same as kg_f). It is not often used these days, but it may still be found. Varieties are ato (gauge pressure, o = 'overpressure') and atü (gauge pressure, German: *überdruck*). It is slightly higher than 736 mm Hg. The term mm Hg as a pressure unit should strictly be the height of a mercury (Hg) column at 0°C; that is, that mercury has the density it has at 0°C.

To round off this discussion of units a table of conversion factors is provided to enable quick conversion of data discussed in the text to make life easier for those that do not have their reference values in metric units (Table 1.2).

It is also worth mentioning that the unit ton is not necessarily unambiguous. In the metric world the ton is 1000 kg while in the Anglo–American units it is 2240 lbs. Often the metric

Table 1.2   *Unit conversion factors.*

From unit to unit	Multiply by	From unit to unit: divide by the same number
ft to m	0.3048[a]	m to ft
lb to kg	0.45359237[a]	kg to lb
lb mol to kmol	0.45359237[a]	kmol to lb mol
°F to °C	Subtract 32, then × 1.8[a]	°C to °F
bar to psi (lb$_f$/sq in.)	14.5037744	Psi to bar
1 mm to micron (μ)	1000	micron to mm
Btu[b] to kJ	1.05435026444	kJ to Btu
kJ to kWh	3600	kJ to kWh
Btu/lb to kJ/kg	2.32444	kJ/kg to Btu/lb
Btu/ft^2·F·h to W/m^2·K	1.751378	W/m^2·K to Btu/ft^2·F·hr
hp (metric) kW	735.49875	kW to (metric) hp
bhp to kW	745.69987	kW to bhp
Imp gallon to m^3	0.00454609	m^3 to imp gallon
US gallon (g) to m3	0.003785412	m^3 to US gallon
gpm (per minute) to m^3/s	0.00006309	m^3/s to US gpm

[a]Implies exact conversion factor, otherwise derived.
[b]This is based on the thermochemical value of BTU, but other definitions range from this value to as high as 1.05987 kJ.

Table 1.3   *Tons, long tons, short tons and tons, and so on.*

Type of ton	Content
1 metric ton (tonne)	1000 kg (approximately 2205 lb)
1 Anglo–American ton (ton, sometimes long ton))	2240 lb (approximately 1016 kg)
1 short ton (in USA and Canada often referred to as 'ton')	2000 lb (approximately 907.2 kg)

Table 1.4   *Various often quoted volumes of gas of given mole mass.*

kmol	Nm^3	Sm^3 (metric)	Sm^3 (US)	SCF (US)
1	22.414	23.645	23.690	836.62

ton is referred to by 'tonne', but if important, this should be verified on a case to case basis. The world of tons is summarised in Table 1.3.

In the air gases industry it is common to talk about plant capacities in tons per day. Clearly it is essential to know which tons are quoted. Ammonia plants are commonly described in tons per day and LNG plant in million tons per year. It is quick to find yourself short-changed.

A very useful summary of conversions between gas volumes and mole contents are given in Table 1.5. Note that the 'normal m^3' is defined at the 'normal conditions', at NTP (normal

Table 1.5 *Values of the gas constant, R.*
(psi is lb$_f$/square in.)

8.31447	J/(mol K)
0.0831447	m^3 bar/(kmol K)
0.0820574	m^3 atm/(kmol K)
8.31447	m^3 Pa/(mol K)
8.31447	m^3 kPa/(kmol K)
1.98721	cal/(mol K)
10.73159	ft^3.psi/(lb mol R)

temperature and pressure). The 'standard m^3' are at STP (standard temperature and pressure). The standard temperature is not the same in Europe and the US.

Another useful compilation is a collection of different values of the gas constant R. These will come in useful as the situation arises.

1.5 Ambient Conditions

Plants have been built in all sorts of places. Some are hot, some are cold and some are to be found at a high altitude where the air is thin. When comparing plant costs and efficiencies, this must be kept in mind. An LNG plant will of course have a better efficiency if the heat sink is at 5°C compared 35°C. On the other hand winterisation may be costly. Special precautions must be made if it is to be operated for weeks on end at −40°C.

1.6 Objective of This Book

The objective of this book is to give the reader a general background for the world of gas processing. It is also a target to provide specialised teaching with respect to the absorption-desorption process in general, and to mass transfer coupled with chemical reaction in particular.

Some topics in this book are treated cursorily and the only justification for including these chapters is to create a starting point for the reader to dive further into those topics. The book that gives specialist in-depth treatment of all you need to know is still to be written.

1.7 Example Problems

Throughout this book we shall need relevant case studies to illustrate the use of the tools and theories developed. The development of these case studies starts here, and they will be based on the problems outlined when discussing typical plant sizes. To a degree reverse engineering will be applied to extract the problems relevant for discussion in this book. Immediate question: Which is the bigger gas processing plant of the following: Flue gas from a 400 MW CCGT, 600 MW coal power, 2000 tonnes per day ammonia plant, 30 MMSCFD natural gas, 3 million Sm^3 per day natural gas plant, or a 7 million tonnes per year LNG plant? After working the example problems, you will know. When gas concentrations are given, they are on a molar (or volumetric) basis unless otherwise specifically stated. Ideal gas is assumed throughout these examples.

1.7.1 Synthesis Gas Plant

A good and well-defined example is an ammonia plant. Here the synthetic gas is eventually converted to ammonia (NH_3). Such a plant is in most situations fed by natural gas at pressure. The gas needs to be treated for sulfur compounds to avoid poisoning of catalysts before processing can proceed. This is followed by 'reforming' the natural gas to H_2 and CO in the first sections of the plant before the CO is converted to H_2 by the help of steam.

$$CH_4 + \tfrac{1}{2} O_2 = CO + 2 H_2$$

$$CO + H_2O = CO_2 + H_2$$

Thereafter the CO_2 must be removed. At this stage we ask ourselves, how much gas must be treated if the ammonia plant has a capacity of 2000 tonnes per day. The eventual reaction is:

$$\tfrac{1}{2} N_2 + \tfrac{1}{2} H_2 = NH_3$$

Now, 2000 tonnes per day work out at:

$$(2000 \, \text{ton/d}) \, (1 \, \text{d/24 h}) \, (1000 \, \text{kg/ton}) = 83\,333 \, \text{kg/h}$$

and since the molecular weight of ammonia is 17, it follows that the ammonia production is:

$$(83\,333 \, \text{kg/h}) \, / \, (17 \, \text{kg/kmol}) = 4902 \, \text{kmol/h}$$

Next it is observed that in the ammonia synthesis reaction $0.5 + 1.5 = 2$ mol of N_2 and H_2 gas are converted to 1.0 mol of ammonia. Hence, the net stream of treated gas after CO_2 removal will be:

$$(2) \, (4902 \, \text{kmol/h}) = 9804 \, \text{kmol/h}$$

Let us assume that the conversion efficiency is 99% and call that $9804/0.99 = 9903$ kmol/h.

It may be worked out from analysis of the ammonia train from the start, but we shall take it as read that the CO_2 content of the gas prior to CO_2 removal is 20% (mol) with a gas pressure of 25 bar and a temperature of 40°C. On this basis the feed to the CO_2 removal unit is:

$$(9903 \, \text{kmol/h}) \, / \, (1 - 20/100) = 12\,378 \, \text{kmol/h}$$

CO_2 to be removed is thus: $12\,378 - 9903 = 2475$ kmol/h.

In this industry it is also quite common to quote flows in Nm^3/h and that works out at:

$$(12\,378 \, \text{kmol/h}) \, \left(22.414 \, Nm^3/\text{kmol}\right) = 277\,440 \, Nm^3/h.$$

With the temperature and pressure given, this means that the actual flow of gas at operating conditions is:

$$\left(277\,440 \, Nm^3/h\right) \, (1.013/25) \, ((273 + 40) \, /273) = 12\,888 \, m^3/h$$

1.7.2 Natural Gas Treatment

Characterising a natural gas treatment plant as small or large is not an exact science. The following example could, for what it is worth, be described as mid-range. A plant is needed to process a stream of 30 MMSCFD.

Using the conversion factor available from Table 1.4, this stream becomes:

$$(30 \text{ MMSCFD}) / (255.002 \text{ MMSCFD/kmol}) = 4902 \text{ kmol/h}$$

This is turn is:

$$(4902 \text{ kmol/h}) (23.645 \text{ Sm3/kmol}) = 115\,908 \text{ Sm}^3/\text{h}$$

With temperature and pressure given as 40°C and 35 bar, the actual gas flow at operating conditions are:

$$\left(115\,908 \text{ Sm}^3/\text{h}\right) (1.013/35) ((273 + 40) /288) = 3645 \text{ m}^3/\text{h}$$

If 8% of this feed is CO_2, then there are $(0.08)(4902) = 392 \text{ kmol } CO_2/\text{h}$ in the feed.

1.7.3 Natural Gas Treatment for LNG

LNG plants are complex and as such their economics thrives on economics of scale. Plant sizes in excess of 10 million tons per year are possible, but we shall look at the implications of a 7 Mton/year capacity.

We shall assume that this capacity is reached by being on-stream for 8600 hours per year. Furthermore, it will be assumed that the average molecular weight of the LNG is 17. Capacity may then be rated as:

$$((7\,000\,000 \text{ ton/y}) (1000 \text{ kg/ton}) / (8600 \text{ h/y})) / (17 \text{ kg/kmol}) = 47\,880 \text{ kmol/h.}$$

If there is 12% CO_2 in the feed, its CO_2 removal plant will receive:

$$47\,880/ (1 - 0.12) = 54\,409 \text{ kmol/h and there will be 5441 kmol } CO_2/\text{h.}$$

Given a temperature of 40°C and a pressure of 50 bar, this implies a flow at operating conditions equal to:

$$(54\,409 \text{ kmol/h}) \left((22.414 \text{ Nm}^3/\text{kmol}) (1.013/50) ((273 + 40) /273) = 26\,120 \text{ m}^3/\text{h.}\right)$$

1.7.4 Flue Gas CO_2 Capture from a CCGT Power Plant

The abbreviation CCGT stands for combined cycle gas turbine (power plant). These plants are often described in terms of CO_2 emission, but we shall approach this from its power rating. A state of the art CCGT will be as big as 440 MW rated power output, and its power efficiency is 58% or more. In this plant gas is burnt under pressure, expanded in the gas turbine and the heat in the hot exhaust is recovered to make steam that is in turn used in steam turbines to boost energy efficiency.

We shall assume a fuel gas feed of 83% (mol) CH_4, 9% C_2H_6, 4% C_3H_8, 1% C_4H_{10}, 2.5% CO_2 and 0.5% N_2. There is also expected to be 3 ppm of H_2S, but this is neglected for the present considerations. Based on heat of combustion data from Perry and Green (1984), the average upper heat of combustion for this gas is 997.06 kJ/mol. The 'upper' value is used since the power process is expected to use a condensing steam turbine at the end. Average molecular weight is estimated to be 18.84. The need for fuel gas is accordingly:

$$(3600)\,((440\,000\,\text{kW})\,/\,(997\,060\,\text{kJ/kmol}))\,/0.58 = 2739\,\text{kmol/h}.$$

The combustion stoichiometry means that 1 mol of CO_2 is formed from 1 mol of CH_4, 2 mol from C_2H_6, and so on. CO_2 itself will of course pass through the combustion unchanged. The CO_2 content of the flue gas may now be estimated:

$$(2739\,\text{kmol/h})\,[1 \times 0.83 + 2 \times 0.09 + 3 \times 0.04 + 4 \times 0.01 + 0.025] = 3273\,\text{kmol/h}.$$

Water in the flue gas from combustion may be similarly estimated based on 1 mol CH_4 giving 2 mol of water, and so on:

$$(2739\,\text{kmol/h})\,[2 \times 0.83 + 3 \times 0.09 + 4 \times 0.04 + 5 \times 0.01] = 5862\,\text{kmol/h}.$$

The oxygen needed for combustion is given from stoichiometry and the need to provide excess oxygen. Here 1 mol O_2 is needed to make 1 mol of CO_2 and $1/2$ mol O_2 is needed to make 1 mol of H_2O. Hence the stoichiometric amount of O_2 needed is:

$$1 \times 3273 + 1/2 \times 5862 = 6136\,\text{kmol/h}.$$

However, air (O_2) is added to the combustion process in excess. An excess factor of 2.8 is quite typical. Using this, the actual O_2 addition will be:

$$(6136\,\text{kmol/h})\,(2.8) = 17\,180\,\text{kmol/h}.$$

The accompanying nitrogen (which here includes the argon) will be:

$$(17\,180\,\text{kmol/h})\,(79/21) = 64\,628\,\text{kmol}\,N_2/\text{h}.$$

Oxygen in the flue gas will be:

$$17\,180\,\text{kmol/h} - 6136\,\text{kmol/h} = 11\,044\,\text{kmol/h}.$$

Assuming a dew point of 10°C for water in combustion air, water in with that is:

$$(0.01227\,\text{bar}/1\,\text{bar})\,(64\,628\,\text{kmol/h}) = 1004\,\text{kmol/h}.$$

Total water from the power plant is then: $1004 + 5862\,\text{kmol/h} = 6866\,\text{kmol/h}$.
   Sum flue gas:

$$64\,628\,\text{kmol}\,N_2/\text{h} + 11\,044\,\text{kmol}\,O_2/\text{h} + 3273\,\text{kmol}\,CO_2/\text{h}$$

$$+ 6866\,\text{kmol}\,H_2O/\text{h} = 85\,811\,\text{kmol/h}.$$

The volumetric flow is:

$$(84\,807\,\text{kmol/h})\,\left(22.414\,\text{Nm}^3/\text{kmol}\right) = 1\,900\,859\,\text{Nm}^3/\text{h}.$$

With a temperature of 100°C and a pressure of 1.05 bar, the actual flow at operating conditions will be:

$$\left(1\,900\,859\,\text{Nm}^3/\text{h}\right)\,(1.013/1.05)\,((273 + 100)\,/273) = 2\,505\,256\,\text{m}^3/\text{h}.$$

1.7.5   Flue Gas CO_2 Capture from a Coal Based Power Plant

The coal technology considered as the basis here is the conventional one where coal is combusted, steam is raised and the steam is used to drive steam turbines to produce electricity. The question raised is: how much flue gas and CO_2 is created from a 500 MW coal fired power plant? Assume a power efficiency of 40% in this case. A further assumption is that the sulfur content of the coal is 1% (which is a low value in this context). We also assume that the final steam turbine in the train is a 'condensing one', which implies that upper heats of combustion should be used. They are 393.550 kJ/mol for carbon and 394.966 kJ/mol for sulfur. On average the heat of combustion is 393.565 kJ/mol.

The amount of coal needed is:

$$((500 \times 1000\,\text{kW}) / (393\,565\,\text{kJ/kmol})) / (40/100) = 3.176\,\text{kmol/s} = 11\,434\,\text{kmol/h}.$$

The amount of CO_2 produced is $(0.99)(11\,434) = 11\,320\,\text{kmol/h}$.

And the amount of SO_2 produced is $(0.01)(11\,434) = 114\,\text{kmol/h}$.

These plants are fired with excess oxygen (or air if you like). We shall assume an excess factor of 1.7. The amount of oxygen needed for combustion is by stoichiometry:

$$11320 + 114 = 11\,434\,\text{kmol/h}.$$

The actual oxygen feed is accordingly:

$$(11,434\,\text{kmol/h})\,(1.7) = 19\,438\,\text{kmol/h}.$$

Excess oxygen is $19\,438 - 11\,434 = 8004\,\text{kmol/h}$.

The amount of nitrogen in the air to combustion is (remembering that there is 21% O_2 and we assume the rest is all N_2):

$$(19\,438)\,(79/21) = 73\,123\,\text{kmol/h}.$$

The total amount of flue gas is the sum of all the components (CO_2, SO_2, N_2, O_2):

When this gas arrives at the CO_2 capture plant, it has been cooled by the steam generation and BFW (boiler feed water) heating. We shall assume that the temperature is 45°C and that the pressure is 1.05 bar. The actual flow of gas neglecting H_2O is then:

$$(11\,320 + 114 + 73\,123 + 8\,004) = 92\,561\,\text{kmol/h}.$$

Its volumetric flow is:

$$(92\,561)\,(22.414)\,(1.013/1.05)\,((45 + 273)\,/273) = 2\,331\,291\,\text{m}^3/\text{h}.$$

1.7.6 CO_2 Removal from Biogas

There are numerous bioreactors about converting organic waste material to methane. The plant sizes cover a very wide range. However, we shall look at a large (in this context) plant that has a nominal capacity 10 million Sm^3 per year. What is the size of gas stream to be treated?

We shall assume that the plant is operating 350 days per year. The temperature is 35°C (assuming mesophilic bacteria), and the pressure is typically 1.1 bar.

Feed gas rate is:

$$\left(10 \times 10^{-6}\, \text{Sm}^3/\text{y}\right) / \left((350\, \text{d/y})\,(24\, \text{h/d})\right) = 1190\, \text{Sm}^3/\text{h}.$$

Actual volumetric feed rate is:

$$\left(1190\, \text{Sm}^3/\text{h}\right)(1.013/0.9)\left((273+15)/288\right) = 1172\, \text{m}^3/\text{h}.$$

Typical CO_2 content of such gas is 35%. The amount of CO_2 in feed is thus:

$$\left(1190\, \text{Sm}^3/\text{h}\right)(0.35)/\left(23.645\, \text{Sm}^3/\text{kmol}\right) = 17.6\, \text{kmol/h}.$$

1.7.7 CO_2 Removal from Landfill Gas

Landfills vary very much in scale, and any gas production from such fills will reflect this. The composition of wellhead gas will also vary, 50% CH_4, 35% CO_2 and 15% N_2 are values in the mid-range for each of these. In addition there will be found H_2S, O_2, and ammonia plus siloxanes. The latter represents a challenging removal problem. One such landfill gas plant produced 13 000 US gal of liquefied CH_4 per day. What gas volume was processed?

The density of liquefied CH_4 at atmospheric pressure is roughly 700 kg/m³. Assuming close to 100% conversion of CH_4, this means that the CH_4 production is:

$$(13\,000\ \text{US gall/d})\left(0.003785\ \text{m}^3/\text{US gall}\right)(1\ \text{d}/24\ \text{h})\left(700\ \text{kg/m}^3\right) = 1435\ \text{kg/h},$$

$$(1435\ \text{kg/h})/(16\ \text{kg/kmol}) = 89.70\ \text{kmol/h}.$$

Since CH_4 constitutes 50% of the feed, the plant feed is:

$$89.70/0.50 = 179.4\ \text{kmol/h}.$$

If we further assume that the pressure is 0.9 bar (abs) and the temperature 20°C, then the actual volumetric feed to this plant is:

$$(179.4\ \text{kmol/h})\left(22.414\ \text{Nm}^3/\text{kmol}\right)(1.013/0.9)\left((273+20)/273\right) = 4857\ \text{m}^3/\text{h}\ :$$

The amount of CO_2 in the feed is: $(179.4\ \text{kmol/h})(0.35) = 62.8\ \text{kmol/h}.$

1.7.8   Summarising Plant Sizes Just Considered

To provide a comparison of plant sizes for background references the previously mentioned capacities are summarised in Table 1.6. Since we have converted all the plants to capacities rated in the same units, such a comparison is now meaningful. It is quickly seen that the relative plant sizes depend on whether they are ranked by actual volumetric flow or the amount of CO_2 to be removed.

Table 1.6 *Summary of plant sizes provided to give the reader a reference frame.*

Type of plant	Actual flow of gas (m³/h)	CO_2 in feed (kmol/h)
Feed to CO_2 treatment in a large ammonia plant	12 888	2 475
Natural gas treatment plant (large)	3 645	392
LNG feed gas	28 327	5 441
Flue gas from 440 MW CCGT gas power station	2 500 000	3 273
Flue gas from 500 MW coal power station	2 331 291	11 320
Biogas reactor treatment feed gas	1 172	17.6
Landfill gas plant feed	4 857	62.8

References

BP(supported by the IGU) (2011) *Guidebook to Gas Interchangeability and Gas Quality* BP in association with the IGU.

Ennis, C.J., K.K. Botros, D. Engler, (2009) On the difference between US Example Supply Gases, European Limit Gases, and respective interchangeability indices. AGA – Operations Conference and Biennial Exhibition, Pittsburgh, PA, May 19–21.

Halchuk-Harrington, R. and Wilson, R.D. (2007) *AGA Bulletin # 36 and Weaver Interchangeability Methods: Yesterday's Research and Today's Challenges*, AGA Publication Ltd.

IGU (2011) *Guidebook to Gas Interchangeability and Gas Quality* BP in association with, IGU.

IGU (International Gas Union) (2013) International Gas, April–September 2013. The Organisation's News Magasin.

IGU (2013) *Wholesale Gas Price Survey – 2013 Edition*, International Gas Union.

Natural Gas Supply Association (2005) White Paper on Natural Gas Interchangeability and Non-combustion End Use.

Perry, R.H. and Green, D. (1984) *Perry's Chemical Engineer's Handbook*, 6th edn, McGraw-Hill.

2

Gas Treating in General

2.1 Introduction

When talking of gas treating, it is most often implied that natural gas is the focus. The natural gas industry is one of the World's largest. However, there is also treatment of gas in the synthesis gas industry, and there are a number of processes used for this that are similar to those used for natural gas treatment. Finally, there is a large industry devoted to separate air to make nitrogen, oxygen and argon, and to an extent, krypton and xenon. Such plants would be cryogenic distillation outfits for large capacities while adsorption and membranes are also in use for smaller units. There is also an emerging interest in CO_2 removal from flue gases caused by the focus on CO_2's role in global warming. This chapter will focus on natural gas.

The term *gas treating* is normally used to cover CO_2 removal, H_2S removal, water removal, hydrocarbon dew pointing and gas sweetening. Gas sweetening is a generic term for sulfur removal. Sometimes the term 'gas conditioning' is used instead of gas treating.

The key question is 'why treat gas?' This is a multi-faceted issue. The gas is produced from a well at a location where there is usually only a negligible market for it. Transport of the gas to the market is the first challenge. Traditionally this has been achieved by pipelines. A lot could be said about pipelines, but here it will suffice to say that these are constructed in some steel material, and the properties of these materials are such that the presence of certain gases must be kept low to ensure the integrity of the pipeline. Hydrogen sulfide is a key component as it may cause stress corrosion cracking. Pipeline specifications may vary, but its content is commonly kept below 3–5 ppm. In the US the number 0.25 grain per 100 SCF is often used, but there is no standard in this matter. It must be remembered that the flow velocity in gas pipelines will be too high to allow integrity for a protective sulfide film on the inner pipe wall. For flow assurance reasons the dew point of the gas must be engineered before entering the pipeline. If the temperature is reduced, both water and hydrocarbons may condense. Water could form hydrate crystals with methane and these could block the flow of gas. Such hydrates are hard and time consuming to get rid of. Clearly no pipeline

Gas Treating: Absorption Theory and Practice, First Edition. Dag Eimer.
© 2014 John Wiley & Sons, Ltd. Published 2014 by John Wiley & Sons, Ltd.

operator would want this to happen. Liquid water could also cause corrosion when acidified by CO_2 that is likely to be present in the gas. This is also undesirable. Finally hydrocarbon condensate could amass to quantities that would cause flow problems if left unchecked. It must be remembered that pipelines follow landscapes where its elevation goes up and down repeatedly. CO_2 is usually also kept below a certain limit, say 1–4%, depending on the local situation.

There is also another reason other than the pipeline considerations to treat gas. Downstream of the transport system, that could be complex, there is a multitude of customers that will use the gas. Their equipment will have been made with certain gas specifications in mind. Here, the gas heating value will an issue, the Wobbe number is often specified and there will be limits on H_2S and CO_2. Corrosion issues apart, H_2S would end up as SO_2 in the flue gas and this would be an environmental problem.

This chapter is a mere primer in the field of gas treating. Further reading includes Kohl and Nielsen (1997), Korens, Simbeck and Wilhelm (2002) and Wagner and Judd (2006). There are also the books by Campbell (1981).

2.2   Process Categories

There is limited attention paid to processing gases in a typical chemical engineering curriculum. This is, however, a huge field where many chemical engineers find employment. In general terms, there are four principally different main methods that may be used to separate gases. They include (in alphabetical order):

- Absorption
- Adsorption
- Cryogenics: liquefaction and distillation
- Membrane permeation.

Ab- and adsorption are often mixed up in write-ups, probably because their spelling is so similar. Process-wise there is a huge difference though. *Adsorption* is essentially a surface phenomenon while *absorption* involves something being dissolved.

Cryogenics involves gases being cooled until they condense after which they may be separated by distillation. Some such processes could also be argued to lean towards absorption and/or desorption. A nitrogen wash unit, sometimes used for synthesis gas treatment, is an interesting case with respect to that kind of discussion.

Membrane technology used for gas separation is in general based on so-called dense membranes that separate gases based on different permeation rates. Small volume niche products within inorganic membranes may be different, but a discussion of this is beyond the present scope.

2.2.1   Absorption

Absorption is a much used process for separating gases, removing undesired gas components or to prevent pollution from stacks. The process is, by its nature, run at supercritical temperature with respect to the main gas component(s). There is no boiling like that seen in distillation columns. The mass transfer process is generally rate controlled. All components are in principle undergoing mass transfer between gas and liquid, but all does not need to be accounted for and/or may be neglected. Mass transfer rates and mass transfer coefficients may differ in different directions for different components.

If a lot of gas needs to be absorbed, large absorbent flows will be needed. This represents an operational cost that, in the end, may be a show stopper for using absorption.

It is a very interesting process, and is in many ways the main focus for this treatise. A separate sub-chapter is dedicated to a preliminary discussion of absorption into alkanolamines in view of these absorbents' commercial importance.

2.2.2   Adsorption

Practical adsorption processes use a granulated material with affinity for the component, or components that are desired to be removed. This material is referred to as the *adsorbent*, while the material adsorbed is referred to as the *adsorbate*. There are four categories of adsorbents commonly used:

- Molecular sieve zeolites
- Activated alumina
- Silica gel
- Activated carbon.

There is also a carbon molecular sieve that is used for making moderate quantities of nitrogen from air, but we shall leave that aside. Also liquids may be treated by adsorption. In gas treating with absorbents, there is usually an adsorption treatment of this absorbent as shall be discussed later under solution management in Chapter 19.

Regeneration of adsorbent may be done by both pressure swing and thermal swing, or a combination. Pressure swing alone is a commercial process that is applied to air separation and at least to hydrogen recovery from streams of synthesis gas. In gas treating contaminant removal by a combination process involving both temperature and pressure variation is mainly used. It could be used to remove water from the gas, and it is used as pretreatment upstream of liquefied natural gas (LNG) trains to ensure sufficiently low dew points. Pressure swing implies that the pressure is changed, and temperature swing implies that the temperature is changed.

Adsorption processes are semi-continuous. By this we mean that they continuously treat the gas without a buffer volume, the discontinuity comes from the need to switch between two or more parallel units. The unit not adsorbing is being regenerated offline at a lower pressure and increased temperature with a heat carrying dilution gas flowing through as illustrated in Figure 2.19 on page 40. This dilution gas may need to be a part of the product gas that most likely will need to be recirculated. In big units more than two parallel columns are often used, sometimes in intricate process stages to emulate some counter-current action while regenerating.

Isotherms for commercial adsorbents are hard to come by. There used to be a couple of companies that handed out leaflets with such content, but such information is certainly not offered on their web sites. In this context it has to be kept in mind that these products are forever being developed such that isotherms may be improved. It is, however, nice to be able to make the odd order of magnitude estimate. This is by no means the result of a thorough

Table 2.1   *References to a few adsorption isotherms and a few extracted data for adsorption levels.*

System	Partial pressure (kPa)	Load (mol/kg)	References
Water (40°C)	0.9	4.6	Kim *et al.* (2003)
	2.1	11.4	
CO_2 (25°C)	2.9	3.7	Cavenati, Grnade and
	125	0.71	Rodrigues (2004)
H_2O–CO_2 binary[a]			Brandani and Ruthven (2004)
C1 (35°C)	216	0.94	Sun *et al.* (1998)
C2	205	2.07	Sun *et al.* (1998)
C3	11	1.67	Sun *et al.* (1998)
C4	0.94	1.4	Sun *et al.* (1998)

Water and CO_2 are for zeolite X while the C1–C4 are for silicalite.
[a] Potential CO_2 adsorbent load could be reduced to one-fifth when water load on the adsorbent is increased from 0 to 10%. This will depend of the adsorbent.

review, but the isotherms used have been published by a number of people and such publications are summarised in Table 2.1 to provide a quick reference as a starting point.

When both CO_2 and water are adsorbed, it should be clear from Table 2.1 that water is significantly more strongly adsorbed and will push CO_2 away as they compete for adsorption sites. In pretreatment of air for air separation plants this means that there will be a CO_2 front moving through the adsorption bed in the direction of flow with a water front pushing from behind. A practical aspect of this is that there is no real need to check for water breakthrough. It is actually easier to handle a CO_2 detector and a water break-through would, in any case, be worse for a cryogenic plant.

On the practical side molecular sieve zeolites could catch 10 g water per 100 g of zeolite in a practical cycle. (Suggested as a quick first approach by a sales engineer a long time ago.) The capacity for CO_2 is less. This has implications for adsorption column design. When the gas quantity to be treated is large and the contaminant concentration is significant, the amount of adsorbent needed could be become very large.

Adsorption is mostly used for trace quantity removal. Specific applications will be discussed as they arise. They could include water removal from gas, and a big application is combined water and CO_2 removal from air feed to ASUs (ASU = Air Separation Unit).

There is a lot of research going on in the hope of finding a solution that may be used for CO_2 removal from flue gas. Recent adsorbents have been reviewed by Hedin and co-workers (2013). They point out that rapid cycling is necessary and believe that some form of structured adsorbent is necessary for success. Treatment of flue gas by vacuum-pressure swing adsorption (VPSA) has been studied (Xiao *et al.*, 2008). They tested three-bed designs using up to 12 steps in the adsorption-desorption cycle to improve CO_2 recovery and purity. Recoveries reported were in the range 70–82%, and purities 82–96%. In the air separation industry VSA is used when oxygen purity does not need to be high, typically 90% although higher can be provided. Argon follows oxygen and is one reason why the purity is that low for reasons of economics. If we now consider the example in Section 1.7.4, the combined cycle gas turbine (CCGT) flue gas example, the following observations may be made:

Amount of CO_2 available for capture: 3273 kmol/h
The partial pressure of CO_2 in feed to capture is 4 kPa (4% of 100 kPa).

Let us assume that the adsorption capacity is 3.7 mol/kg adsorbent based on the numbers in Table 2.1. This is equivalent to 0.0037 kmol/kg. Note that this is theoretical adsorption capacity at 25°C. Actual temperature is higher, such the estimate here is too low unless the gas is cooled. The actual capacity is lower than the theoretical. The estimate of adsorbent bed size(s) next is for these reasons too low.

This means that the need for adsorbent is:

$$(3273 \, \text{kmol CO}_2/\text{h}) \, / \, (0.0037 \, \text{kmol CO}_2/\text{kg adsorbent}) = 884\,595 \, \text{kg adsorbent/h}.$$

The bulk density of such adsorbents is in the order of 700 kg/m^3. The volume of adsorbent needed, or saturated, is:

$$(884\,595 \, \text{kg/h}) \, / \, (700 \, \text{kg/m}^3) = 1264 \, \text{m}^3/\text{h}.$$

For a TSA (temperature swing adsorption) on-stream time is typically 8 h, but let us consider 4 h. The size of bed is then:

$$(1264 \, \text{m}^3/\text{h}) \, (4 \, \text{h}) = 5056 \, \text{m}^3.$$

In a VSA cycle adsorption times are typically 4–10 min, perhaps 2 min is possible. Let us consider 6 min, in which case the size of adsorbent bed is:

$$(1264 \, \text{m}^3/\text{h}) \, (6/60 \, \text{h}) = 126 \, \text{m}^3 \text{ which is still a big bed.}$$

A number of beds in parallel would be needed to enable a cycle as referenced previously. Next, we would need to consider the size of valves and piping but this is left to the reader.

2.2.3 Cryogenics

Cryogenics is often defined as process engineering at temperatures below 150 K. This is merely to provide a definition. It is essentially about liquefying so-called permanent gases to enable separation by distillation. The biggest application is in the field of separating air to make gaseous or liquid nitrogen, oxygen and argon, which is a very mature and highly developed process (Haussen and Linde, 1985). However, interesting new concepts are still forthcoming more than 100 years after Linde introduced his famous double column for air separation (Fu and Gunderson, 2013). There are many applications of low temperature processes, often referred to as *cryogenic*, even if they do not always go as low as 150 K. The process thinking and need for special refrigeration are common denominators for this 'class' of processes.

Since the temperature will have to be reduced below the point where H_2O and CO_2 in the gas will start to condense or freeze out, these components are usually removed prior to lowering the temperature. There are process concepts where this effect is exploited, however. In the past it was normal to use so-called reversing heat exchangers to cool air before cryogenic distillation. Here one heat exchanger would be cooling the air and freeze water and CO_2 onto its surface, another parallel unit would be heated slightly to let the water and CO_2 sublimate. It is noticed that a variation of this concept has recently been proposed for CO_2 removal (Pan, Clodic and Toubassy, 2013). In this concept the thawing is done under pressure such that CO_2 is recovered as liquid. This should make the CO_2 compressor superfluous.

Cryogenic based processes are used in natural gas processing for recovering ethane and heavier gases from natural gas. The rationale is that ethane, propane and higher alkanes have a higher market value on their own, since they are valuable as raw materials for synthesis. Natural gas is mainly sold at a lower price reflecting its fuel value. When recovering the heavier components, natural gas is cooled until the heavier components condense and a certain recovery has been achieved. How much of that can be economically recovered will vary from situation to situation. Methane is, however, the dominant raw material for making ammonia.

It should not be forgotten that there are applications also for cryogenic separations in synthesis gas context. There are nitrogen wash, the Braun purifier and ammonia loop bleed gas hydrogen recovery, plus more.

When considering mixtures for cryogenic distillation, it is useful to list the components in the order of increasing boiling point. This gives the order of which they would be distilled over the top, and as such is useful information when sketching a distillation train. Relevant data are listed for selected gases in Table 2.2.

A further help to get a feel for relative volatilities and temperatures to be met at various pressures in the processes to be discussed may be had from Figure 2.1 where vapour pressures of selected compounds are shown. Argon, mentioned later for synthesis gas separation, is omitted but lies close to oxygen and is slightly more volatile.

Table 2.2 *Critical points and normal boiling points of a selection of gases.*

Gas	Symbol	Molecular weight	Melting point (K)	Boiling point (K)	Critical temperature (K)	Critical pressure (bar)
Helium	He	4.00	2.2	4.2	5.2	2.29
Hydrogen	H_2	2.02	13.9	20.4	33.2	12.9
Nitrogen	N_2	28.01	63.2	77.3	126.1	33.9
Carbon monoxide	CO	28.01	68.1	81.6	132.3	35.0
Argon	Ar	39.95	83.8	87.3	150.7	48.6
Oxygen	O_2	32.00	54.4	90.2	154.8	50.8
Methane	CH_4	16.04	90.7	111.7	190.6	46.0
Ethane	C_2H_6	30.07	89.9	184.6	305.5	48.8
Carbon dioxide	CO_2	44.01	216.6	[a]	304.2	73.7
Hydrogen sulfide	H_2S	34.08	187.6	212.9	373.6	90.1
Propane	C_3H_8	44.10	85.4	231.1	369.8	42.5
i-Butane	i-C_4H_{10}	58.12	113.6	261.5	406.9	37.0
n-Butane	n-C_4H_{10}	58.12	134.9	272.7	425.2	38.0

Melting points given are the triple points.
[a] CO_2 solidifies when the pressure is reduced below 5.18 bar. It has no 'normal' boiling point that may be compared. Its melting point is at 5.18 bar.

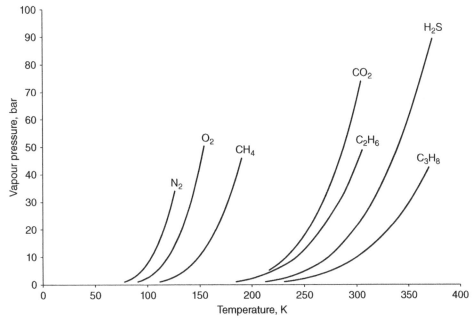

Figure 2.1 *Vapour pressures of selected components. Based on data from L'Air Liquide (1976).*

The principle of a refrigeration process is shown in Figure 2.2. It is akin to a heat pump or refrigerator in what it does. Essentially energy is picked up from a source (that has to be a little warmer) at low temperature implying that the refrigerant must attain a temperature below the lowest process temperature. A so-called Joule–Thomson valve (JT valve) is usually the last step in achieving this. (It is just a throttle valve, but it is commonly referred to as a Joule–Thomson valve for this application). The refrigeration step is generally done by letting the refrigerant evaporate at low pressure. The ensuing cold vapour is heat exchanged against, for example (as shown) warm refrigerant being cooled and condensed. Thereafter the (spent) refrigerant, that has been warmed, is compressed before it is recycled for cooling, condensation and new energy pick-up. This refrigeration cycle is shown with an expander after some cooling of the gas, then further cooling and a final temperature reduction over a JT valve. The expander takes work out of the process and provides energy efficient cooling while the JT valve is an isenthalpic expansion that just leads to a lower temperature. Note that an ideal gas would see no temperature reduction in a JT–valve, and also that gases being warm enough will actually see a temperature increase on such a pressure let-down. Hydrogen at room temperature is a case in point. The process shown is merely to illustrate the overall principle of providing cooling (refrigeration) at a temperature below the available heat sink. To develop such a process the heat exchanger network in combination with compressors and pressure let-downs must be studied. The theoretical basis for this has recently been discussed by Wechsung *et al.* (2011), and they propose new formal procedures for approaching this problem. For a big industrial scale process it is essential to make the most out of optimising the process flowsheet while exploiting

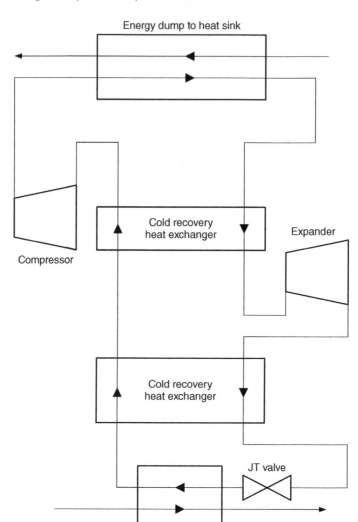

Figure 2.2 *The principle of a refrigeration process. How energy is picked up at a low temperature and disposed of at a higher temperature.*

available machinery and heat exchangers intelligently. There is no point in a fantastic process if the necessary machines cannot be bought. Sometimes compromises must be made between optimum process features and available hardware.

2.2.3.1 The Expander Process and Gas Fractionation

The expander process for recovering C_2+ is illustrated in Figure 2.3. Here the temperature reduction is achieved by expanding the gas in an expansion turbine. Enough liquid is formed to provide 'reflux' for the separation tower. A part of the feed gas may be used to reboil

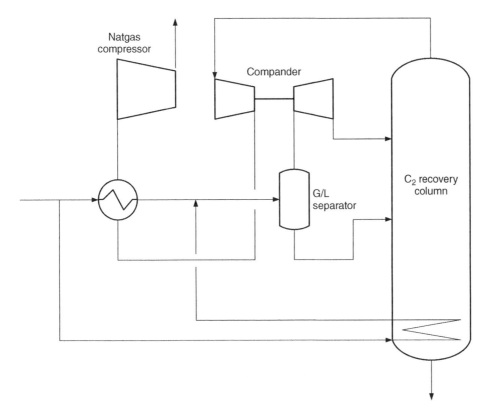

Figure 2.3   *Simple sketch of the expander based C$_2$+ recovery process.*

the recovery column. The expander is used to boost the pressure of the natural gas before it is further compressed for export in a dedicated compressor. Such a process will give more efficient separation of these gases than a straight partial condensation process. It must be understood that the flowsheet sketched in Figure 2.3 is a simplified version for illustration of the principle only. This also goes for the ensuing illustrations. In cryogenics the trick of the trade is to exploit the 'householding' of the 'cold' to the full to squeeze the last bits of efficiency out of the process.

The expander process may be taken a step further as shown in Figure 2.4. Here, it is illustrated in a rough way as to how natural gas may be fractionated into a number of products. How the refrigeration arrangements are made are not shown. Some variations may be proprietary, and others are dependent on the gas composition and what is required in the commercial situation under which the plant is designed. Relative market prices of the components will change over time and also locally since transport costs mean that there is no *one* market as such, but rather a fragmented market pattern.

2.2.3.2   *Ryan–Holmes Process (or Technology)*

The so-called Ryan–Holmes process makes use of liquefied gases to remove CO_2 from natural gas (Holmes *et al.*, 1983). This process provides a solution to break the CO_2-ethane

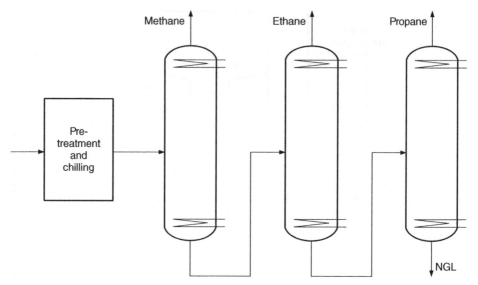

Figure 2.4 *A generic distillation train to recover methane, ethane and propane as overhead products from the 'demethaniser', 'deethaniser' and 'depropaniser' respectively. Residue from the latter is an NGL stream of C_4+.*

azeotrope when desorbing the CO_2. There will be cases when it is the preferred process. It is essentially an add-on feature to the separation trains for natural gas described previously.

Developed in 1979 by Ryan and Holmes, this process has been installed in a number of applications, for example the Chevron Buckeye CO_2 plant (Garner, 2008). The process is described in US patents 4318723 and 4428759 (Holmes and Ryan, 1982; Ryan and O'Brien, 1984) and also by Ryan and O'Brien (1986). It is more correct to refer to the Ryan–Holmes process as a technology rather than a process. This is because many process variations can be made based on the fundamental idea. The fundamental idea is the use of a liquid additive to the process to enable low temperature separation of CH_4 and CO_2 without the latter precipitating as a solid. This additive also helps to break the azeotrope between CO_2 and C_2H_6 and the equilibrium pinch in the CO_2 rich region of the binary CO_2-C_3H_8. To understand the beauty of this it has to be realised that the phase behaviour of CO_2 is rather special in that its liquid form solidifies if its pressure (also partial) falls below 5.18 bar. This will obviously happen if CH_4 is to be purified to a typical pipeline specification of less than 2% CO_2. The use of the additive makes CO_2 much more soluble and prevents such freeze-out. There is a deep understanding of thermodynamics behind the formulation of the Ryan–Holmes technology. They were the first to go this way and there have been an appreciable number of occasions when this technology has been the preferred solution. Being the preferred solution in an application is essentially about giving the best economic result in the end and this includes taking care of local potential problems that could cause downtime or expensive waste handling. Figures 2.5 and 2.6 show a couple of variations of how this technology can be used. These examples were given at a time when there was little interest in recovering C_2H_6 as such, but propane and natural gas liquid (NGL) were interesting products as money makers.

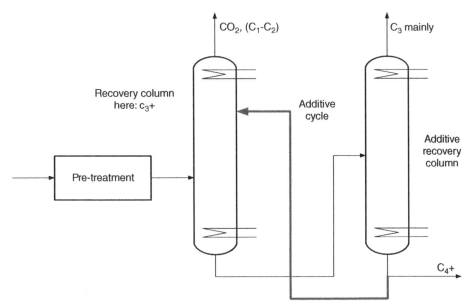

Figure 2.5 *Ryan–Holmes process with C$_3$H$_8$+ recovery and CO$_2$ separated and recovered for EOR purposes. The figure is based on information from Ryan and O'Brien (1986).*

The pretreatment alluded to in Figure 2.5 comprises compression as necessary and water removal. Water is a perpetual problem for low temperature processes since it will precipitate as ice on the process surfaces not the least in heat exchangers. Their patents can be referred to provide insight to the process itself. The additive is typically C$_4$+, and adding in the order of 10–20% of this to the liquid at or near the top of propane recovery (C$_3$+) column gives most of the CO$_2$ solubility effect. Process pressure discussed is in the order of 35 bar. This pressure allows subcritical operation with respect to CH$_4$ and will provide the highest possible temperature to improve CO$_2$ solubility. This implies a temperature of around 185 K in the presence of nearly pure CH$_4$ but if the overhead was nearly pure CO$_2$ the temperature would be more like 273 K. The second, additive recovery column would be operated at a pressure and temperature such that the recycled additive is fed to the first column at a temperature as close as possible to that in this column in order to avoid excessive evaporation.

A more complex variation of the same technology is shown in Figure 2.6. Here, it has also been desirable to separate the CH$_4$ for export as sales gas. This is the main difference from the process variation described previously.

The provision of refrigeration for this process has not been discussed in the patents and presentations found. The columns are shown with just indicated condensers and reboilers. Since the condensers will mostly operate at temperatures below a normal 'heat sink', the cooling must be provided by a heat pump cycle akin to that shown in Figure 2.2 that can transfer the condenser heat duty to the end cooling medium that will be air or cooling water based depending on the location. It must be understood that the refrigeration cycle used in practice will be integrated with the process to maximise energy efficiency. The refrigeration cycle does not need to be a closed one. Reboilers should be driven by heat sources close to

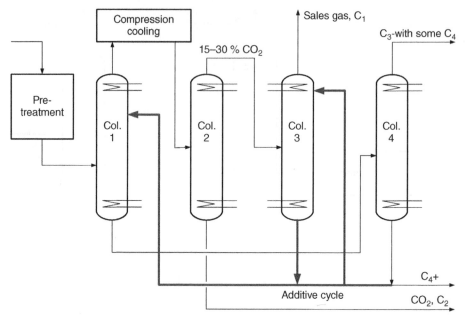

Figure 2.6   *Ryan–Holmes process with sales gas (CH_4) and C_3H_8+ recovery as well as CO_2 separated and recovered for EOR purposes. The figure is based on information from Ryan and O'Brien (1986).*

the process temperature for two reasons. Heat integration energy issues represent the one, and reboiler temperature driving force is the other. A pinch analysis of the process will give the first pointers.

2.2.3.3   Nitrogen Rejection

The rejection of nitrogen from natural gas is necessary when the content of nitrogen becomes too high. This has to do with the sales gas specification that would typically include the gas' heating value and Wobbe number. Sometimes nitrogen is present in the natural gas because it is just there, and other times it is present as a consequence of Enhanced Oil Recovery (EOR) schemes based on nitrogen injection. The processing needs are in principle the same although the relative streams in the process may be different. There are four different process variations normally considered. As usual the choice is dictated by the local conditions and perhaps preferences. Bidding for contracts could also influence the choice since the differences between processes are not all that big. This technology has many similarities to the distillation of air and it is not a surprise to find that the big technology providers from this field are also active in nitrogen rejection, but there are also others. In nitrogen based EOR schemes it is not unusual that a company will offer to provide the nitrogen as an 'over-the-fence' contract and also offer a similar deal for the nitrogen rejection. It allows the field operator, who is unlikely to have experience with such plants, to leave this part of the processing to others. The processing has been discussed in a number of presentations (Alvarez, Hilton and Vines, 1984; Alvarez and Vines, 1985; Browne and Aberle, 1983; Davis and Kindt, 1986; Klotz *et al.*, 1983; Vines, Newton and Peeples, 1982; Vines, 1986). However, for those interested in process details

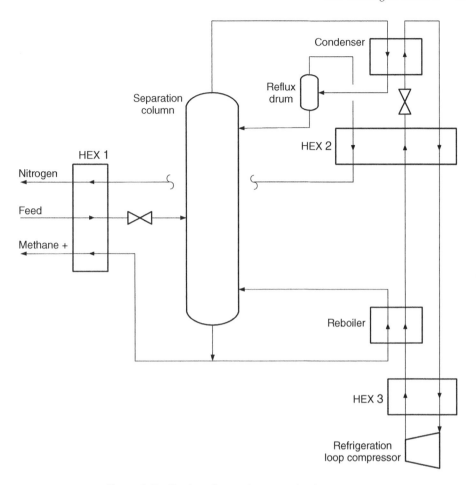

Figure 2.7 *Single column nitrogen rejection process.*

it is better to go to the patent literature where actual examples are usually quoted (Agrawal, 1997; Brostow, Roberts and Geoffrey, 2005; Clare and Oakey, 2008; Davis *et al.*, 1985; Pahade, Maloney and Handley, 1987).

The easiest variation of a nitrogen rejection process to understand conceptually is chosen in Figure 2.7. The heat exchangers are shown as generic blocks, but are likely to be based on plate-fin heat exchangers that represent the dominating technology for this in the air separation industry. The most likely alternative would be spiral wound heat exchangers. Descriptions of these are provided in the books by Kays and London (1984), Hesselgreaves (2001) and Reay (1999).

In the single column process, as is referred to, the feed is first cooled to such a temperature that it is partially liquefied on pressure let-down, and then fed to some middle point in the column. Nitrogen is more volatile than CH_4 and is concentrated in the top of the column while CH_4 and heavier components will be recovered from the bottom. A refrigeration loop is shown, and some energy house-holding of the energy involved is also indicated. The fluid in the refrigeration process is matched to the refrigeration needs. It could be methane, but the choice is open. The refrigeration cycle is shown with a JT valve to provide the lowest temperature, but an expansion turbine may also be integrated as an option.

As in the separation of air, a so-called double column arrangement is also used, and this is illustrated in Figure 2.8. Also here the feed gas is cooled to a very low temperature and let down in pressure to provide a two-phase feed to the bottom of the lower column. Nitrogen will rise to the top of this column where it will be condensed while providing heat to reboil the bottom liquid of the upper column. This is made possible by operating the two columns at different pressures such that the dew point of the nitrogen in the top of the lower is higher than the boiling point of the CH_4+ in the sump of the upper column. The pressure implications may easily be seen from the vapour pressures shown in Figure 2.1. Nitrogen is recovered from the top of the upper column while CH_4+ is coming out from the bottom of the bottom. Both streams recovered are used for chilling the feed gas in the main heat exchanger. There is a need to provide a refrigeration loop to provide reflux in the top of the upper column as depicted in Figure 2.8. This is just one of many ways of doing this and is only included to provide a basic understanding for the principle. The choice of refrigeration fluid is a parameter that may be used to optimise the process. As always, the costs involved, both investment and operational, provide the object function for this exercise. The advantage of the double column technique is a reduction in energy. For this reason these double columns are always used in the air separation industry, except for smaller units when only nitrogen is wanted as a product.

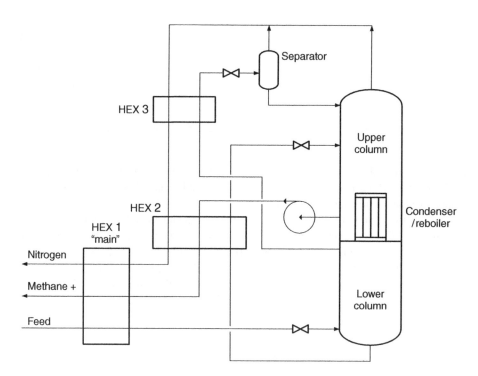

Figure 2.8 *Double column nitrogen rejection process.*

The double column principle has also been further developed to a triple column process. Such a configuration gives further advantages with respect to flexibility and energy. The fourth approach to nitrogen rejection is a dual column system (Air Products, 2009).

2.2.3.4 Controlled Freeze Zone Process

This is a process developed by Exxon that has yet to reach commercialisation. Patent was granted in 1985, which means that development work must have been undertaken prior to this (Valencia and Denton, 1985). It was revealed that it was to be tested in a demonstration plant at Shute Creek in Wyoming in 2010–2011 (Kelley *et al.*, 2011). No information on the results has been found. The process is illustrated in Figure 2.9. It is elegant in its simplicity. In this process CO_2 is allowed to freeze and precipitate under controlled conditions.

According to the example given by Valencia and Denton (1985) the feed is first pre-cooled and expanded somewhat in a JT valve to a pressure of 38 bar (550 psi) and 226 K ($-62°F$). It is then fed to the lower section a couple of trays down from the top. In the bottom part of the column there is a rectification taking place where the CO_2, being the less volatile, will

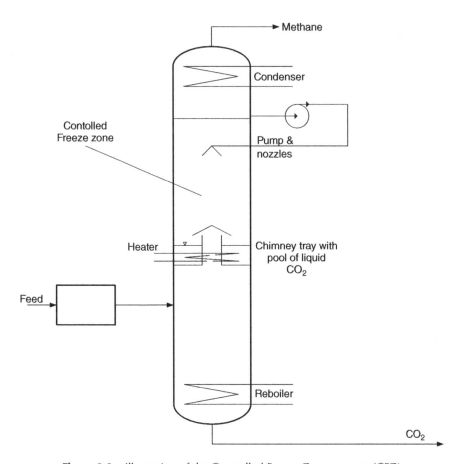

Figure 2.9 *Illustration of the Controlled Freeze Zone process (CFZ).*

predominantly go to the bottom while the methane goes into the next section with some CO_2. Section top temperature is 226 K while the bottom is 275 K. The reboiling will render the CO_2 essentially free of methane, but any heavier components like ethane, and so on will follow CO_2. In the middle spray section the gas meets cold liquid being sprayed into the column from above at a temperature of 186 K. This cold liquid will cool the gas going up to the extent that CO_2 will freeze and precipitate into the pool at the chimney tray that is at 187 K. This pool is heated by the gas coming up from below and by a control heater to ensure that the pool is essentially free of solid CO_2. The gas from the freeze section rises to the tray section above where methane is distilled off. The temperature profile in this section is fairly constant at 184–186 K. Reflux is provided by a condenser to render the methane essentially CO_2 free. Any lighter components like nitrogen will follow the methane.

2.2.3.5   Synthesis Gas Purification for Ammonia Train

The classic process here is the so-called nitrogen wash column. Its use has traditionally been in ammonia trains where the hydrogen synthesis starts with a gasifier where oxygen is used. The nitrogen thus made available from distillation of air is used in due turn in the process when the gas and the nitrogen is cooled to 100 K or so to allow nitrogen to be liquid at the process pressure. Synthesis gas containing methane and argon is fed to the bottom of the cryogenic absorber while liquid nitrogen is fed to the top. The liquid nitrogen washes out the methane and argon from the gas while a suitable amount of nitrogen is evaporated to make the gas leaving the top of the absorber stoichiometrically suitable for ammonia synthesis.

A later development is the so-called Braun purifier (Grotz, 1969). It is associated with an ammonia train variation built for a long time by the company C.F. Braun. Figure 2.10 shows the principle. Here, the synthesis gas is partially cooled before being extracted to do work in an expansion turbine and in this way being used to provide refrigeration for the process. The penalty is that all the synthesis gas is let down in pressure, meaning that the downstream ammonia synthesis compressor gets a lower suction pressure. Before that the synthesis gas is fed to the bottom of an absorber, where the gas is washed by a reflux provided by the absorber bottom stream that is let down in temperature by a JT valve before it is used to cool the absorber top and provide the reflux. Both streams are thereafter piped back to the main heat exchanger for 'cold recovery'.

It has been found that this could be further improved by using the waste stream to provide the refrigeration in the expander (Eimer and Øi, 1992; Øi and Eimer, 1997). This variation is shown in Figure 2.11. It saves energy and to a degree also investment in the ammonia synthesis gas compressor since the suction pressure is higher and suction volume smaller.

There is further use of cryogenics in the ammonia synthesis loop itself. It is desirable to recover hydrogen, and that may be done by cooling the gas down until the accumulated inerts in the form of argon and methane plus excess nitrogen condenses.

2.2.4   LNG Trains

The LNG market in 2012 was 240 million tonnes with about 30% being a spot market. Liquefaction of natural gas has been commercially practised for 50 years in 2014. Capacities in

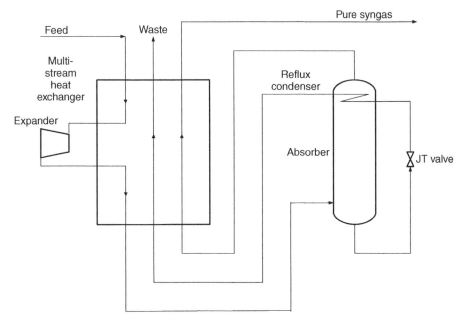

Figure 2.10   *The 'Braun purifier'.*

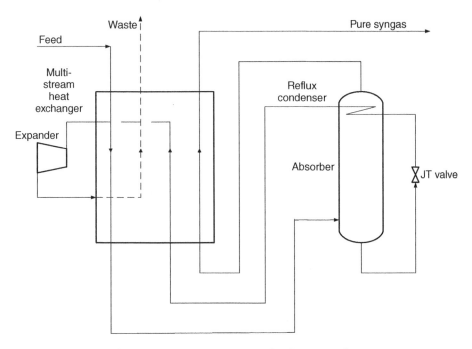

Figure 2.11   *Improvement on the 'Braun purifier'.*

the early days quickly rose to 1 Mt/year and established itself at 3 Mt/y for a period. However, it is a very cost focused business and economics of scale has to be considered. There was a leap in the number of projects in the late 1990s with the Middle East region making its way into the LNG market based on huge availability of gas that had little alternative usage. In this process bigger capacity trains were sought, and trains of 7 Mt/y were proposed. A number of trains for 7.8 Mt/y were built in QATAR, but apart from these sizes have been in the range 3–5.5 Mt/y over the last 10 years, some much smaller (IGU, 2013a). The chase for larger sizes seems to have died. The case for 10–12 Mt/y was also argued but never realised. Implications and feasibilities of such big liquefiers have been discussed by Kaart *et al.* (2007) and Sawchuk *et al.* (2004). Tanker sizes in the last 15 years have been 125 000–150 000 m^3 as the 'normal', up to 180 000 m^3 as large and a few over 200 000 m^3 (IGU, 2013a).

The feed gas to the LNG trains needs to be treated to allow it to be condensed without flow channels being blocked by frozen or very viscous fluids. LNG will eventually be cooled to 111 K (−162°C). This demands upstream removal of water as far as possible, CO_2 is usually removed to 50 ppm or less, while the heavier hydrocarbons can be drained out as they condense on the initial cooling. The actual LNG process is mainly about refrigeration and how to make that as economic as possible. Eventually the liquefied gas must be cooled to its normal boiling point to avoid the need for pressurised LNG tanks. The temperature will be close to the boiling point of methane.

The focal point in a refrigeration process, that an LNG train really is, is to match the temperature–enthalpy curves on the cold and warm side of the heat exchanger system: see Figure 2.12. Thereafter, it is a case of designing the refrigeration system which is really a heat pump dumping the heat released on liquefaction to either cooling water or

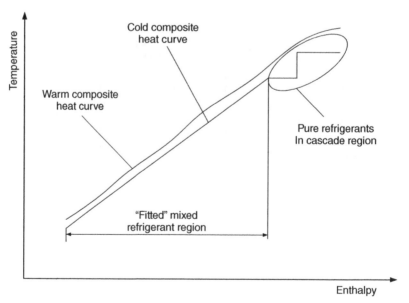

Figure 2.12   *The composite heat curves and effects of 'fitted' mixed refrigerants compared to a cascade of pure refrigerants.*

air like illustrated in principle in Figure 2.2. The natural gas stream to be liquefied is multi-component and as such it does not have a defined condensation temperature. It condenses over a temperature range. If a mixed refrigerant (MR) is made up such that its temperature behaviour matches that of the gas to be liquefied, it is possible to get a good match between the warm and cold composite heat curves. It also helps that the gases are at or above their critical points such that there is no pronounced heat of condensation. Pressures of the process and refrigerant streams are also important. This problem complex has recently been put into a mathematical system as discussed by Wechsung *et al.* (2011).

The dominating liquefaction process is Air Products' so-called C3-MR accounting for almost two-thirds of the name plate capacity (IGU, 2013a). It uses propane as a first stage refrigerant for so-called pre-cooling before the gas is cooled and liquefied by a MR cooling cycle. A key point in this process is a coil-wound heat exchanger (CWHE) that can be made very large, up to 5 m diameter and 55 m long (Air Products, 2013). The C3-MR process is illustrated in Figure 2.13 (Pillarella *et al.*, 2007). The propane is expanded at a number of pressures to provide refrigeration at a number of temperature levels in the pre-cooling stage. This is essentially a cascade system although in one loop with one refrigerant. Both the natural gas to be liquefied and the MR is cooled down by propane. The MR is split into a liquid and a gas stream downstream of the pre-cooling. The liquid is further cooled in the CWHE, but taken aside before the really cold end, expanded and used as refrigerant in the lower section of the CWHE. The MR gas fraction is cooled along with natural gas all the way to the top of the CWHE where it is expanded and used as refrigerant in the top section. The MR is collected from the bottom of the CWHE and returned to the MR compressor. The natural gas coming out of the top of the CWHE must be flashed and stabilised before being piped to a storage tank. Normal boiling point of propane is 231 K, and that is as low as the propane cycle can take the chilling. In the top of the CWHE the gas must be closing in on 111 K that is the end target. The CWHE is really the focal point in this process. It represents a means of making a really big capacity heat exchanger, and a multistream one

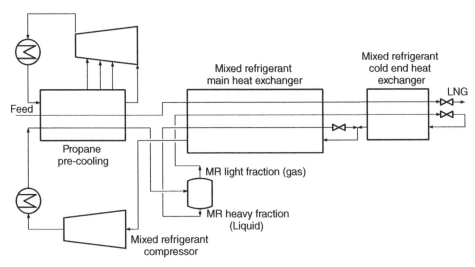

Figure 2.13   *A stylised process flow diagram of Air Products' C3-MR liquefaction process. (Top of the CWHE is to the right).*

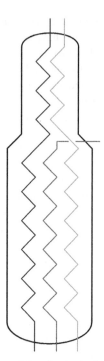

Figure 2.14 *A sketch of a CWHE used for the both the main and cold end heat exchanger in Figure 2.13. The process streams from left to right are natural gas, heavy mixed refrigerant and light mixed refrigerant. The cold and 'less cold' liquefied refrigerants that are used for cooling are fed at the top and at the 'neck' respectively. The zigzag pattern denotes coil-wound tubes. The cooling streams flow downwards in the free space counter-currently to the rising gas streams.*

at that, and is a bit of a cornerstone in the C3-MR process. There is a figure sketch of a CWHE in Figure 2.14.

When a market appeared for bigger capacities, it became clear that there would be bottle-necks in increasing the liquefaction capacity of the process described. One was the size of the CWHE, another aspect was the compressor and its driver for these refrigeration loops. At this time compressors tended to be driven by gas turbines since this was found to give the best economics. Insofar as bigger compressors being available, there would be very few who could offer the necessary size, and in terms of gas turbines from General Electric, there would be a need to go from frame 7 size to frame 9 that had at that stage not been used for such a purpose. ConocoPhillips introduced a cascade process with three refrigeration loops. Air Products went for unloading the C3-MR process by producing 'warm LNG' that was subsequently cooled by an add-on nitrogen liquefaction cycle and could then spread the work load on three compressors as well. The Air Products process is referred to as AP-X and is illustrated in Figure 2.15. The refrigerant of the final cooling loop is nitrogen, a process well proven from many air separation plants. Nitrogen is compressed, chilled by heat exchange, partly expanded in an expander and a JT valve, then used to cool the gas before the still cold nitrogen is sent for 'cold recovery' before being returned to the compressor (Pillarella *et al.*, 2007).

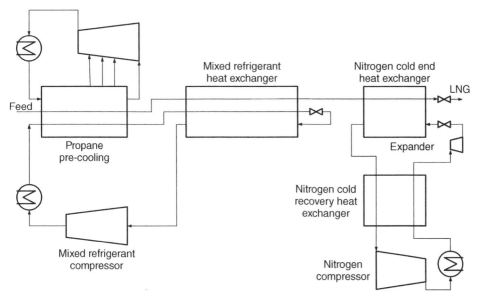

Figure 2.15 *A stylised process flow diagram of Air Products' AP-X liquefaction process. Both the mixed refrigerant and nitrogen cold end heat exchangers are based on the spiral wound technology.*

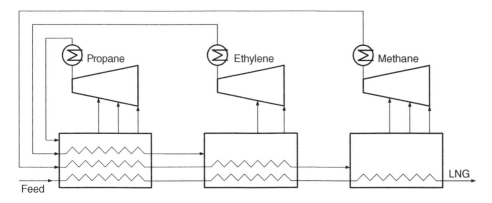

Figure 2.16 *A cascade liquefaction process.*

A cascade process is a little more stringently laid out. There are variations, but in principle, it can be illustrated as in Figure 2.16. The warmest refrigeration loop is used to cool both natural gas and the other refrigerants, then the second warmest is used for cooling both natural gas and refrigerants and so on. The figure shows a simplified flow diagram to just illustrate the principle. No JT valves or expanders are shown. More extensive discussion of such processes may be found in the literature (Meher-Homji *et al.*, 2007; Diocee *et al.*, 2004).

There are a few more processes about, but it is left to the reader to find descriptions of these in the literature. A good starting point is the review article by Lim, Choi, and Moon

(2012). Good sources of information on LNG technology include the regular proceedings of the LNG conferences known as LNG-17 for the last one (IGU, 2013b), proceedings of the World Petroleum Conferences (but less so), and regular LNG sessions at AIChE meetings (American Institute of Chemical Engineers).

The CWHEs have been mentioned. An alternative is the use of multiple blocks of aluminium plate-fin heat exchangers. They are the work horses of the air separation industry and are also used in LNG.

LNG plants built on ships or barges have been targeted for at least 30 years. At long last this seems to come about although nothing is yet built and in operation but a few projects have reached the finance investment decision stage (IGU, 2013a). For such projects technical features like a floating base in motion and offshore transfer of LNG to a tanker are additional challenges.

2.2.5 Membranes

The class of membranes that has been used at a scale interesting for gas treating is polymeric asymmetric membranes with a 'gas tight' skin. By a 'gas tight' skin it is meant that there will be no volumetric flow as such through it. This skin can separate gases because the permeabilities of the gases are different. Using such a selective, permeable membrane may seem to be the ultimate solution to gas separation problems. Unfortunately nature is not that simple. The difference in permeabilities is generally too small to provide absolute separation and the material properties may not tolerate the process conditions.

The driving force for permeation is a difference in partial pressures (or fugacities) between the two sides of the membrane for the gas in question. The side where the gas enters and leaves is the retentate side and the gas held back is referred to as the retentate. The other side where a gas permeates to is the permeate side, and the gas leaving from there is the permeate. The pressure of the permeate limits how much of the gas that may be removed. This may be illustrated by the following simplified case: When removing CO_2 from natural gas, the gas could be fed at 50 bar with a CO_2 concentration of 10%. If the permeate side is 2 bar, the maximum CO_2 removal would be:

$$(50\,\text{bar} \times 0.10 - 2\,\text{bar}) / (50\,\text{bar} \times 0.10) = 0.6 \text{ or } 60\%.$$

In practice the recovery would be significantly lower. Unfortunately CH_4 also permeates the membrane, and its partial pressure on the retentate side would be much higher than that of the CO_2. Hence CH_4 is lost to the permeate. This loss is significant, and often a permeate recycle scheme is needed to improve the process' selectivity and methane yield as shown in Figure 2.17. Such a recycle would imply the use of a compressor and usually a second membrane unit. There is nothing wrong with that and it all boils down to a question of economics, but the simplicity of the membrane principle is lost. The higher the CO_2 content in the feed, and the higher the allowed CO_2 in the retentate, the more competitive membrane technology becomes. The dilution of the permeate with CH_4 would of course improve the driving force for the CO_2 permeation and could provide an element of 'counter-current processing'.

Membrane units also have a limited capacity for feed gas. Scale-up is normally done using parallel units. Hence, its competitiveness is better for smaller plants than for big ones.

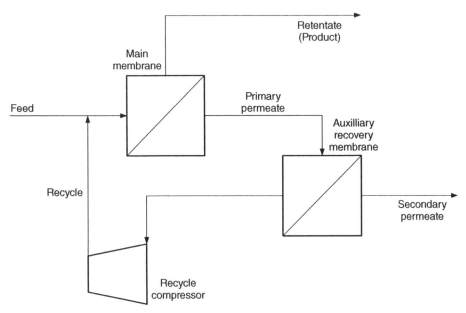

Figure 2.17 *Typical membrane cell arrangement for removing CO₂ from natural gas. Similar configuration is also used to recover hydrogen from an ammonia plant synthesis bleed stream.*

When the natural gas pressure becomes high, it is a challenge to find membrane materials that do not swell prohibitively. Membranes can also plasticise in the presence of CO_2. It is usually required to place a unit upstream of the membrane unit to filter the gas and remove any condensate droplets from the gas. Hydrogen sulfide has so far seemed to be a problem for membranes. A lot of research is going on to bring out forever better membrane materials. There are as yet no permanent truths about limits of membranes. The technology is moving forwards.

Membrane technology is used in gas treating, but it has not become the panacea hoped for by some.

2.3   Sulfur Removal

As already explained there are many reasons why sulfur components need to be removed from natural gas before the gas may be 'shipped'. Even when further processing is to be done at site, sometimes sulfur removal up front will be required. This is very much the case in downstream plants where catalysts are used for processing.

How sulfur components are removed will be subject to local conditions and requirements. Engineers must always strive to make the processing as cost efficient as possible under the constraints imposed. The latter point is important but represents a challenge for the inexperienced, and sometimes even for the experienced engineer. Some guidance may be found is overview papers like that of Tennyson and Schaaf (1977).

When discussing technology, it is essential to define the problem at hand. In this case it involves deciding on what size of field and extent of H_2S problem to analyse. To put

Table 2.3   *Definition of the present 'Reference Case'.*

Gas quantity	1.5 million Sm^3/d or 62 500 Sm^3/h
H_2S content	100 ppm (mol)
Pressure	60 bar

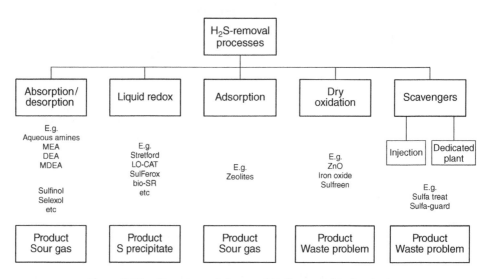

Figure 2.18   *Overview of classes of H_2S removal technologies.*

this into perspective a 'Reference Case' is defined in Table 2.3 and this will be used in the discussions to follow. The size of field would correspond to one of the Gullfaks platforms, or Brage (all North Sea fields), in order of magnitude. However, the choice of H_2S level or any other data is not based on these fields.

There are many technologies available for removing H_2S from natural gas. They may be grouped in four main classes, as illustrated in Figure 2.18. Which option is picked for any *one* situation is influenced mainly by economics, but not without considering past experience, technological back-up and local preference. The choice is seldom clear-cut.

2.3.1   Scavengers

Scavengers are simply chemicals that react irreversibly with the sulfur compounds targeted for removal. When the scavenger is spent, it must be disposed of in an appropriate manner. This usually means that somebody else must be paid to take care of the environmental aspects. The well-known scavenger ZnO will typically be sent back to the vendor for recycling of the zinc. The vendor will also take care of the sulfur.

Scavengers are chemicals that may react either reversibly or irreversibly with H_2S to remove it from the gas. According to the dictionary, all these methods could be referred to as scavengers but this is not done by the practitioners in the field. There is no precise definition of scavengers but the term is used to describe chemicals injected into the gas stream to react with the H_2S, mainly without side reactions. These chemicals may be injected into pipe-lines or applied in special scavenging towers. They are typically used when the H_2S level is below 200 ppm (Dalrymple, Trofe, Leppin, 1994). From the present author's experience in this business, North Sea operators would like to use an alternative method well below this figure, probably an order of magnitude lower. However, it is usually difficult to retrofit new equipment, especially on an offshore platform. Hence the use of scavengers will be stretched beyond their up-front considered normal operating range.

Dalrymple, Trofe, Leppin (1994) divide H_2S scavengers into three categories, which are:

1. Caustic/sodium nitrite solutions
2. Non-regenerable amines/triazines
3. Iron-based.

Schaack and Chan (1989) review the use of such chemicals by Dome Petroleum in Canada. Seeing the variety used by Dome and their comments, there is no single choice that represents an omni-preference. The choice is always situation dependent. Scavengers are particularly interesting locally (in Norway) because they seem to fit the requirements in the North Sea region quite well. They are also known to be in use in the region already.

2.3.2 Adsorption

It is possible to remove H_2S from natural gas by adsorption on a variety of materials, for example zeolite molecular sieves. These are generally used to remove the last traces of water, CO_2 and H_2S upstream of LNG condensation trains. One North Sea platform had a molecular sieves plant to treat for water prior to H_2S appearing. This plant was successfully adjusted to handle the H_2S as well. Adjustments would typically involve modifying the adsorption/desorption cycle. A general process concept is shown in Figure 2.19. Desorption of H_2S, and anything else adsorbed, will involve the use of heat and a stripping gas. This gas is also used as a heat transfer medium. Hence, a low pressure, sour gas is produced as a byproduct.

2.3.3 Direct Oxidation–Liquid Redox Processes

The best known of these processes is probably the Stretford process which uses vanadium, valence state 5, as oxidant. Many problems have been associated with this process in the past (Andrews, 1989). Even if the process may now be understood, the use of vanadium poses environmental problems. The process needs a large absorption tower and a large oxidation tank to regenerate the solution. Sulfur, which is produced from the H_2S, is precipitated directly. It is not an immediate choice for platform use. Newer concepts using the redox principle make use of iron as oxidant. The most publicised is the LO-CAT (presently owned by Merichem). A LO-CAT II has also appeared (Hardison, 1991). SulFerox is a similar technology available from Shell (Buenger and Kushner, 1988). The LO-CAT technology uses iron at the 1500 ppm level while SulFerox uses 2–3%. Both SulFerox and LO-CAT use, for example, KOH to keep the solution slightly alkaline. CO_2 would react with this to some degree. A version of the LO-CAT process is shown in Figure 2.20. The solution is contacted with the sour feed gas in the absorber. Spent absorbent is pumped to the oxidizer for regeneration where oxygen in air is used as oxidant. There is a newer version where the absorber is included in the oxidizer unit. Also a newer concept is the Japanese process

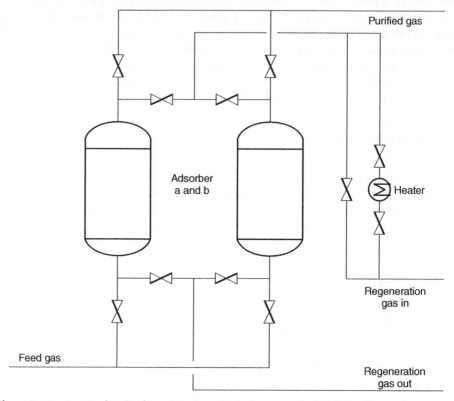

Figure 2.19 *A typical H₂S adsorption plant. The gas assumed used for regeneration could be part of the purified gas, and most likely will be.*

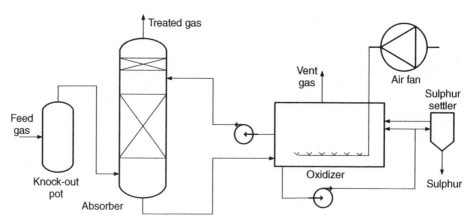

Figure 2.20 *LO-CAT II in so-called conventional configuration. There is an autocirculation version where the absorber is included as part of the oxidizer.*

Bio-SR, which uses an un-chelated form of iron which is regenerated microbially. Shell also offers a microbiological process called Paques (O'Brien *et al.*, 2007). A large bioreactor needed for regeneration reduces its potential. Some aspects of chelation chemistry are discussed by Bedell *et al.* (1988). Successful operation of the liquid redox processes will depend on a proper understanding of the fundamentals. The reference list for installations of these technologies runs into three digits. The present development efforts may well lead to improved processes in the future. Once their increased competitiveness opens a larger market, more development money is likely to be spent giving, hopefully, even better processes. The application range for these redox processes is generally 1–20 tons of sulfur per day and they are applicable to a vast range of gases. A deeper discussion of these processes could easily merit a book of their own. More information is available from the web pages of the licensors and the book by Kohl and Nielsen (1997). There are also the annual conference proceedings from the annual meeting of the GPA and the Laurance Reid Gas Conditioning Conference.

2.3.3.1 Dry Oxidation Processes

A variety of such processes is available, and the book by Kohl and Nielsen (1997) may be referred to for an initial overview. Wood chips impregnated with iron oxide may be used, but could pose a safety problem. A more modern technology is zinc oxide on some carrier. ICI has developed an improved version that shows activity even at 40–50°C, but the efficiency is much higher at elevated temperatures (ICI Katalco, personal communication, 1994; Schaack and Chan, 1989). It features routing of flows to keep vessels in series with the freshest material downstream. One vessel will handle the flow while the other is emptied and refilled. This way of operating will saturate the material fully. Dry oxidation plants can be quite big, in which case handling of large amounts of solids is involved. If this technology were to be used for the typical field defined in Table 2.3, it would mean two vessels 50–100 m^3, and the use of 500 m^3 ZnO per year. Those 500 m^3 must be handled twice, once in and once out, and ZnO represents a waste problem. Three applications based on ICI's Puraspec process are known. One was an onshore treatment of gas from the offshore field, Morecambe Bay, in the Irish Sea region. The H_2S content was reduced from 5 to 2.75 ppm, and some 400 m^3 of the adsorbent had then to be changed annually. The amount of gas treated was in the order of 780 000 Nm^3/h. Another was at the North West Button field where the H_2S in 50 000 Nm^3/h was removed from 10 ppm to spec. The third is installed on the Scott platform, originally used as a temporary measure, to lower H_2S from 11 to 2 ppm while waiting for onshore facilities to come on-stream, but was kept on to allow production of sour wells.

2.3.4 Claus Plants

A Claus plant is not used directly for treating the main gas stream. It is a plant where H_2S is converted to elemental sulfur, and the feed stream is most likely the desorber overhead from an absorption–desorption plant.

2.3.4.1 Classic Claus Plant

The Claus process may be explained with reference to Figure 2.21. The H_2S containing feed gas is fed to the Claus furnace along with air. In this furnace about one-third of the

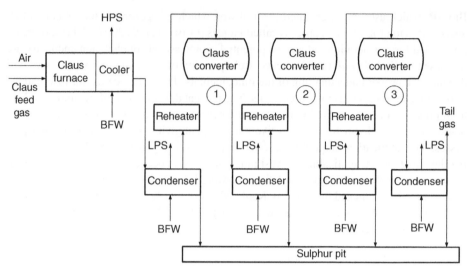

Figure 2.21   *A classic Claus plant configuration. (BFW = Boiler Feed Water, HPS = High Pressure Steam and LPS = Low Pressure Steam.)*

H_2S is converted to SO_2.

$$H_2S + 3/2\, O_2 = H_2O + SO_2. \tag{2.1}$$

This is an exothermic reaction and the temperature is in the region of 1100–1400°C depending on the amount of 'inert gas' present. The heat evolved is recovered to make High Pressure Steam (HPS). There is also the reaction:

$$2\, H_2S + SO_2 = 2\, H_2O + 3\, S. \tag{2.2}$$

Elemental sulfur is condensed as the gas is cooled. It is further cooled while making Low Pressure Steam (LPS). The condensed sulfur is allowed to flow to the sulfur pit where it is collected.

The gas is then reheated and fed to the catalytically based Claus Converter where H_2S reacts with SO_2 to form more elemental sulfur according to Reaction 2.2. As the gas from the converter is cooled, sulfur condenses and is routed to the pit. In a modern plant there are three such converters in series. These are operated at successively lower temperatures to recover more and more sulfur.

It makes economic sense to recover as much sulfur as possible although there is a point of diminishing returns. However, environmental considerations dictate that the recovery must be close to 100%. The 'Tail Gas' is a key point here. There is unconverted H_2S in this tail gas and that must be taken care of. The tail gas treatment is a process field of its own and is discussed further next.

2.3.4.2   Split-Flow Claus Plant

The classic Claus process needs to have a very high concentration of H_2S to make the plant work. The clue is stable operation of the Claus Furnace at a high enough and controllable temperature. A minimum of 50% H_2S is usually seen as the minimum. If the H_2S

concentration is below this, it has become standard practice to let about two-thirds of the feed gas bypass the furnace and let the remaining one-third be fully converted to SO_2 when combusted with air. The two streams are subsequently mixed before the gas goes to the catalytic converters. With the split-flow technology Claus plants have been operated down to 10% or so H_2S.

2.3.4.3   Oxygen Blown Claus Plant

Spiking pure oxygen into the air to increase oxygen concentration is an alternative to a split-flow process. When the oxygen concentration becomes higher than ~28%, special burners are needed. There are several companies offering such technology and there are many installations in operation.

Blowing the furnace with oxygen is also a way to debottleneck a Claus plant because it means that less gas will flow through the equipment.

2.3.4.4   Claus Plant Tail Gas

The most well-known tail-gas treatment technology is the so-called SCOT-plant (Kohl and Nielsen, 1997). (SCOT = Shell Claus Off-gas Treatment). What it essentially does is to convert any sulfur compound like COS, CS_2 to H_2S by using a catalyst and a reducing gas. When this is accomplished, the H_2S is absorbed in an alkanolamine solution from where it is desorbed and recycled to upstream of the absorption plant that feeds the Claus plant. The SCOT absorption process is designed to absorb as little CO_2 as possible. The final tail gas from the SCOT plant may be incinerated or routed to a convenient sink in the gas treating plant. There are alternatives, but the SCOT process serves to illustrate the challenge at hand.

2.3.5   Novelties

Proposals have been made to remove H_2S from gas by microbial means (Sublette, 1990), but no evidence of tests using conditions similar to those existing in gas processing has been found. Some research has been done into manipulating conditions in the reservoir to prevent H_2S from forming. Neither approach will be discussed here since information is lacking for quantifying their potential technical and economic benefits. Alexander and Winnick (1994) discuss a concept where an electrochemical membrane separator is used to treat natural gas. It is claimed that it gives treatment costs around 50% of that projected for aqueous methyldiethanolamine (MDEA) followed by a Claus plant with SCOT tail gas clean-up. It may be an interesting concept for the future, but the reference to Claus/SCOT technology implies that this is targeted at higher H_2S levels than the discussion here. Membranes are also discussed by Meldon and Dutta (1994). Their discussion concerns limitations on membrane technology, an important topic if future research is considered.

2.4   Absorption Process

The absorption/desorption processes based on amines are those most widely used, even for relatively low H_2S levels in the gas to be treated. The rest of the book is dedicated to this process. This is merely a primer for those without prior knowledge.

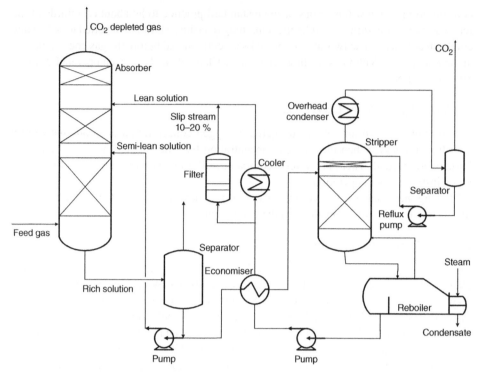

Figure 2.22 *A high pressure absorber with a split-stream configuration where the absorber is fed separately by both a lean and semi-lean absorbent solution. The bulk of the CO_2 is removed by the semi-lean solution.*

There are many such processes available, and they may be divided into three categories according to the absorbent used. They are: (1) physical absorption processes, (2) absorption combined with reversible reaction with carbonate solutions and (3) amine solutions. Kohl and Nielsen (1997) give an introduction to the commonly used processes. Further developments may be traced by studying the annual proceedings from the Laurance Reid Gas Conditioning Conference and the Annual Convention of the GPA. A typical process based on an amine solution is shown in Figure 2.22. These processes have the ability to handle large quantities of H_2S economically, and CO_2 may also be controlled if necessary. The use of co-current absorption using static mixers is an interesting variation when the demand for separation stages is low (see, e.g. Baker and Rogers, 1989). The H_2S is eventually desorbed from the circulating solution. The vent gas is normally treated to recover the sulfur. This is dictated both by economics and environmental constraints. A minimum of sulfur is required to make this exercise profitable. This criterion is not met in the North Sea region, and a sulfur plant with product handling on a platform 100 miles offshore is nobody's dream. Removal of CO_2, which may also be achieved in such plants, is only undertaken occasionally so far. Hence, the conventional large absorbers and desorbers are not immediate choices for H_2S removal.

References

Air Products (2009) Increased Production Through Enhanced Oil Recovery (EOR). Nitrogen Injection and Nitrogen Rejection. Leaflet.

Air Products (2013) Air Products' MCR Coil-Wound Heat Exchangers. Leaflet.

Agrawal, R. (1997) Separation of fluid mixtures in multiple distillation systems. US Patent 5,692,395.

Alexander, S.R. and Winnick, J. (1994) Removal of H_2S from natural gas through an electrochemical membrane separator. *AIChE J.*, **40**, 613–620.

Alvarez, M.R., Hilton, M.F. and Vines, H.L. (1984) Dome's NRU is successfully treating gas from an EOR project. *Oil Gas J.*, **82**, 95–99.

Alvarez, M.R. and Vines, H.L. (1985) Nitrogen rejection/NGL recovery for EOR projects. *Energy Prog.*, **2**, 67–72.

Andrews, E.M. (1989) Stretford Chemistry and Analytical Methods Handbook. GRI Report, Project 5086-253-1275, PB89-173595, GRI/GTI (Gas Technology Institute), Des Plaines, IL.

Baker, J.R., J.A. Rogers, (1989) High efficiency co-current contactors for gas conditioning operations. Proceeding of the Laurance Reid Gas Conditioning Conference, Norman, OK, pp. 249–256.

Bedell, S.A., Kirby, L.H., Buenger, C.W. and McGaugh, M.C. (1988) Chelates' role in gas treating. *Hydrocarbon Process., Int. Ed.*, **2**, 63–66.

Brandani, F. and Ruthven, D.M. (2004) The effect of water on the adsorption of CO_2 and C_3H_8 on type X zeolites. *Ind. Eng. Chem. Res.*, **43**, 8339–8344.

Brostow, A.A., M.J. Roberts, C. Geoffrey, (2005) Nitrogen rejection from condensed natural gas. US Patent 6,978,638.

Browne, L.W., J.L. Aberle, (1983) Flexible, integrated NGL recovery/nitrogen rejection systems. Annual Gas Processors Associated Mtg.; (United States); Conference: 62, Proceedings of the GPA Ann Meeting, pp. 167–175.

Buenger, C.W., D.S. Kushner (1988) The SulFerox process – Plant design considerations. Proceeding of the 67th Annual Convention GPA, Dallas, TX, March 14–16, pp. 56–62.

Campbell, J.M. (1981) *Gas Conditioning and Processing*, Campbell Petroleum Series, vol. 1 and 2, Campbell Petroleum Corporation, Norman, OK.

Cavenati, S., Grnade, C.A. and Rodrigues, A.E. (2004) Adsorption equilibrium of methane, carbon dioxide, and nitrogen on zeolite 13X at high pressures. *J. Chem. Eng. Data*, **49**, 1095–1101.

Clare, S.R., J.D. Oakey, (2008) Nitrogen rejection method and apparatus. US Patent 7,373,790.

Dalrymple, D.A., T.W. Trofe, D. Leppin, (1994) Gas industry assesses new ways to remove small amounts of "H_2S" and "H_2S" scavenger selection index updated. *Oil Gas J.*, **92**, 54–60; 56.

Davis, R.A., D.M. Herron, J.W. Pervier, H.L. Vines, (1985) Nitrogen rejection from natural gas integrated with NGL recovery. US Patent 4,504,295.

Davis, R.A., J.T. Kindt, (1986) Compression fit for nitrogen rejection in EOR projects. AIChE Annual Meeting, November, Paper 3e, Preprint.

Diocee, T.S., P. Hunter, A. Eaton , A. Avidan, (2004) Atlantic LNG train 4 – The world's largest LNG train. LNG-14, Proceedings 14th International Conference and Exhibition on Liquefied Natural Gas, QATAR. Paper PS2-1.

Eimer, D., L.E. Øi, (1992) Method for purification of synthesis gas. PCT Patent WO/1992/012927.

Fu, C. and Gundersen, T., (2013) Recuperative vapour recompression heat pumps in cryogenic air separation processes. *Energy*, **59**, 708–718.

Garner, M. (2008) Chevron Buckeye CO_2 Plant Treating of Natural Gas Using the Ryan/Holmes Separation Process. BSc Senior Project Report, University of Texas at the Permian Basin.

Grotz, B.J., (1969) Hydrocarbon reforming for production of a synthesis gas from which ammonia can be prepared. US Patent 3,442,613.

Hardison, L.C., (1991) Recent developments in acid gas treatment using the autocirculation LO-CAT H_2S approach. AIChE Spring National Meeting, Houston, TX, April 7–11, Session 16c.

Haussen, H. and Linde, H. (1985) *Tieftemperaturtechnik*, 2nd edn, Springer-Verlag.

Hedin, N., Andersson, L., Bergström, L. and Yan, J. (2013) Adsorbents for the post-combustion capture of CO_2 using rapid temperature swing or vacuum swing adsorption. *Appl. Energy*, **104**, 418–433.

Hesselgreaves, J.E. (2001) *Compact Heat Exchangers. Selection, Design and Operation*, Pergamon Press.

Holmes, A.S., Price, B.C., Ryan, J.M. and Styring, R.E. (1983) Pilot tests prove out cryogenic acid-gas/hydrocarbon separation processes. *Oil Gas J.*, **81** (26), 85–86.

Holmes, A.S., J.M. Ryan, (1982) Cryogenic distillative separation of acid gases form methane. US Patent 4,318,723.

IGU (2013a) *International Gas Union World LNG Report, 2013 Edition*, International Gas Union, www.igu.org (accessed 24 March 2014).

IGU (2013b) *17th International Conference and Exhibition on Liquefied Natural Gas (LNG 17)*, International Gas Union, Gas Technology Institute (GTI), International Institute of Refrigeration (IIR), Houston, TX Hosted by American Gas Association April 16–19.

Kaart, S., W. Elion, B. Pek , R.K. Nagelvoort, (2007) A novel design for 10-12 MTPA LNG trains. LNG-15, Proceedings 15th International Conference and Exhibition on Liquefied Natural Gas, Barcelona, Spain, Paper PS2-3.

Kays, W.M. and London, A.L. (1984) *Compact Heat Exchangers*, 3rd edn, Krieger Publishing, Malabar, FL..

Kelley, B.T., Valencia, J.A., Northrop, P.S. and Mart, C.J. (2011) Controlled freeze zone for developing sour gas reserves. *Energy Proc.*, **4**, 824–829.

Kim, J.H., Lee, C.H., Kim, W.S., Lee, J.S., Kim, J.T., Suh, J.K. and Lee, J.M. (2003) Adsorption equilibria of water vapour on alumina, zeolite 13X, and a zeolite X/activated carbon composite. *J. Chem. Eng. Data*, **48**, 137–141.

Klotz, H.C., Copeman, T.W., Vines, H.L. and Miller, E.J. (1983) Gas processing developments. *Hydrocarbon Process., Int. Ed.*, **62**, 84–87.

Kohl, A. and Nielsen, R. (1997) *Gas Purification*, Gulf Publishing.

Korens, N., D. Simbeck, D.J. Wilhelm, (2002) Process Screening Analysis of Alternative Gas Treating and Sulphur Removal for Gasification. Revised Final Report to DOE, Task Order No 739656-00100, Task 2, SFA Pacific, Inc., Mountain View, CA.

L'Air Liquide (Division Scientifique) (1976) *Encyclopedie des Gaz* Gas Encyclopaedia, Elsevier.

Lim, W., Choi, K. and Moon, I. (2012) Current status and perspectives of liquefied natural gas (LNG) plant design. *Ind. Eng. Chem. Res.*, **52**, 3065–3088.

Meher-Homji, C.B., Yates, D., Weyermann, H.P., Masani, K., Ransbarger, W. and Gandhi, S. (2007) Aeroderivative gas turbine drivers for the ConocoPhillips optimized cascade LNG process – world's first application and future potential. LNG-15, Proceedings 15th International Conference and Exhibition on Liquefied Natural Gas, Barcelona, Spain, Paper PS2–6.

Meldon, J.H. and Dutta, A. (1994) Analysis of ultimate permselectivity for H_2S over CO_2 in alkaline solutions. *Chem. Eng. Sci.*, **49**, 689–697.

O'Brien, M., C. Wentworth, A. Lanning, T. Engert, (2007) Shell-Paques bio-desulfurization process directly and selectively removes H_2S from high pressure natural gas – start-up report. Proceeding of the Laurance Reid Gas Conditioning Conference, Norman, OK.

Øi, L.E. and Eimer, D. (1997) New separation opportunities in ammonia train, in *Ammonia Plant Safety and Related Facilities*, vol. **37**, AIChE, New York(based on papers from the 1996 Symposium in Boston)..

Pillarella, M., Y.-N. Liu , J. Petrowski , R. Bower, (2007) The C3MR liquefaction cycle: versatility for a fast growing, ever changing LNG industry. LNG-15, Proceedings 15th International Conference and Exhibition on Liquefied Natural Gas, Barcelona, Spain, Paper PS2-5.

Pahade, R.F., J.J. Maloney, J.R. Handley, (1987) Process to separate nitrogen and methane. US Patent 4,664,686.

Pan, X., Clodic, D. and Toubassy, J. (2013) CO_2 capture by antisublimation process and its technical economic analysis. *GHG Sci. Technol.*, **3**, 8–20.

Reay, D. (1999) *Learning from Experiences with Compact Heat Exchangers*, CADDET, IEA, Analysis Series, vol. **25**, CADDET.

Ryan, J.M., J.V. O'Brien , (1984) Distillative separation employing bottom additives. US Patent 4,428,759.

Ryan, J.M., J.V. O'Brien (1986) Practical application of Ryan/Holmes technology to EOR (CO_2) gas processing. Proceeding of the Laurance Reid Gas Conditioning Conference, OK.

Sawchuk, J., R. Jones, C. Durr , K. Davis , (2004) BP's big green train: Benchmarking next generation LNG plant designs. LNG-14, Proceedings 15th International Conference and Exhibition on Liquefied Natural Gas, QATAR. Paper PS2-4.

Schaack, J.P. and Chan, F. (1989) H_2S scavenging-1. Formaldehyde-methanol, metallic oxide agents head scavengers list processes. *Oil Gas J.*, **87**, 51–55; 81–82; 45–48; 90–91, 4 parts..

Sublette, K.L. (1990) Microbial treatment of sour gases for the removal and oxidation of H_2S. *Gas Sep. Purif.*, **4**, 91–96.

Sun, M.S., Shah, D.B., Xu, H.H. and Talu, O. (1998) Adsorption equilibria of C_1 to C_4 alkanes, CO_2, and SF_6 on silicalite. *J. Phys. Chem. B*, **102**, 1466–1473.

Tennyson, R.N., R.P. Schaaf, (1977) Guidelines can help proper process for gas-treating plants. *Oil Gas J.*, **10**, 78–80 and 85–86.

Valencia, J.A., R.D. Denton, (1985) Method and apparatus for separating carbon dioxide and other acid gases from methane by the use of distillation and a controlled freezing zone. US Patent 4,533,372.

Vines, H.L. (1986) Upgrading natural gas. *Chem. Eng. Prog.*, **82** (11), 46–50.

Vines, H.L., Newton, C.L. and Peeples, C.L. (1982) Nitrogen rejection facilities for inert-gas enhanced recovery projects. *Adv. Cryog. Eng. (US)*, **27**(CONF-810835-), 991–1000.

Wagner, R., B. Judd, (2006) Fundamentals – gas sweetening. Laurance Reid Gas Conditioning Conference, Norman, OK.

Wechsung, A., Aspelund, A., Gundersen, T. and Barton, P.I. (2011) Synthesis of heat exchanger networks at subambient conditions with compression and expansion of process streams. *AIChE J.*, **57**, 2090–2108.

Xiao, P., Zhang, J., Weley, P., Li, G., Singh, R. and Todd, R. (2008) Capture of CO_2 from flue gas streams with zeolite 13X vy vacuum-pressure swing adsorption. *Adsorption*, **14**, 575–582.

3

Rate of Mass Transfer

The objective of this chapter is to provide an introduction to the concept of rate of mass transfer (RMT) and column design. The material will be familiar to chemical engineers but will represent new territory to chemists. This analysis will be discussed in more depth in Chapter 7.

3.1 Introduction

Distillation and absorption processes are separation processes with several stages in series. When dealing with distillation, we are used to assuming that equilibrium is reached on each stage. To compensate for this not being true the term *stage efficiency* is introduced. This will be 50–80% in most cases implying that the initial assumption is at most out by a factor of 2. In absorption of CO_2 this stage efficiency could easily be 10% or even less. Hence the estimate might be for three stages, but 30 would need to be installed. This borders on the ridiculous, but based on empiricism this is somehow made to work. It might seem that an equally good approach would be to say that all CO_2 absorption columns shall be a fixed height.

The reason for these low stage efficiencies is simply the great capacity for CO_2 in chemical absorbents coupled with the fact that this CO_2 must be transferred by a mix of convection and diffusion. This is a slower process than the condensation/boiling seen in distillation.

On this basis it would seem that absorption column design would benefit from a more mechanistically oriented approach. Basing the design on RMT coupled with transfer driving force and contact area seems to be a good start. There will still be empirical parameters involved, but these may be measured directly or indirectly from dedicated experiments, or may be available from information collected from operating columns.

Gas Treating: Absorption Theory and Practice, First Edition. Dag Eimer.
© 2014 John Wiley & Sons, Ltd. Published 2014 by John Wiley & Sons, Ltd.

3.2   The Rate Equation

There will be no transfer of mass between the phases unless there is a driving force to push it. That driving force will be proportional to the deviation from equilibrium between gas and liquid. The mass transfer will also be proportional to the contact area between the two phases. Based on this we can write:

$$(Rate\ of\ mass\ transfer)\ \alpha\ (contact\ area)\,(driving\ force) \tag{3.1}$$

To turn this into an equation we introduce the mass transfer coefficient:

$$(Rate\ of\ mass\ transfer) = (mass\ transfer\ coefficient)\,(contact\ area)\,(driving\ force) \tag{3.2}$$

It is more convenient to write this using standard chemical engineering symbols but first we need to discuss the driving force. Basic knowledge of two-film theory and values at the interface between the two phases are assumed. (This will be discussed in detail in Chapter 7.) In this text all driving forces shall be in terms of concentrations. Other ways of doing it shall be treated as deviations. It will be convenient to make such deviations in many situations. The key to the notation used for various concentrations used in this text is given in Figure 3.1. The component i will be omni-present, but we will need its concentrations in the gas phase in particular, both in the gas bulk and at its interface with the liquid. Furthermore its concentrations will also be needed at the liquid side of the interface and in the liquid bulk.

Using these symbols, we can now write two equivalent equations to describe the RMT of species i.

$$RMT_i = k_G a \left(C_i^{Gb} - C_i^{Gi} \right) \tag{3.3}$$

and

$$RMT_i = k_L^0 a \left(C_i^{Li} - C_i^{Lb} \right) \tag{3.4}$$

where RMT_i is the rate of mass transfer of 'i' per unit volume (mol·m^{-3}·s^{-1}), k_G and k_L are the gas side and liquid side film mass transfer coefficients respectively (m·s^{-1}) and a is the contact area per unit volume (m^2·m^{-3}).

There is nothing fundamentally new to this compared to other textbooks, most of which tend to work with pressures in the gas phase. However, it is emphasised that driving forces shall be in units of concentrations in this text. This is to keep mass transfer coefficients in

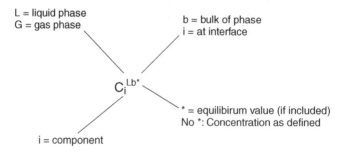

Figure 3.1   *Key to notation for concentrations as used in this text.*

units of velocity (m·s^{-1}), which is in line with the values and units that logically materialise from the dimensionless Sherwood number when this is used. Further discussion of this is deferred to later chapters.

3.3   Co-absorption and/or Simultaneous Desorption

Most textbooks concentrate on one species being absorbed and this is fine for teaching the basics of absorption. In real life all components present will be subject to mass transfer between the phases to some degree. Fortunately a number of the components may be ignored in mass transfer analysis because they contribute little to the mass transfer process and because we are not particularly interested in those components. That may change if we, for example, become interested in studying a trace component.

Some components are likely to be desorbed rather than absorbed. This is usually the case with the absorbent itself since it is not likely to be present in the feed gas. Absorbents tend to have limited volatility and desorption of them can thus most often be ignored. Another case is water that has considerable volatility. If the gas enters with a water content below the column dew point, water will evaporate (i.e. desorb); and since this is an endothermic process, it will influence the temperature profile through its influence on the local heat balance.

Care should be taken in choosing which components are to be studied.

3.4   Convection and Diffusion

In simple distillation analysis it is customary to assume equimolar overflow. This implies that the number of moles leaving the liquid matches the number of moles entering the liquid from the gas. The implication is that the mass transfer is limited to diffusion. There is no convection of gas or liquid towards or from the interface. In absorption this is not the case. The process is arranged such that there will be a net flux of gas to the interface from where one or more of the components will be absorbed in the liquid. Those components that are negligibly absorbed will need to make their way away from the interface back into the bulk of the gas by diffusion. This diffusion is driven by the concentration difference set up by the 'convective sweeping' of these components towards the interface. An effect of this is that the species absorbed will be present in a lower concentration at the interface than in the bulk of the gas. This will slow down the rate of absorption.

3.5   Heat Balance

Absorption is exothermic and desorption is endothermic. In this sense these processes are related to condensation and evaporation, although the mechanisms are different.

It may, or may not, be necessary to run a heat balance when making design calculations for an absorption column. If the absorption is purely physical and there is little of it, it may well be that the heat effect is negligible. It is the designer's call to decide the extent of the design calculations.

For absorption of CO_2 into alkanolamines, there could easily be a 20 K change in temperature along the column, usually with both ends being cooler. Such a temperature change merits the use of a heat balance.

3.6   Axially along the Column

The height of column needed to carry out the separation or treatment at hand is found by integrating the RMT from one end of the column to the other. Focus will be on the key component, or components, if more than one is specified. However, it could be that all the components must be considered. Typically the effect of water mass transfer may have its influence due to the associated heat effect.

An evaluation must be made of which parameters may be considered constant and which parameters must be varied along the column. The specific interfacial area is probably not very dependent on the various axial profiles of parameters. The mass transfer coefficient on the other hand, is likely to be influenced by a number of parameters. Only experience will help making such decisions. However, if the analysis has to be done by computer anyway, which is likely, there is little reason to simplify given that the program code at hand can handle it all within an acceptable time. If the code has limitations, these limitations must be understood. Given that the code is to be custom made, there is a lot of incentive to keep it simple. If a virgin code has to be written, it should be kept to a minimum at a start with code planned for extensions. Designating more parameters as variables axially along the column could be added one by one. Experience has shown that this is the quickest way to the ultimate target.

3.7   Flowsheet Simulators

Flowsheet simulators have column simulators included but most default versions are equilibrium based. ASPEN Plus and a couple of others offer rate based column simulators as an option at considerable extra cost. ASPEN HYSYS is equilibrium based. ProTreat has a rate based absorber. It is not clear from the BR&E's web page what the standard offer with TSWEET is, and it now comes in a package referred to as ProMax. ChemShare is another. There are more on the market, but these are probably the most well known. It is also possible to look for specialised software without full flowsheeting capability. The advantage must be weighed against the extra expense. It must be decided as to how critical the use of a rate based model is to reach the target at hand. If it is a question of establishing a heat and mass balance only, the column models may be played around with until the column designs look reasonable in comparison with what is known about the column size. The equilibrium stages approach should not be discarded all together, in spite of earlier arguments. It might be the only option available in a given situation. The key is to know what you are doing (always).

If it is vital to analyse the column profiles in detail, however, a rate based simulator should be considered. It is also possible to buy these simulations on a consultancy basis. That may be a better approach if the analysis is a one off and there is little or no experience in using such simulators in the organisation.

3.8 Rate versus Equilibrium Approaches

It may be argued that rate based column models only introduce a new empirical parameter in the form of $k_G a$ or $k_L a$ to replace the stage efficiency. That is correct, but it is not the full story. When modelling we should always strive towards models that are based on the underlying and fundamental mechanisms of the processes we are trying to simulate. The stage efficiency does not represent a mechanism, it is more like a symptom. The interfacial area introduced in the rate based model is sound enough but difficult to precisely determine. The mass transfer coefficient on the other hand is a proportionality constant derived from empirical experience although it is possible to develop this concept from diffusion principles and a few assumptions regarding hydrodynamics. In any case it is not possible to separate the mass transfer coefficient and the interfacial area from column data since they are multiplied with each other in the model. However, the mass transfer coefficient may be broken down into further factors that may be related to mass transport phenomena existing in the process. On this basis it is argued that a rate based analysis is to be preferred and it is certainly the modern trend in chemical engineering.

Design of heterogenic catalytic reactors has, for many years, been based on a rate approach, albeit an approach with a lot of empiricism built into it. Such designs are commonly based on the so-called space velocity which may simply be defined as Nm^3 (Nm^3 is normal m^3) of gas per volume of catalyst per unit time, typically $Nm^3 \cdot m^{-3} \cdot h^{-1}$. Since the flow of gas is in 'normal volume', the value to be used will depend on the pressure. Temperature at feed is often specified independently. Different catalysts used for different applications will have different space velocities associated with them. It is not exactly based on process mechanisms but is rate based.

Further Reading

Cussler, E.L. (2009) *Diffusion: Mass Transfer in Fluid Systems*, 3rd edn, Cambridge University Press, Cambridge.

Richardson, J.F., Harker, J.H. and Backhurst, J.R. (2002) *Coulson and Richardson's Chemical Engineering*, 5th edn, vol. **2**, Butterworth-Heinemann.

4.5 Rate versus Equilibrium Approaches

For Further Reading

4

Chemistry in Acid Gas Treating

4.1 Introduction

To understand systems where chemical reactions play an important role, it is essential to have a fundamental understanding of the chemical reactions involved. Since theories tend to evolve over time, and CO_2, H_2S and amines have been studied fundamentally for over 80 years, it is also necessary to know early theories in order to interpret and use old data correctly. This chapter aims to provide this insight.

In gas treatment both gases and liquids are involved. The liquid, often referred to as *absorbent*, is often a mixture. This mixture could have any number of components. Usually one or perhaps two are chemically active toward the acid gases. The rest is (usually) referred to as the *solvent*. Both aqueous and nonaqueous solvents may be encountered. Chemicals used like alkanolamines are also in principle solvents and they influence the properties of the liquid when mixed with other solvents including water. For concentrated solutions this must be understood and accounted for.

The reader of this book is more likely to be a chemical engineer than a chemist. We chemical engineers (including myself) do not normally have very extensive training in organic chemistry and are therefore a little lost when intricate considerations need to be made. Hence, it is a good idea to start the discussion of gas treating chemistry with a small primer in organic chemistry. If you happen to be a chemist, feel free to skim or skip this material.

Absorbents where base like molecules are present are common in acid gas removal. Hence, an understanding of what goes on with regard to chemistry is a must unless the processes are to be treated like black boxes. Maybe such boxes represent a nice idea at first consideration but you would be helpless when there is a need for the least bit of troubleshooting. So here goes.

The field of chemistry of acid gases and amines follow the same principles as the organic chemistry at large. The laws of nature are no different. It is therefore a good idea to discuss the relevant basic chemical principles as a start to set the scenery where the acid gas and the basic absorbents must fit in. More treatises of this, or these, theme(s) may

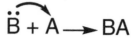

Figure 4.1 *How electron movement in reaction mechanisms is depicted. Note the arrow starting from where the electrons are and pointing to where they are set to move.*

be found in good books on organic chemistry. (The books by Carey and Sundberg (1990) and McMurry (1992) are good examples.) There are four types of reactions normally met in organic chemistry. Addition reactions are where something is being added to a molecule. This typically covers CO_2 reacting with an amine. There are elimination reactions where a part of a molecule is shed. An example would be CO_2 splitting off from the carbamate when CO_2 is desorbed. There are also substitution reactions and rearrangements but these are not important to the basic absorption–desorption process. The field of degradation reactions may be different. Chemical bonds can split symmetrically with one electron going each way (homolytic split) or it may split asymmetrically with both electrons going one way leaving charged groups of atoms (heterolytic split). It is also useful to know when presented with reaction mechanisms that the convention is to draw arrows starting from the site where the electrons are and make the arrow point to the target site, as demonstrated in Figure 4.1.

As will be remembered from basic organic chemistry, the electrons will not be distributed evenly between two atoms in a bond unless they exert the same attraction to them. In the chemical systems met in gas treating there will be differences in the attraction of electrons from the different atoms like C, N and O when present. They exhibit different degrees of electronegativity. We talk about nucleophiles (that are inherently negatively charged) and electrophiles (that are inherently positively charged) when we move to the world of molecules. The CO_2 molecule has both C and O atoms in it, and in this case there is a tendency for the electrons to flock to the Os which are stronger electrophiles than the C. This makes the molecule polarised. This polarity is important for how this molecule engages in chemical reactions. The magnitude of such charges in a molecule is also influenced by other atoms, or groups of atoms, in a molecule. This will become evident when we discuss the alkanolamines used in absorption a little later. Most organic molecules are actually electrically neutral, it just so happens that the chemistry of absorption of acid gases are mainly about polar reactions. This is, of course, no chance happening. It is due to the polar nature of the acid gas molecules.

In polar reactions, not to be confused with ionic reactions, charged centres will approach each other. Negative charges will seek positively charged parts of a molecule and positive charges will seek sites with a surplus of electrons, that is, a negatively charged site. The obvious example is a CO_2 molecule being attracted by the electron surplus associated with the N in the amine group in an alkanolamine. In organic chemistry this is referred to as 'functional groups'. Such groups may be an alcohol group like –OH, an amine group like –NH_2, or a methyl group like –CH_3. These are the main groups we will meet in gas treating, but there are many more. The acid gases themselves are looked at individually. Figure 4.2 shows how these functional groups are polarised. It is the negatively charged N-atom in the amine that is the site of attraction for the acid molecules

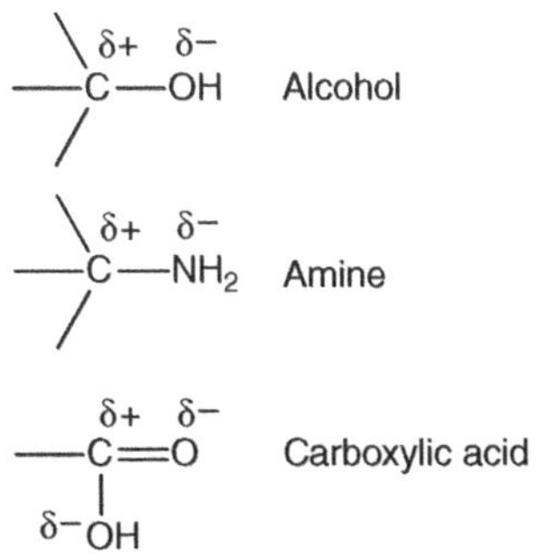

Figure 4.2 *Polarities of selected functional groups.*

when they are absorbed. The OH-group in the alkanolamine is there to make it more soluble in water and to lower its vapour pressure, but in principle it could also be placed on a site in the amine such that it will influence the negative charge on the N-atom. Properties of groups are influenced by others in the vicinity. This is exploited in so-called sterically hindered amines.

It may be timely at this point to remind ourselves about the concept of acids and bases. In the 1880s Svante Arrhenius (Swedish chemist) developed his theory of ionisation of aqueous solutions. He allocated acidic properties to the presence of H^+ ions and basic/alkaline properties to the presence of OH^- ions. This may be referred to as the *water-ions definition*, often thought of as the classic definition of acid and bases.

A further development in 1923 was the Brønsted–Lowry definitions of acids and bases (Johannes Nicolaus Brønsted was a Danish chemist, 1879–1947; Thomas Martin Lowry was British, 1874–1936):

- An acid is any molecule or ion that can act as a proton donor.
- A base is any molecule or ion that can act as a proton acceptor.

However, many molecules do not have a proton to donate because there is no 'H' in the molecule. Many of these molecules are still acidic in character. The acid–base picture was extended by Lewis (Gilbert Newton Lewis was an American chemist, 1875–1946):

- An acid is a molecule or an ion that can accept one or more electron pairs.
- A base is a molecule or an ion that can donate one or more electron pairs.

This last definition is very much a key to organic chemistry. Lewis' definition also came in 1923. Both the Brønsted–Lowry and the Lewis definitions are in active use today.

4.2 'Chemistry'

What makes a chemical reaction take place? There has to be a reduction in Gibbs free energy of a system if a chemical reaction is to proceed. On a stoichiometric level the sum of Gibbs free energy of formation for all the reactants and products must show a negative value for a reaction to proceed as summarised by Equation 4.1

$$\Delta G = \sum_{products} \upsilon \Delta G^f - \sum_{reactants} \upsilon \Delta G^f \qquad (4.1)$$

Here, G is the Gibbs free energy, υ is the stoichiometric coefficient and superscript f indicates the value of formation. The higher the difference, the more the reaction will proceed before it is halted by its equilibrium condition being approached. The reactions between acid gases and alkanolamines will proceed spontaneously (not to be confused with 'instantaneously') at room temperature. Other reactions will not proceed until heated up to a higher temperature. This is fortunate as most amine degradation reactions that could be very damaging for the process will not take place to a significant extent until we are well above the temperatures normally met in this process. The traditional way of depicting the extent of the reaction and the Gibbs free energy relation is shown in Figure 4.3(a,b). That extra energy that needs to be overcome to get under way is referred to as the *activation energy*. There are often intermediates in a reaction path. Figure 4.3(b) shows how this is depicted in the Gibbs free energy versus reaction progress diagram.

The energy needed to break a bond, or that made available when a bond is formed, is referred to as the *bond dissociation energy*. Tabulations of such energies form part of the toolbox for the organic chemist and provide some guidance as to the reactivity of molecules. This is, however, not a fully predictable method, merely a guide. A few relevant bond energies are shown in Table 4.1. The energies are those needed to break the bond indicated. The higher the bond strength, the more energy it takes to break it. How bonds appear in commonly used amines may be seen from Figure 4.4.

These considerations say nothing about how fast the reactions will proceed. On a macroscopic level it is common to describe reaction rates by:

$$r = k[reactant\ 1]^\varsigma [reactant\ 2]^\varsigma \qquad (4.2)$$

where k is a constant, referred to as the *kinetic constant* for the reaction, the brackets denote the concentrations of the reactants specified inside, and ς is the reaction order with respect to the reactant. The orders need not be the same for all reactants. Two reactants are indicated

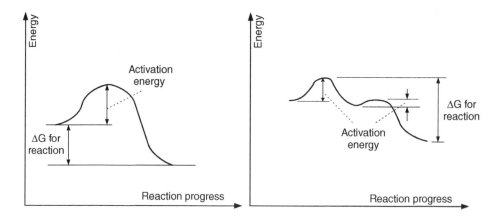

Figure 4.3   *(a,b) Relation between reaction progress and Gibbs free energy.*

Table 4.1 Bond dissociation energies.

Bond	Strength (kJ/mol)	Bond	Strength (kJ/mol)
C–C	339	CH_3–H	435
C–H	410	C_2H_5–H	410
C–O	331	CH_3–OH	381
C=O	724	C_2H_5–OH	381
C–N	276	CH_3–CH_3	368
N–H	385	C_2H_5–CH_3	356
O–H	456	CH_3–NH_2	335
		HO–H	498
		NH_2–H	431

Extracted from McMurry (1992) and Carey and Sundberg (1990).

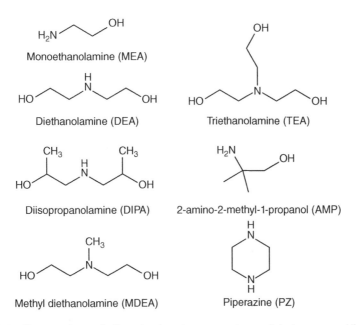

Monoethanolamine (MEA)

Diethanolamine (DEA)

Triethanolamine (TEA)

Diisopropanolamine (DIPA)

2-amino-2-methyl-1-propanol (AMP)

Methyl diethanolamine (MDEA)

Piperazine (PZ)

Figure 4.4 *Commonly used alkanolamines in gas treating and their structural formulae.*

here, but any number might be involved. However, if more than two molecules are directly involved, the reaction is likely to be slow as the probability of three or more molecules meeting in the right configuration is much lower than for just two molecules. It is possible to determine reaction orders from measurements of chemical kinetics. Such a reaction order might be 'broken', that is, not a whole number. Clearly this is not physical in the sense that a fraction of a molecule participates. In such situations there will be a reaction path involving more than one step, either in series, in parallel or both. To determine the reaction mechanism on a molecular level further measurements must be made. Such measurements would typically involve determination of intermediates. Molecular modelling could also play a role.

Moving on from the core fundamentals of chemical reactions it is quickly realised that reactions do not in practice take place without other molecules around. A gas phase reaction may do this in theory but reactions in the liquid phase take place with a solvent surrounding the reactants. This adds another layer of complexity. The traditional amine absorbent is present in an aqueous solution in a concentration in the order of 10% (mol), that means that 1 in 10 molecules in the solution is an amine. Water as a liquid is polar, more so than most solvents.

In the water molecule Hs are slightly positively charged while the Os are negatively charged. It is also split into ions, a process that is referred to as *autoprotolysis*,

$$H_2O = H^+ + OH^- \tag{4.3}$$

and its equilibrium constant (i.e. its autoprotolysis constant) is relatively high compared to other solvents.

$$K'_{ap} = \frac{[H^+][OH^-]}{[H_2O]} \tag{4.4}$$

It is customary to report this as:

$$K_{ap} = K'_{ap}[H_2O] \tag{4.5}$$

$K_{ap} = 10^{-14}$, or as often reported: $pK_{ap} = -\log(10^{-14}) = 14 = pK_w$ in the notation more familiar to most (see Figure 5.2). The amine molecules are also polar and they acquire, to an extent, a proton in water. Alkanolamines will acquire positive charges from water that is protonated. This protonation reaction may be written with monoethanolamine (MEA) as an example. It is customary to write it in the form of:

$$MEAH^+ = MEA + H^+ \tag{4.6}$$

The protonated MEA is here an acid that can shed an H^+. This way of writing the protonation reaction allows logical reporting of its basicity as 'acid strength'.

$$K_a = \frac{[MEA][H^+]}{[MEAH^+]} \tag{4.7}$$

It may seem a little illogical at first, but it has become the established practice. The basicity relation would be, based on reaction:

$$MEA + H_2O = MEAH + +OH- \tag{4.8}$$

$$K'_b = \frac{[MEAH^+][OH^-]}{[MEA][H_2O]} \tag{4.9}$$

The fundamental information is the same. There is an easy relation between acid and basic strengths that is simply quoted here, whereas above $K_b = K'_b[H_2O]$:

$$K_{ap} = K_a K_b \tag{4.10}$$

Since *pK* notation is logarithmic, this could also be written:

$$pK_w = pK_{ap} = pK_a + pK_b \qquad (4.10a)$$

In this environment there is plenty of opportunity for engaging in hydrogen bonding. The amine molecules are solvated, which means that such a molecule is surrounded by solvent molecules that are attached to the amine molecule by hydrogen bonds such that a 'shell' of solvent molecules exists around the amine molecule. When the solvent is water, this solvation may be referred to as *hydration*. This is illustrated qualitatively in Figure 4.5. The solvation shell must be broken before the amine molecule is available to react with another molecule. It would be expected that the rate of reaction is different when the solvent is changed. This is confirmed by investigations made by a number of research teams (Alvarez-Fuster *et al.*, 1981; Kadiwala, Rayer and Henni, 2012; Sada, Kumazawa and Han, 1985; Versteeg, van Dijk and van Swaaij, 1996). There is also energy associated with this solvation process. As always in chemistry a process will not spontaneously happen unless the system moves to a level of lower energy and some heat evolvement would be expected when a molecule is solvated. To break the solvation shell, energy must by analogy be provided. This process also takes time and its kinetic constants will differ from solvent to solvent as the strength of the solvation bonding will change (Table 4.2).

Steric effects are common in organic chemistry. Stereochemistry is a field of its own. There are also steric effects associated with the amine molecules. In this corner of the

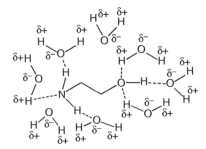

Figure 4.5 *Solvation (or hydration) of MEA as a qualitative example. Illustrated in two dimensions (2D) whereas the solvation is, of course, active in all three dimensions.*

Table 4.2 *Heats of solution in infinitely dilute solutions. (Maham et al., 1997)*

Amine	Amine in water (kJ/mol)	Water in amine (kJ/mol)
MEA	−12.3	−6.4
DEA	−14.5	−4.4
TEA	−14.1	−0.7
MDEA	−20.3	−5.4

Values used are averages of their experimental and literature values.

organic chemistry this is exploited to engineer the amine group's ability to engage in reactions with CO_2. The examples described have to do with atoms or functional groups being such positioned on the amine molecule that it gets in the way of CO_2 molecules that would like to react with the amine. This is usually done by placing a big functional group like methyl or ethyl, and so on, on the C-atom (often referred to as the α-atom or α-carbon) that is bonded to the amine's N-atom. There is a class of absorbent amines referred to as sterically hindered amines that shows this phenomenon. Members of this group of amines are not able to form stable carbamates with CO_2 and need, like the tertiary amines, to hydrolyse CO_2 to make bicarbonates (or carbonates) to bind CO_2. The amino methyl propanol (AMP) molecule in Figure 4.4 is an example of such an amine.

These groups will have properties like ions, and are indeed ions, when it comes to reacting with H_2S. H_2S, being a Brønsted acid, will ionise and provide a negatively charged ion in the form of HS^- to react with the amine ion. This is an ionic reaction, and such reactions take place at an infinite rate. Further discussion of this, and its implications, will be returned to.

The foregoing is sufficient to deal with the major acid gases CO_2 and H_2S. To deal with the other (minor) sulfur containing gases like CS_2, COS and mercaptans (thiols) a further approach to reaction basics is necessary. The ideas of Lewis acids and bases are still key concepts, but on top of that there is a concept introduced by Pearson (1963) that has found its way into the chemistry literature (Carey and Sundberg, 1990). A molecule can be polarised, as already discussed. This polarisability deals with how electrons in a molecule react to charges near them. It is useful to introduce two semi-qualitative terms referred to as *hardness* and *softness*. They describe the ease or difficulty involved with attaining the charge polarisation in a molecule. It also deals with electron density. We could, in the Lewis sense, talk about hard and soft acids and bases. The concept as such is introduced here as an introduction to the basic concepts but it will be further discussed when we discuss the chemistry of the minor sulfur containing gases.

To explain the different chemical behaviour of the minors from the majors it is useful to go into further detail with respect to Lewis acids and bases. Pearson (1963) introduced the concept of hard and soft acids and hard and soft bases. A base or an acid is considered hard if the polarised 'charges' are small, as in being close together. They are soft if they are big with 'charges' spread thinly. The majors, CO_2 and H_2S, are rated as hard acids. The minors are seen as soft (Richardson and O'Connell, 1975). Pearson also put forward the general principle that soft acids prefer to coordinate with soft bases while hard acids prefer to coordinate with hard bases and vice versa (Pearson and Songstad, 1967). The standard alkanolamines are rated as hard bases and so is water to the extent that water is involved.

Basically the absorption of CO_2 and H_2S in alkaline solutions (based on amines) is mostly about acid–base chemistry. This approach is emphasised in the discussions that follow, first by discussing the acid character of these compounds, and then by discussing their reactions with amines.

Zwitterions as a concept was brought into the gas treating chemistry by Danckwerts (1979) based on work by Caplow (1968). There is an unsettled discussion whether these take part in the reaction mechanism between CO_2 and the alkanolamines or not. Zwitterions as such are certainly an item in organic chemistry. The word has German roots and simply refers to a molecule being a hybrid one, in this case a molecule that has both acid and basic groups (ionised) in the same molecule. Such a situation is common with amino acids.

Although the relevant absorption systems were used earlier to provide data and concrete information in the previous material, it is now time to become really specific and discuss the researched and industrially used systems. Included in this discussion are various concepts and hypotheses/theories described in the literature. There is no universally agreed reaction mechanism for the amine absorbents except for H_2S. For chemical engineering purposes the important information comprises the stoichiometry of reactions to estimates of mass balances and the reaction kinetics because the latter can influence the mass transfer rates. The various ways of relating to reaction kinetics including explanations of reaction mechanisms are described one by one. It is important to keep track of the relation between measured data and the kinetic expressions where they belong.

4.3 Acid Character of CO_2 and H_2S

When delving into the chemistry of these compounds, it is necessary to make use of the concepts for acids due to Lewis and Brønsted respectively. (The Brønsted concept is often referred to as the *Brønsted–Lowry theory*, but here only Brønsted will be used, for convenience.)

H_2S is a Brønsted acid, which means it engages in direct proton transfer reactions. The chemistry is quite simply $H_2S + B = BH^+ + HS^-$ where B is any base, for example a Brønsted base like an amine or even H_2O. This reaction is extremely fast.

In solution H_2S is diprotic and could therefore undergo two dissociation steps:

$$1st : H_2S \leftrightarrow H^+ + HS^- \tag{4.11}$$

$$2nd : HS^- \leftrightarrow H^+ + S^{2-} \tag{4.12}$$

These reactions do not involve the solvent as such since the solvent does not take part in the reactions. Reactions 4.11–4.12 will, in principle, take place in any solvent. The nature of the solvent will, however, influence the reaction equilibria through the activity coefficients. It is worth noting that HS^- is also a Brønsted acid in Reaction 4.12. This dissociation of HS^- will, however, only be significant in very strongly basic solutions.

The dissociation equilibrium constant of Reaction 4.12 is of the same order as the equilibrium constant for the dissociation of water. This implies that HS^- dissociation will only be significant in strong alkali solutions and may be neglected when dealing with the normal alkanolamine solutions. The dissociation constants of H_2S and CO_2 are shown in Figure 4.6 as a function of temperature. The minimum values of pK_a with respect to temperature are as reported in the reference.

CO_2 is a Lewis acid, which means it accepts electron pairs from an electron donor; for example, a Lewis base. On absorption in water CO_2 is hydrolysed in a reaction where H_2O acts as a Lewis base:

$$CO_2 + H_2O = H_2CO_3 \tag{4.13}$$

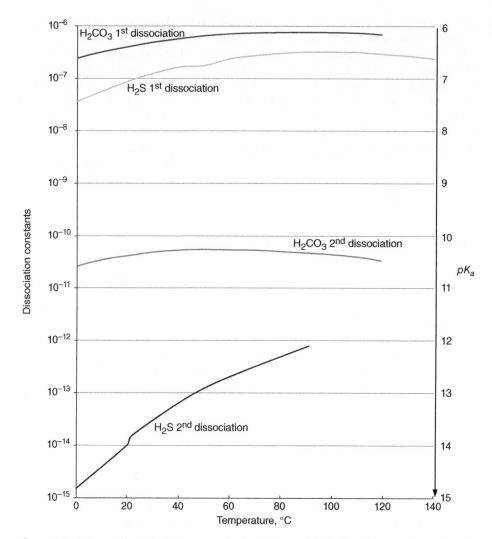

Figure 4.6 *Dissociation equilibrium constants of CO_2 and H_2S. Based on data from Astarita et al. (1983).*

or

$$O :: C :: O + H : O : H = O :: C : O : H$$

$$,,$$

$$O : H \qquad (4.14)$$

The equilibrium of this reaction is such that more than 99% is present as free CO_2 (i.e. $CO_2(aq)$), leaving less than 1% bound in the form of H_2CO_3 (Mahan, 1965, p. 494). The diprotic carbonic acid dissociates according to:

$$1\text{st} : H_2CO_3 \leftrightarrow H^+ + HCO_3^- \tag{4.15}$$

$$2\text{nd} : HCO_3^- \leftrightarrow H^+ + CO_3^{2-} \tag{4.16}$$

The second dissociation only takes place in strongly basic solutions (just like H_2S). Dissociation constants are given in Figure 4.6. Both H_2CO_3 and HCO_3^- are Brønsted acids. Any Brønsted acid is also a Lewis acid, but not vice versa.

Water is a Lewis base also in the molecular form, H_2O. It is known that the reaction rate even between OH^- and CO_2 is slow. The reaction between CO_2 and molecular water is even slower (Astarita *et al.*, 1983, p. 210). Actually H_2O can behave as both a base and an acid since it is an amphiprotic neutral solvent, see Figure 5.1.

Having discussed the basic principles of chemistry and the acid character of CO_2 and H_2S, it is time to look at what is said in the literature with respect to the specific chemistry. There are a few concepts about. It is elected to start with the chemistry of the alkanolamine solutions.

4.4   The H_2S Chemistry with any Alkanolamine

The chemistry of H_2S has essentially already been dealt with when describing its acidic character. Since it is a Brønsted acid and can split off a proton directly, it will react directly with any base according to:

$$H_2S + B = BH^+ + HS^- \tag{4.17}$$

As already stated. B could be any alkanolamine RNH_2, R_2NH or R_3N. It is an ionic reaction and it is, hence, instantaneous. The reaction is the same no matter what type of amine it meets, or any other base for that matter. H_2S reacts directly also with tertiary amines, or amines where the zwitterion formation would be hindered.

4.5   Chemistry of CO_2 with Primary and Secondary Alkanolamines

The chemistry of CO_2 with primary, secondary and tertiary amines in aqueous solutions has been reviewed in the literature a number of times, notably Versteeg and van Swaaij (1988a,b; 1996), Crooks and Donnellan (1989) and Aboudheir *et al.* (2003). It is necessary to treat the chemistry of primary and secondary amines separately from the chemistry of tertiary amines since the reaction mechanisms are different, as implicit in the discussion of steric effects earlier, and as will become evident from the following discussion. The reason for the difference stems from the lack of hydrogen atoms attached to the nitrogen in tertiary amines. There is an unresolved conflict of views with respect to the reaction mechanisms for primary and secondary amines and CO_2. It is reasonable to say that most of the work done has been carried out by chemical engineers. Most of the emphasis has thus been on reaction kinetics and whatever these can tell us. Although chemists have probably been consulted, a rigorous study of the reaction mechanisms by chemists would be desirable. As the situation is, it is prudent not to draw categorical conclusions regarding the chemical mechanisms involved. Options for interpretation should be kept open. The present text aims to give accounts of proposed mechanisms and also to summarise older work in order to make the use of these older papers easier.

A zwitterion mechanism proposed by Caplow in 1968 and revitalised/extended to alkanolamines by Danckwerts in 1979 is the currently accepted mechanism. This mechanism is, for instance, supported in the Versteeg and van Swaaij (1988a). However, Aboudheir *et al.* (2003) claim that only the termolecular mechanism due to Crooks and Donnellan (1989) can explain their kinetic data. The ability of the reaction mechanism to allow a derivation of chemical kinetics that may explain fractional reaction orders with respect to the amine is a key issue in this discussion. From a chemistry point of view spectroscopic identification of intermediates will also be needed before this discussion may be landed.

4.5.1   Zwitterion Mechanism

According to the zwitterion mechanism, the reaction, or reactions, start with a direct reaction between CO_2 and the amine to form the zwitterion, an electrophilic addition reaction (McMurry, 1992), and this zwitterion reacts with a base B to lose its proton as shown below. Zwitterions are further explained by Albert and Serjeant (1984).

$$CO_2 + R_1R_2NH \leftrightarrow R_1R_2N^+HCOO^- \tag{4.18}$$

$$R_1R_2N^+HCOO^- + B \leftrightarrow BH + +R_1R_2NCOO^- \tag{4.19}$$

Let k_2 be the forward rate constant and k_{-1} (signifying second and first order reactions) be the reverse rate constant of Reaction 4.18, and furthermore, let k_b be the forward rate constant of Reaction 4.19. The base B can be any base in the solution, in aqueous solution the bases are R_1R_2NH, OH^- and H_2O. (Hence, there really ought to be three Reactions 4.20a–c.) The zwitterion concept leads to the following expression for the reaction rate:

$$r_{CO_2} = \frac{k_2\left[CO_2\right]\left[R_1R_2NH\right]}{1 + \dfrac{k_{-1}}{\sum k_b\,[B]}} \tag{4.20}$$

The mathematics behind this expression are straightforward once a pseudo steady state for the zwitterion concentration is assumed. The starting point is that:

$$r_{CO_2} = k_2\left[CO_2\right]\left[R_1R_2NH\right] - k_{-1}\,[Z] \tag{4.21}$$

And since the rate of formation through Equation 4.18 and consumption through the Equation 4.19(a–c) of the zwitterion (Z) is equal, it follows that:

$$k_2\left[CO_2\right]\left[R_1R_2NH\right] - k_{-1}\,[Z] = [Z]\left\{\sum k_b\,[B]\right\} \tag{4.22}$$

Equation 4.22 may be rearranged to:

$$[Z] = \frac{k_2\left[CO_2\right]\left[R_1R_2NH\right]}{\left\{\sum k_b\,[B]\right\} + k_{-1}} \tag{4.22a}$$

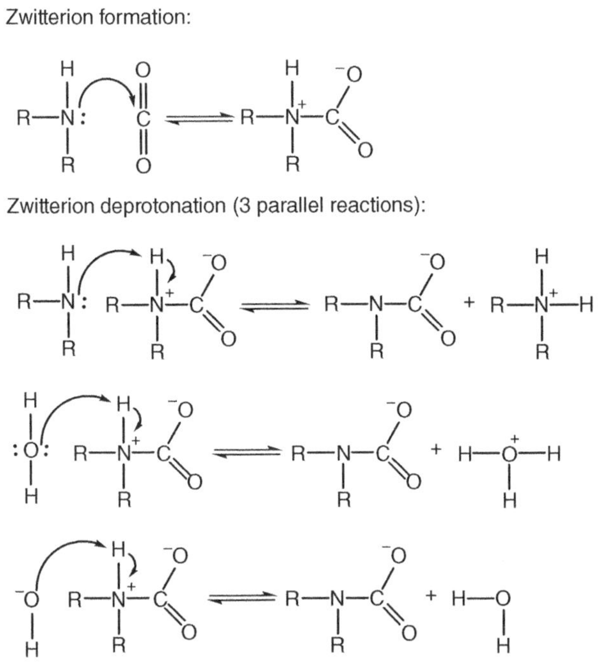

Zwitterion formation:

Zwitterion deprotonation (3 parallel reactions):

Figure 4.7   *The zwitterion mechanism illustrated.*

Inserting the value of [Z] from Equation 4.22(a) into Equation 4.21 and rearranging will yield Equation 4.20.

There will also be a direct, albeit slow, reaction between CO_2 and OH^- ions in the solution. The observed reaction rate is thus a little higher than predicted by Equation 4.20:

$$r_{CO_2,obs} = r_{CO_2} + k_{OH} [CO_2] [OH^-] \qquad (4.23)$$

Both reaction paths are open to OH^-, but its direct reaction with CO_2 is considered to be slow, whereas the reaction with the zwitterion is fast. Hence, it is reasonable to assume that this path will dominate the OH^- consumption. Since little OH^- is present in the solutions, this reaction is of less importance kinetically.

Figure 4.7 shows the collected zwitterion reaction mechanisms involved. It should be pointed out that the present approach to the zwitterion theory involves 'a unified base approach' (i.e. all bases react similarly with the zwitterion).

4.5.2   Termolecular Mechanism of Crooks and Donnellan

Crooks and Donnellan (1989) questioned the validity of the zwitterion mechanism. They argue that the relative sizes, of kinetic rate – and equilibrium constants, implied for the rate expression obtained from the zwitterion theory, are unreasonable. Crooks and Donnellan (op. cit) claim that the 'termolecular reaction mechanism' shown in Figure 4.8 is more realistic than the zwitterion mechanism.

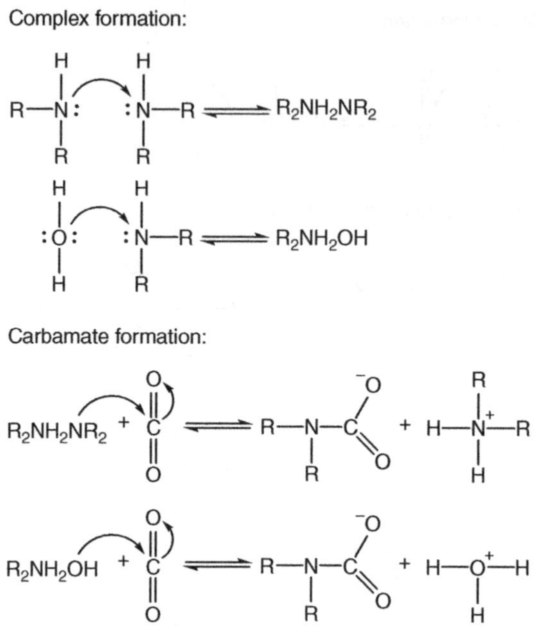

Complex formation:

Carbamate formation:

Figure 4.8   *The termolecular mechanism postulated by Crooks and Donnellan.*

Littel, Versteeg and van Swaaij (1992a) discussed the Crooks–Donnellan work. They concluded that their approach is unable to explain the fractional reaction orders observed in some instances. On this basis they rejected the hypothesis. Aboudheir *et al.* (2003) claimed that only the termolecular mechanism could explain their kinetic data. There is evidently disagreement in the literature with respect to this. As will be shown next, reaction kinetic data are not suitable to distinguish between these mechanisms since they both give rise to similar kinetic rate expressions.

Crooks and Donnellan do not discuss directly how this mechanism is developed into a reaction rate expression. It seems reasonable, however, that a pseudo equilibrium is reached by the following two reactions describing their 'complex formation':

$$R_1R_2NH + H_2O = \{(R_1R_2NH)(H_2O)\} \tag{4.24}$$

$$R_1R_2NH + R_1R_2NH = \{(R_1R_2NH)(R_1R_2NH)\} \tag{4.25}$$

where the right-hand sides represent the intermediate association products, the adducts.

Assuming Reactions 4.24 and 4.25 reach a pseudo equilibrium state, involves assuming that these reactions are much faster than the reactions:

$$CO_2 + \{(R_1R_2NH)(H_2O)\} \rightarrow \{(H_3O^+)(R_1R_2NCOO^-)\} \tag{4.26}$$

and

$$CO_2 + \{(R_1R_2NH)(R_1R_2NH)\} \rightarrow \{(R_1R_2NH_2{}^+)(R_1R_2NCOO^-)\} \tag{4.27}$$

The equilibria of the Reactions 4.24–4.25 are described by:

$$K_{15} = \frac{[(R_1R_2NH) \cdot (H_2O)]}{[R_1R_2NH][H_2O]} \tag{4.28}$$

$$K_{16} = \frac{[(R_1R_2NH) \cdot (R_1R_2NH)]}{[R_1R_2NH][R_1R_2NH]} \tag{4.29}$$

The rate of CO_2 consumption is proportional to the concentrations of both solvent complexes, $\{(R_1R_2NH)(H_2O)\}$ and $\{(R_1R_2NH)(R_1R_2NH)\}$. Their concentrations may be calculated from the equilibrium expressions (4.28) and (4.29). The following rate expression then ensues:

$$r_{CO_2} = k_a K_1 [CO_2][R_1R_2NH][H_2O] + k_b K_2 [CO_2][R_1R_2NH]^2 \tag{4.30}$$

This is in practice the same expression as the one derived by Crooks and Donnellan (op. cit) purely from an inspection of their observed rate data. A direct reaction between CO_2 and water is not considered by them. The rate expression in Equation 4.30 may also be derived from Equation 4.20 if $k_{-1} \gg \Sigma(k_b B)$ and OH^- is neglected. Both first and second order reaction kinetics may be derived from Equation 4.30, depending on the value of the constants. If both terms are significant, a fractional reaction order with respect to the amine would follow. Hence both mechanisms may explain a fractional reaction order.

4.5.3   Australian Approach

This is a recent addition to the way of viewing how CO_2 reacts with amines in aqueous solutions. It has come out of collaboration between Marcel Maeder at the University of Newcastle (NSW) and the CSIRO Research Institute. Their way of looking at the chemistry of this system is summarised in Figure 4.9. In the ensuing discussion their original nomenclature for constants are kept to make reference to the original for actual data easier.

Their approach is to list all reactions that can explain the connection between the various compounds, assign kinetic constants where relevant or equilibrium constants where the reactions are instantaneous. Constants are eventually fitted to experimental data by regression.

To relate water and CO_2 they list the following reactions:

$$CO_2 (aq) + H_2O = H_2CO_3 \; (k_1 \text{ is forward}, k_{-1} \text{ backward kinetic constant}) \tag{4.31}$$

$$CO_2 (aq) + OH^- = HCO_3{}^- \; (k_2 \text{ is forward}, k_{-2} \text{ backward kinetic constant}) \tag{4.32}$$

$$CO_3{}^{2-} + H^+ = HCO_3{}^- \; (K_3 \text{ as equilibrium constant}) \tag{4.33}$$

$$HCO_3{}^- + H^+ = H_2CO_3 \; (K_4 \text{ as equilibrium constant}) \tag{4.34}$$

$$OH^- + H^+ = H_2O^- \; (K_5 \text{ as equilibrium constant}) \tag{4.35}$$

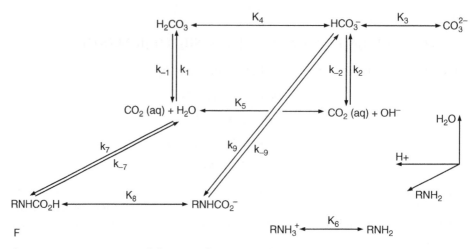

Figure 4.9 *An overview of the Australian approach to explain the chemistry of CO_2 and amines. The figure is recreated based on a figure in one of their papers (Conway et al., 2012). Subscripts on constants follow their convention, K = equilibrium constants and lower case k = kinetic constants. Single line, double arrows are ionic/instant reactions. Double lines are kinetically controlled reactions. Adapted with permission from Conway et al. (2012). Copyright (2012) American Chemical Society.*

Reactions involving amines are:

$$RNH_2 + H^+ = RNH_3^+ \ \left(K_6 \text{ as equilibrium constant}\right) \tag{4.36}$$

$$RNH_2 + CO_2 \text{ (aq)} = RNHCO_2H \ \left(k_7 \text{ is forward}, k_{-7} \text{ backward kinetic constant}\right) \tag{4.37}$$

$$RNHCO_2^-H+ = RNHCO_2H \ \left(K_8 \text{ as equilibrium constant}\right) \tag{4.38}$$

$$RNH_2 + HCO_3^- = RNHCO_2^- \ \left(+H_2O\right) \ \left(k_9 \text{ is forward}, k_{-9} \text{ backward kinetic constant}\right) \tag{4.39}$$

A difference from the preceding proposals is that there is no amalgamated kinetic expression for kinetics. Here, all reactions are considered. The rest is a question of mathematics. From this point of view it is not possible to compare this approach with those of Danckwerts–Caplow and Crooks–Donnellan by visual inspection for reaction orders.

Gaining a deeper insight in reaction mechanisms necessitates further investigations other than reaction kinetics. An example of this would be the use of spectroscopic techniques, NMR as an example, to identify species in the solutions (Perinu *et al.*, 2014).

4.5.4   Older Representations

It may seem superfluous to discuss old ways of seeing the reactions involved with CO_2 absorption into amines. There is a vast amount of literature available from years prior to Danckwerts' proposal of the zwitterion mechanism, however, and this literature is worth using. Hence it is useful to acquire at least a cursory knowledge of the older representations.

4.5.4.1   Traditional Concepts

This is not a concept to explain the reaction mechanism involved, but even so may be convenient for analysing the kinetics and stoichiometry. Three reactions are considered. These are known as carbamate formation (CF), bicarbonate formation (BF) and the so-called carbamate reversion (CR). All of them will be discussed next.

It is customary to represent the net reaction between CO_2 and an amine (the CF reaction) by:

$$CO_2 + 2\,R_1R_2NH = R_1R_2NH_2^+ + R_1R_2NCOO^- \tag{4.40}$$

Historically, this was based on a reaction mechanism where CO_2 reacts with the amine to form an intermediate that immediately reacts with another amine molecule to form carbamate and a protonated amine:

$$CO_2 + R_1R_2NH = R_1R_2NCOOH \tag{4.41}$$

$$R_1R_2NCOOH + R_1R_2NH = R_1R_2NCOO^- + R_1R_2NH_2^+ \tag{4.42}$$

The CF could also be the zwitterion mechanism by the amine, explained in terms of R_1R_2NH, being the base B. The hypothesis of Crooks and Donnellan will yield the CF via Reaction 4.27. There is practical conflict between these approaches.

The intermediate, R_1R_2NCOOH, may be referred to as a carbamic acid. These are unstable, but the implied adduct on the right-hand side of Equation 4.42 is the carbamate that is a corner stone in CO_2 removal by primary and secondary amines.

The BF reaction is another reaction assumed to take place. Its overall concept may be written:

$$CO_2 + R_1R_2NH + H_2O = R_1R_2NH_2^+ + HCO_3^- \tag{4.43}$$

BF cannot be explained in terms of the zwitterion mechanism. As is seen from Figure 4.7, water taking the role of the base B does not lead to bicarbonate. Nor can OH^- as base in the zwitterion mechanism explain Reaction 4.43. However, OH^- could, as pointed out when discussing the zwitterion mechanism, also react directly with CO_2 according to:

$$CO_2 + OH^- = HCO_3^- \tag{4.44}$$

Astarita, Savage and Bisio (1983) explain that for high loads of CO_2 in the solution (i.e. from $\alpha > 0.5$ mol CO_2/mol amine) there is a third reaction that dominates (in the sense that it is a bottle neck), the CR, which may be written as:

$$R_1R_2NHCOO^- + H_2O = R_1R_2NH + HCO_3^- \tag{4.45}$$

This 'frees' amine to re-enter into the CF reaction. The CR is slower than the CF, so this conversion to bicarbonate (CR) becomes limiting. In practice, CR will start at lower loadings, α, than 0.5. There is no real discontinuity at $\alpha = 0.5$ although that is the point where, stoichiometrically, all the amine molecules are spoken for.

4.5.4.2   The 'Astarita Representation'

This approach is basically a very useful mathematical representation. In the book *Gas Treating with Chemical Solvents* (Astarita, Savage and Bisio, 1983) a few simplifying concepts and limiting solutions are presented. The chemistry fits the zwitterion mechanism, but a few special terms are introduced. They are well worth noticing.

Astarita, Savage and Bisio (1983) simplified the reaction system to make it more compact and easier to handle. By mathematical manipulation of the previous equations, they arrived at *one* reaction involving the volatile component, CO_2:

$$CO_2 + 2\,R_1R_2NH = R_1R_2NH_2{}^+R_1R_2NCOO^- \tag{4.46}$$

and two equilibria in the solution involving non-volatile components only:

$$R_1R_2NH + HCO_3{}^- = H_2O + R_1R_2NCOO^- \tag{4.47}$$

$$R_1R_2NH + HCO_3{}^- = R_1R_2NH_2{}^+ + CO_3{}^- \tag{4.48}$$

The latter reaction is only significant at very high pH, and may be neglected in the amine systems and conditions usually employed.

The method for solving Equations (4.46–4.48) using 'extent of reaction' is known also from traditional literature on multiple equilibrium reactions. The book by Astarita, Savage and Bisio (1983) gives the details for solving these equations.

4.6   The Chemistry of Tertiary Amines

The current concept of the reaction between CO_2 and tertiary amines is usually attributed to Donaldson and Nguyen (1980). The proposed mechanism is referred to as base catalysis where the amine acts as a base, and the chemistry may be described as follows (see also Figure 4.10). The lone electron pair on the amine nitrogen attacks at the incremental positive charge on water hydrogen to give a protonation reaction:

$$R_3N : + H - O - H = R_3H^+ + OH^- \tag{4.49}$$

OH^- then reacts with CO_2 according to:

$$CO_2 + OH^- = HCO_3{}^- \tag{4.50}$$

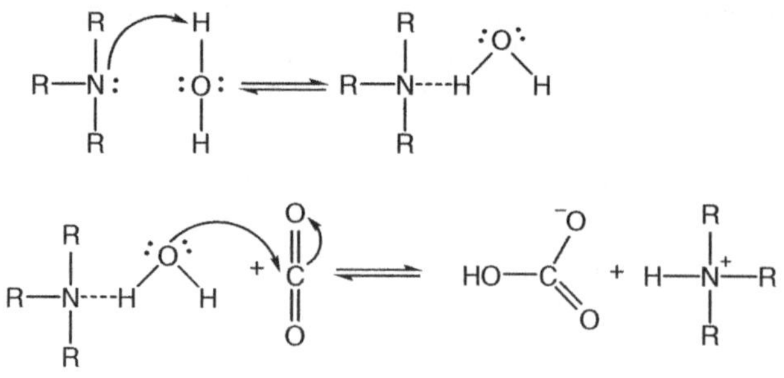

Figure 4.10   *Mechanism for CO_2 and tertiary amines.*

which is the rate determining step. The overall reaction becomes:

$$CO_2 + H_2O + R_3H = HCO_3^- + R_3NH^+ \tag{4.51}$$

Although R_3N and R_3NH+ are not the same, the term base catalysis is still justified since the amine protonation reaction may be added to close the 'amine catalysis' loop. Donaldson and Nguyen (op cit.) discuss other conceivable reaction mechanisms. One possibility is a reaction involving the alcohol OH-group on the alkanolamine with H^+ being released 'acid-fashion', but this is only significant at very high pH (Jørgensen and Faurholt, 1954). Blauwhoff (1982) found that this did not influence the reaction kinetics up to at least pH = 10.7. Crooks and Donnellan (1990) support the proposed mechanism and point out that the reverse reaction agrees with an accepted mechanism in organic chemistry.

It is pertinent to raise the question of why the zwitterion does not play a role in the reaction between CO_2 and tertiary amines. Since CO_2 is a Lewis acid and the tertiary amine is a Lewis base, there ought to be a reaction as with the other amines:

$$O :: C :: O + R_3N := R_3N^+COO^- \tag{4.52}$$

With primary and secondary amines this zwitterion is $R_2N^+HCOO^-$ and here the 'H' disappears when reacting with a base. This second reaction cannot occur with tertiary amines since any NR bond is more stable (or balanced in charges). Hence, there is no appreciable direct reaction between CO_2 and R_3N since the zwitterion represents a 'dead end'. The zwitterion is unstable so the concentration stays low and does not represent a significant potential for CO_2 storage. Hence, it may be neglected.

The 'Astarita representation' (see Section 4.5.4.2) may be modified to apply also for the reaction between CO_2 and tertiary amines (More on this by Astarita, Savage and Bisio, 1983).

4.7 Chemistry of the Minor Sulfur Containing Gases

Since the gases carbonyl sulfide (COS), carbon disulfide (CS_2) and mercaptans (RSH) are less abundant and often not present, we may group them under the label 'minor sulfur containing gases', or maybe just 'minor gases' for short. They all behave in a different way chemically speaking from the CO_2 and H_2S which we could refer to as the 'majors' for short. Their removal cannot be accomplished by the standard aqueous alkanolamine solutions that are successful in removing the two majors. These standard solutions include MEA, diethanolamine (DEA), methyldiethanolamine (MDEA) and triethanolamine (TEA) (Richardson and O'Connell, 1975). They argue that to capture COS and CS_2 it is necessary to look for absorbents where soft bases are available. There are a number of such absorbents, and they are summarised in Table 4.3. It is further known from Wölfer (1982) that the solubilities of COS and CH_3SH are on level with H_2S and CO_2. The order of solubilities in a solvent like Sepasolv is $CH_3SH > H_2S > COS > CO_2$ and they are all an order of magnitude higher than CH_4. Furthermore Huffmaster (1997) points out that the presence of Sulfolane in the Sulfinol process aids the absorption of organic sulfur compounds.

The alcohols are amphiprotic and considered as soft bases (or acid as the case might be). Also the other solvents are soft in this respect. The more complex molecules actually have both hard and soft base spots. All the absorbents listed can also remove CO_2 and H_2S, but

Table 4.3 *Absorbents that include soft bases to capture COS and CS_2.*

Absorbent	Solvent	Chemical base
Fluor solvent	Propylene carbonate	None
Rectisol	Methanol (at a −70°C level)	None
Purisol	n-Methyl-α-pyrrolidone	None
Selexol	Dimethyl ether of polyethylene glycol	None
Amisol	Methanol	MEA
Sulfinol	Sulfolane (tetrahydrothiphene dioxide)	DIPA or MDEA
Sepasolv	Oligoethylene glycol and methyl isopropyl ethers	None

Richardson and O'Connell (1975), Wölfer (1982) and Kohl and Nielsen (1997).

where there is no chemical base such removal is achieved through physical solubility. This can be, and is, done but more absorbent will need to be circulated. A process like Rectisol that necessitates a very low temperature due to CO_2 solubility and own volatility has its competitive edge when the process stream needs to be cooled like in synthesis gas plants with a cryogenic process step.

Looking further at the literature, it appears that the real world is, however, a little more complicated. There is extensive evidence that COS reacts also with the common alkanolamines, and quite strongly so. In fact, this reaction represents a problem in such gas treating because the reaction products are not reversed when using the normal desorption conditions for these processes. This is particularly the case for MEA where the reaction with COS is also quite fast. It has, for this reason, been stated that the common alkanolamines MEA, DEA, MDEA are unsuitable for treating gases where COS is present. A significant amount of research has been published in this area as shall become clear in the ensuing discussions.

4.7.1   The COS Chemistry

The carbonyl sulfide molecule has received much less attention than CO_2 and H_2S, but it has been widely studied. Its structural formula is shown in Figure 4.11. The normal boiling point is −50°C, which makes it a gas at ambient conditions.

Its reactions with alkanolamines have been studied by Sharma (1965), Al-Ghawas, Ruiz-Ibanez and Sandall (1989), Al-Ghawas and Sandall (1991) and Littel, Versteeg and van Swaaij (1992a,b). Sharma proposed an analogy to the standard reaction for CO_2 at the time. Littel, Versteeg and van Swaaij based on work by Ewing, Lockshon and Jencks (1980) and Millican *et al.* (1983), proposed that the reaction between COS and primary and secondary amines is analogous to that of CO_2 and goes via the zwitterion mechanism:

$$COS + R_1R_2NH \leftrightarrow R_1R_2NH^+COS^- \tag{4.53}$$

$$R_1R_2NH^+COS^- + B \leftrightarrow BH^+ + R_1R_2NCOS^- \tag{4.54}$$

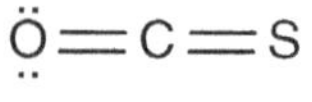

$$\ddot{O} = C = S$$

Figure 4.11 *Structural representation of COS.*

The end product is referred to as a thiocarbamate. B is any base, H_2O, OH^- or R_1R_2NH. Kinetic data of Al-Ghawas, Ruiz-Ibanez and Sandall and Littel *et al.* support this proposal. So far, so good, but in chemistry an explanation of kinetic data is necessary but not conclusive evidence for a reaction mechanism. Rahman, Maddox and Mains (1989) made a study of the reaction between CO_2 and several alkanolamines using NMR as a tool. They found support for the formation of the thiocarbamate.

Al-Ghawas and Littel also made kinetic studies of the reaction between COS and tertiary amines. It must be pointed out that the rates measured by Littel, Versteeg and van Swaaij are much lower than those reported by Al-Ghawas, Ruiz-Ibanez and Sandall (1989, 1991) and Alper (1993). Disagreements between rate measurements aside, their data support the proposal that there is an analogous reaction scheme to CO_2 also for tertiary amines. The reaction is then referred to as *base catalysed*, and the reaction may be written:

$$R_3N : +H - O - H = R_3NH^+ + OH^- \qquad (4.55)$$

OH^- then reacts with COS according to:

$$COS + OH^- = HCO_2S^- \qquad (4.56)$$

which is the rate determining step. The overall reaction becomes:

$$COS + H_2O + R_3NH = HCO_2S^- + R_3NH^+ \qquad (4.57)$$

Kohl and Nielsen (1997) mention irreversible reactions between COS and various alkanolamines in their book. For primary and secondary alkanolamines Rahman *et al.* (1989) suggest that the thiocarbamates undergo further reaction according to a scheme that was proposed by Pearce, Arnold and Hall (1960). This is shown in Figure 4.12. Littel, Versteeg and van Swaaij (1992a) carried out NMR analysis of the loaded solutions of some amines. Their analysis shows that there are significant amounts of degradation products in the MEA solution while hardly any were found in DEA and AMP solutions. MEA also reacts much faster with COS. This supports the general wisdom that MEA should be avoided in favour of DEA for gas treating if COS is present in the gas.

Vaidya and Kenig (2009) have given a comprehensive overview of past studies made of the kinetics between COS and alkanolamines. All the standard alkanolamine amines have been studied. Details of who studied what and determined rate coefficients may be found in their work. However, they overlooked the 2007 study of COS and MDEA by Rivera-Tinoco and Bouallou (2007).

When COS is present in the gas, it is customary to let this influence the choice of amine used for gas treating. Such choice is based on experience from relevant process plants. Richardson and O'Connell suggested that hybrid absorbents using both water and an organic solvent and optionally an alkanolamine could be a good choice. The picture is not clear and it would be a good idea to seek the advice of suppliers of gas treating technology where this is a problem. They would either have actual plant experience to help or you would most likely go to another supplier. There is too much ambiguity in the information available to make clear conclusions.

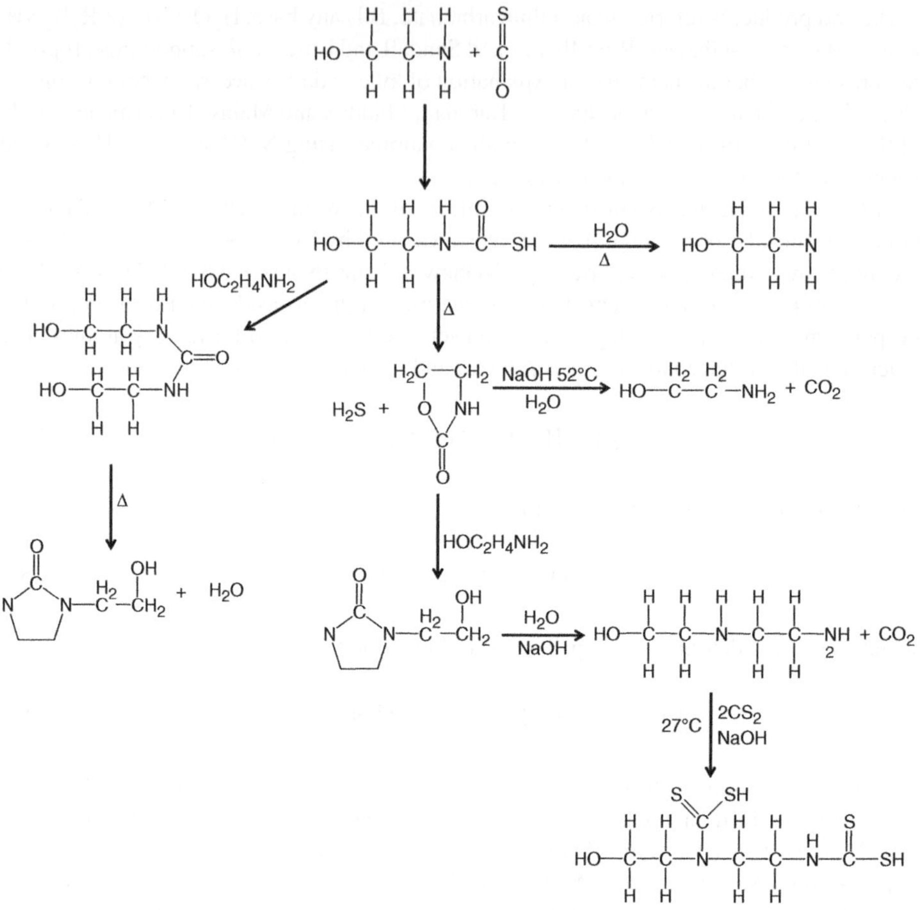

Figure 4.12   *Reactions following the absorption of COS. (Pearce, Arnold and Hall, 1960)*

4.7.2   Chemistry of CS₂

CS_2, carbon disulfide, is used as a solvent for some chemical reactions, for example making rayon from cellulose. Its boiling point is 46°C. Hence, it is a liquid at room temperature and pressure. Its structural formula is sketched in Figure 4.13.

Little attention has been given to CS_2 in relation to alkanolamines in the chemical engineering literature. Kohl and Nielsen (1997) point out that CS_2 is not removed to a great degree in processes based on primary and secondary amines, and that CS_2 does not react with tertiary amines. To the extent that it is absorbed, it seems it may reappear as COS in the stripper. Pearce, Arnold and Hall (1960) also studied CS_2 in their work, but passed this over

$$S = C = S$$

Figure 4.13   *The structural formula of CS₂.*

in their paper with a reference to this part of their work not being ready to share. (A search for further papers by these authors drew a blank.) Umbreit (1961) points to the reaction between CS_2 and secondary amines being used to determine the latter spectroscopically via the reaction product dithiocarbamic acid. Dawodu and Meisen (1996) concluded that CS_2 react with DEA to form dithiocarbamates but the reaction rate was orders of magnitude lower than that for CO_2 and DEA. They also argue that there are reasons to expect also other degradation products. There is little to be found when searching the literature on this subject.

Quantitative determination of COS and CS_2 may be made by a method given by Schwack and Nyazi (1993).

If CS_2 is known to be present in the feed gas to an absorber, a dedicated literature search addressing also the circumstances should be made. And as for COS, it is prudent to look for advice.

4.7.3 Chemistry of Mercaptans (RSH)

In modern organic nomenclature, mercaptans are known as thiols. If 'R' is a methyl group, the IUPAC name is 'methanethiol'. The older name would be 'methylmercaptan'. The term mercaptan comes from their ability to react with mercury and thus capture it. In gas treating 'R' is typically C_1–C_4.

There seems to be a limited amount of information on mercaptans (RSH) available in relation to its chemistry with alkanolamines. This is not a big surprise when the most outstanding feature of RSH is an extremely obnoxious smell, even at extremely low concentrations. They are also toxic. It is not tempting to volunteer to work with these compounds. Special care would need to be exercised in the laboratory.

The structural formula is shown in Figure 4.14. It is like an H_2S molecule with an H substituted by an alkyl group.

It is known that mercaptans will react with strong alkali like NaOH.

$$2\,RSH + 2\,NaOH = 2\,NaSR + 2\,H_2O \tag{4.58}$$

which may be reversed by oxidation according to:

$$4\,NaSR + O_2 + 2\,H_2O = 2\,RSSR + 4\,NaOH \tag{4.59}$$

This is the basis for the Merox and Merichem processes (Wizig, 1985). Kohl and Nielsen (1997) point out that RSH will react with bases like alkanolamines. However, the reaction product is not very stable in the sense that only limited amounts of RSH will be removed from the gas by an alkanolamine absorption process. They also refer to an investigation where it was looked for a few specific irreversible degradation products with alkanolamines, and the investigator found none. Rahman, Maddox and Mains (1989) claim that

Figure 4.14 *Structural formula of a generic thiol or mercaptan.*

Table 4.4   *Selected data for mercaptans.*

	Molecular weight	Boiling point, °C	Solubilities, mol (l·Mpa)		
			H_2O	n-Hexane	Toluene
Methyl mercaptan	48	6	2	6	17
Ethyl mercaptan	62	35	1	8	40
n-Propyl mercaptan	76	67–68	0.7	13	55
n-Butyl mercaptan	90	98	–	14	55
Methane (for comparison)	16	−162	0.007	0.3	0.2

Solubilities are at 50°C. (Bedell and Miller, 2007).

MEA is too weak a base to ionise a mercaptan. Richardson and O'Connell (1975) argue that mercaptans should be removed by the use of a 'soft acid'. They list mercaptans under both soft acid and soft base categories. It is further argued that solvents such as Selexol and Sulfinol have such organic properties within the molecules of the solvent. Similar claims are made by Bedell and Miller (2007) and Bedell (2012). They review solubilities of CH_3SH (methyl mercaptan, or methane thiol) and show that the solubility is greater in hexane than water and greater still in toluene. In the more recent work the term 'inclusion compound' is used. As chemical evidence this is circumstantial, but it is a starting point. There is certainly chemistry available for removing mercaptans from gas. A few data to help describe a few mercaptans are summarised in Table 4.4.

4.8   Sterically Hindered Amines

This has emerged as a field of its own after it was introduced to gas treating by Sartori and Savage in 1983 in what is now a much quoted paper. Their main contribution was to discover that these amines offered advantages with respect to CO_2 capacity and reaction rates, which meant that loadings akin to tertiary amines could be achieved while providing much higher reaction rates. Sterically hindered amines had, however, already been described earlier by Frahn and Mills (1964). They had shown that certain primary and secondary alkanolamines did not form carbamates and explained that by steric hindrance when groups are attached to the α-carbon. In their 1983 paper Sartori and Savage defined sterically hindered amines structurally as amines in which the amino group is attached to a tertiary carbon atom. This is best described by the examples shown in Figure 4.15.

Sartori and Savage did not state that sterically hindered amines could not form carbamates, but pointed out that any carbamate formed would be much less stable than the carbamates formed by the standard amines discussed hitherto. Following this argument, any carbamate formed would quickly undergo the CR reaction (i.e. hydrolysis by water) to form bicarbonate. Chakraborty, Astarita and Bischoff (1986) made a study of CO_2 absorption into aqueous solutions of hindered amines where they infer a direct reaction between CO_2 and the amine without making any speculation as to what kind of adduct was formed as an intermediate. The end product was bicarbonate. The same research team has also made a

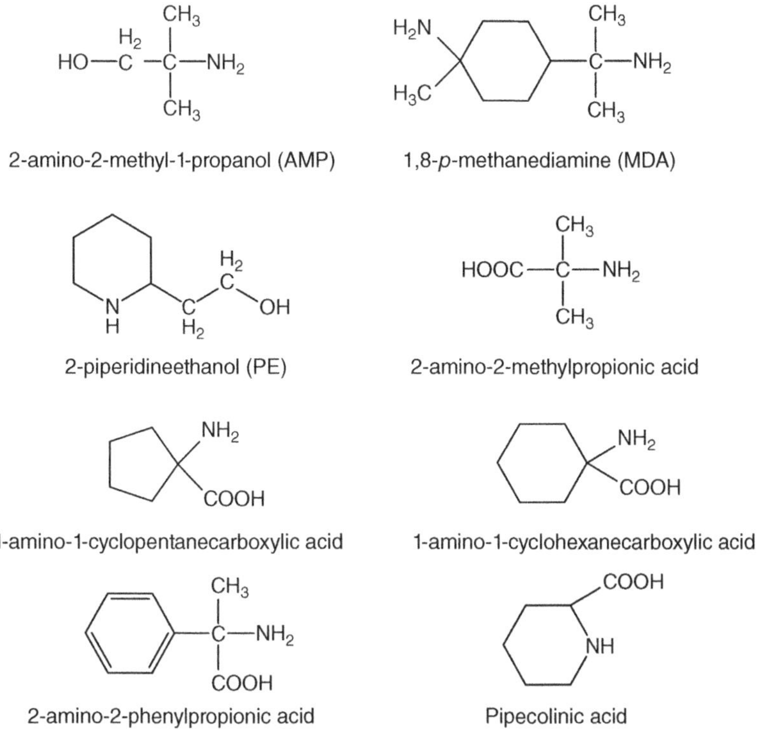

2-amino-2-methyl-1-propanol (AMP)

1,8-*p*-methanediamine (MDA)

2-piperidineethanol (PE)

2-amino-2-methylpropionic acid

1-amino-1-cyclopentanecarboxylic acid

1-amino-1-cyclohexanecarboxylic acid

2-amino-2-phenylpropionic acid

Pipecolinic acid

Figure 4.15   *Examples of sterically hindered amines.*

much more fundamental study to throw light on the issue of steric hindrance (Chakraborty *et al.*, 1988). They made use of molecular modelling as well as spectrometric methods. It appears that the introduction of methyl groups and bigger to replace hydrogen atoms on the α-carbon influences the electron donor properties of the N-atom such that the amine as a base becomes 'softer'. This makes it a poorer match for CO_2. It is concluded that such fundamental studies support the more macroscopic observations made in measuring absorption equilibria and kinetics. Values for some carbamate stability constants are shown in Table 4.5. There is clearly a difference as to how stable the carbamates of different amines are, and it is not difficult to see that there could be a soft transition between 'stable' and 'unstable' carbamates. Although DIPA is not listed as a sterically hindered amine, an inspection of the structural formulas will quickly suggest that DIPA is more hindered than DEA as the table also reflects.

Concerning the chemistry, it is clear from the literature that the kinetics of sterically hindered amines' reaction with CO_2 is first order with respect to the amine. This may be accommodated both by assuming that no carbamate is formed at any time like in the reaction with tertiary amines, but from a kinetic point of view the formalism of either mechanism applied with primary and secondary amines could also be used. There is only the one

Table 4.5 *Carbamate stability constants of selected amines. (Astarita et al., 1983; Sartori and Savage, 1983)*

Amine	K_c at 40°C
MEA	12.5 (Sartori); 27 (Astarita)
DEA	2.0 (Sartori)/0.4 (Astarita)
DIPA	0.4 (Astarita)
AMP	<0.1 (Sartori)

Astarita's values for DEA and DIPA cross over at around 40°C.

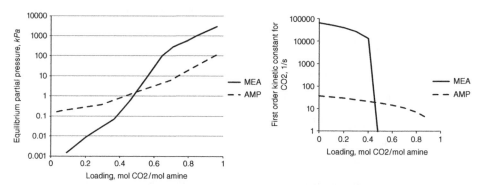

Figure 4.16   *(a,b) CO_2 absorption equilibria and first order rate constant for CO_2 of the sterically hindered AMP versus the 'normal' MEA. These figures are based on data from Jou, Mather and Otto (1995), Teng and Mather (1990), Ying and Eimer (2013) and Saha, Bandyopadhyay and Biswas (1995).*

paper by Chakraborty *et al.* that has gone beyond the kinetics. The conclusion regarding reaction mechanism is not absolute and is open for research.

Significant efforts have been made to research sterically hindered amines alone or in mixtures since their introduction by Sartori and Savage. Research efforts have been directed at equilibrium determinations and measurement of reaction kinetics.

As shown in Figure 4.16 sterically hindered amines have an advantage over primary and secondary amines in that they have both a lower equilibrium pressure of CO_2 and higher reaction rates when the loading approaches 0.5 mol/mol and for higher loadings. This behaviour is partially attributable to the reaction stoichiometry.

4.9   Hot Carbonate Absorbent Systems

Leaving the organic base systems behind, there are also processes based on inorganic chemistry to capture CO_2 and H_2S. The typical alkali like NaOH and KOH are too strong as bases to allow for reversing the reaction when desorption is desired. However, their salts turn out to be very suitable and form a useful buffer at the right pH between the first and second dissociation of the two acids. Written in terms of KOH, the reaction equations are:

$$K_2CO_3 + CO_2 + H_2O = 2\ KHCO_3 \qquad (4.60)$$

for CO_2, and for H_2S absorption:

$$K_2CO_3 + H_2S = KHCO_3 + KHS \qquad (4.61)$$

These reactions are straightforward, and the buffer systems and the inherent equilibria are discussed in Chapter 5. However, if no CO_2 was absorbed, the second reaction would proceed one step further little by little as CO_2 was lost from the system:

$$2\ KHCO_3 = K_2CO_3 + CO_2 + H_2O \qquad (4.62)$$

$$2\ KHS = K_2S + H_2S \qquad (4.63)$$

The bisulfide is virtually non-regenerable (Tosh *et al.*, 1960). This is an issue to bear in mind when deciding on an appropriate absorption–desorption system. Since these reactions involve salts, clearly their solubilities become an issue. Even if these systems are operated at elevated temperatures; that is, from 70°C and upwards for absorption, the solubilities put a limit on the concentration of the solution. The chemistry and operation of these systems were discussed by Benson and co-workers in a number of papers (Benson and Field, 1969; Benson, Field and Haynes 1956; Benson, Field and Jimeson 1954). (The work reported in these papers lead to the well-known Benfield process.)

These reactions are nice and simple. However, the reaction kinetics are slow and along with this the mass transfer kinetics also become slow relative to the best organic systems discussed earlier. In the past both arsenic and vanadium salts have been use to speed up the kinetics but as these catalysts have been abandoned for environmental reasons it is elected not to discuss the chemistry involved here.

These days, alkanolamines are the promoters of choice but there are several of them. Claims are made at intervals that this or that new promoter is better than previous systems, and probably they are, given the process conditions and challenges quoted in that presentation. Since there are several chemicals being used, it is reasonable to conclude that there is no one chemical that outperforms the others in all respects. DEA has a long history as an additive to the carbonate process to speed up absorption rates. Since the advent of the sterically hindered amines at least both AMP and PE have been studied. The Catacarb process is said to have a used non-toxic catalyst, but the nature of such chemical promoter has not been revealed (Eickmeyer, 1962).

The hot carbonate solutions also have the ability to hydrolyse COS such that this sulfur may thereafter be removed as H_2S:

$$COS + H_2O = H_2S + CO_2 \qquad (4.64)$$

When DEA is used to speed up the reactions and mass transfer, the chemistry of the kinetically dominating reaction is like the zwitterion mechanism but the carbonate salts represent extra bases for the reaction Equation 4.19 and the kinetics in Equation 4.20.

4.10   Simultaneous Absorption of H_2S and CO_2

This situation is discussed by Astarita, Savage and Bisio (1983). Essentially, H_2S reacts faster with the amine (particularly tertiary amines) than CO_2 since H_2S reacts instantaneously, but CO_2 will partially replace some of the H_2S already absorbed, or reacted, when CO_2 in due time has completed its reaction. This situation will, in principle, be the same in both aqueous and nonaqueous solutions. The relative strengths of the two acids may change from one solvent to another. In water the carbonic acid has a lower pK_a value and is the stronger. Simultaneous absorption of these gases will be further discussed in Chapter 13. The effects are more rate-oriented than chemistry based and discussion is therefore deferred.

4.11   Reaction Mechanisms and Activators–Final Words

The reactions associated with H_2S in these absorption systems are not debated. Since H_2S is a Brønsted acid the chemistry seems straightforward and it happens instantaneously.

The chemistry of CO_2 with respect to the primary and secondary amines is not settled. Kinetic data alone cannot identify the mechanisms, it can only give pointers. On the other hand, any proposed mechanism must be able to explain the kinetic measurements. Further investigations involving spectroscopy and other suitable techniques are required to settle this discussion. From an engineering point of view, however, the exact nature of the reaction mechanism is seldom all that important. When analysing mass transfer kinetics, the chemical kinetics are really all that is needed. It is, however, good to be precise when discussing these issues. Maybe a better insight into the actual reaction mechanism would point to improved absorption chemicals?

'Activators' is an interesting subject. What does it really mean when a so-called activator is added to a solution? There has, for example been a lot of attention given to piperazine added to say MDEA or AMP. The reaction between CO_2 and piperazine is much faster than with the other amines, and its natural end reaction is 'carbamate'. Obviously the 'activation' will cease when all piperazine is converted to carbamate. Further activity will demand that the carbamate is hydrolysed to bicarbonate to recover piperazine for further reaction. No evidence has been found in the literature suggesting any other path. A similar argument could be made for say MEA in MDEA. Having said this, it certainly looks as if amine mixtures have interesting properties for CO_2 capture.

4.12   Review Questions, Problems and Challenges

- **Problem 1:** Claims are made that only the termolecular or the zwitterion reaction mechanism can explain measured data for reaction kinetics. Often this is due to finding fractional reaction order with respect to the amine. Discuss such a claim in view of the reaction kinetic equations derived. Question any stated or implicit assumption made in the derivation of them.
- **Problem 2:** Why is it that, say Selexol, is seen as a good absorbent for COS?
- **Problem 3:** What are the merits of mixing an absorbent system like Sulfinol (DIPA, Sulfolan and water)?

References

Aboudheir, A., Tontiwachwuthikul, P., Chakma, A. and Idem, R. (2003) Kinetics of the reactive absorption of carbon dioxide in high CO_2-loaded, concentrated aqueous monoethanolamine solutions. *Chem. Eng. Sci.*, **58**, 5195–5210.

Albert, A. and Serjeant, E.P. (1984) *The Determination of Ionization Constants*, 3rd edn, Chapman & Hall, London.

Al-Ghawas, H.A., Ruiz-Ibanez, G. and Sandall, O.C. (1989) Absorption of carbonyl sulphide in aqueous methyldiethanolamine. *Chem. Eng. Sci.*, **44**, 631–639.

Al-Ghawas, H.A. and Sandall, O.C. (1991) Simultaneous absorption of carbon dioxide, carbonyl sulphide and hydrogen sulphide in aqueous methyldiethanolamine. *Chem. Eng. Sci.*, **46**, 665–676.

Alper, E. (1993) Comments on kinetics of reaction of carbonyl sulphide with aqueous MDEA. *Chem. Eng. Sci.*, **48**, 1179–1180.

Alvarez-Fuster, C., Midoux, N., Laurent, A. and Charpentier, J.C. (1981) Chemical kinetics of the reaction of CO_2 with amines in pseudo m-nth order conditions in polar and viscous organic solutions. *Chem. Eng. Sci.*, **36**, 1513–1518.

Astarita, G., Savage, D.W. and Bisio, A. (1983) *Gas Treating with Chemical Solvents*, John Wiley & Sons, Inc., New York.

Bedell, S.A., (2012) Method and composition for removal of mercaptans from gas streams. US Patent 2012/0280176 A1.

Bedell, S.A. and Miller, M. (2007) Aqueous amines as reactive solvents for mercaptan removal. *Ind. Eng. Chem. Res.*, **46**, 3729–3733.

Benson, H.E. and Field, J.H. (1969) New data for hot carbonate processes. *Petrol. Refiner*, **39**, 127–132.

Benson, H.E., Field, J.H. and Haynes, W.P. (1956) Improved process for CO_2 absorption uses hot carbonate solutions. *Chem. Eng. Prog.*, **52** (10), 433–438.

Benson, H.E., Field, J.H. and Jimeson, R.M. (1954) CO_2 absorption employing hot potassium carbonate solutions. *Chem. Eng. Prog.*, **50** (7), 356–364.

Blauwhoff, P.M.M. (1982) Selective absorption of H_2S from sour gases by alkanolamines solutions. Dr IR, thesis. University of Twente, Twente, NL.

Caplow, M. (1968) Kinetics of carbamate formation and breakdown. *J. Am. Chem. Soc.*, **90**, 6795–6803.

Carey, F.A. and Sundberg, R.J. (1990) *Advanced Organic Chemistry; Part A: Structure and Mechanisms*, 3rd edn, Plenum, New York.

Chakraborty, A.K., Astarita, G. and Bischoff, K.B. (1986) CO_2 absorption in aqueous solutions of hindered amines. *Chem. Eng. Sci.*, **41**, 997–1003.

Chakraborty, A.K., Bischoff, K.B., Astarita, G. and Damewood, J.R. Jr., (1988) Molecular orbital approach to substituents in amine-CO_2 interactions. *J. Am. Chem. Soc.*, **110**, 6947–6954.

Conway, W., Wang, X., Fernandes, D., Burns, R., Lawrance, G., Puxty, G. and Maeder, M. (2012) Toward the understanding of chemical absorption processes for post-combustion capture of carbon dioxide: electronic and steric considerations from the kinetics of reactions of CO_2(aq) with sterically hindered amines. *Environ. Sci. Technol.*, **47**, 1163–1169.

Crooks, J.E. and Donnellan, J.P. (1989) Kinetics and mechanism of the reaction between CO_2 and amines in aqueous solution. *J. Chem. Soc., Perkin Trans. 2*, 331–333.

Crooks, J.E. and Donnellan, J.P. (1990) Kinetics of the reaction between CO_2 and tertiary amines. *J. Org. Chem.*, **55**, 1372–1374.

Danckwerts, P.V. (1979) The reactions of CO_2 with ethanolamines. *Chem. Eng. Sci.*, **34**, 443–446.

Dawodu, O.F. and Meisen, A. (1996) Degradation of aqueous diethanolamine solutions by carbon disulfide. *Gas. Sep. Purif.*, **10**, 1–11.

Donaldson, T.L. and Nguyen, Y.N. (1980) CO_2 reaction kinetics and transport in aqueous amine membranes. *Ind. Eng. Chem. Fundam.*, **19**, 260–266.

Eickmeyer, A.G. (1962) Catalytic removal of CO_2. *Chem. Eng. Prog.*, **58** (4), 89–91.

Ewing, S.P., Lockshon, D. and Jencks, W.P. (1980) Mechanism of cleavage of carbamate anions. *J. Am. Chem. Soc.*, **102** (9), 3072–3084.

Frahn, J.L. and Mills, J.A. (1964) Paper ionophoresis of amino compounds. Formation of carbamates and related reactions. *Aust. J. Chem.*, **17**, 256–273.

Huffmaster, M.A. (1997) Stripping requirements for selective treating with Sulfinol and amine systems. Laurance Reid Gas Conditioning Conference, Norman, Oklahoma.

Jørgensen, E. and Faurholt, C. (1954) Reactions between CO_2 and amino alcohols. II. TEA. *Acta Chem. Scand.*, **3**, 1141–1144.

Jou, F.-Y., Mather, A.E. and Otto, F.D. (1995) The solubility of CO_2 in a 30 mass percent monoethanolamine solution. *Can. J. Chem. Eng.*, **73**, 140–147.

Kadiwala, S., Rayer, A.V. and Henni, A. (2012) Kinetics of carbon dioxide (CO_2) with ethylenediamine, 3-amino-1-propanol in methanol and ethanol, and with 1-dimethyl-amino-2-propanol and 3-dimethylamino-1-propanol in water using stopped-flow technique. *Chem. Eng. J.*, **179**, 262–271.

Kohl, A. and Nielsen, R. (1997) *Gas Purification*, Gulf Publishing.

Littel, R.J., Versteeg, G.F. and van Swaaij, W.P.M. (1992a) Kinetics of COS with primary and secondary amines in aqueous solutions. *AIChE J.*, **38**, 244–250.

Littel, R.J., Versteeg, G.F. and van Swaaij, W.P.M. (1992b) Kinetic study of COS with tertiary alkanolamine solutions. 1. Experiments in an intensely stirred batch reactor. *Ind. Eng. Chem. Res.*, **31**, 1262–1269.

Maham, Y., Mather, A.E. and Hepler, L.G. (1997) Excess molar enthalpies of (water + alkanolamines) systems and some thermodynamic calculations. *J. Chem. Eng. Data*, **42**, 988–992.

Mahan, B. (1965) *University Chemistry*, Addison-Wesley, Reading, MA.

McMurry, J. (1992) *Organic Chemistry*, 3rd edn, Brooks/Cole, Pacific Grove, CA.

Millican, R.J., Angelopoulos, M., Bose, A., Riegel, B., Robinson, D. and Wagner, C.K. (1983) Uncatalyzed and general acid catalysed decomposition of alkyl xanthates and monothiocarbonates in aqueous solutions. *J. Am. Chem. Soc.*, **105**, 3622–3630.

Pearce, R.L., J.L. Arnold, C.K. Hall, (1960) A study of carbonyl sulphide in natural gas processing. Laurance Reid Gas Conditioning Conference.

Pearson, R.C. (1963) Hard and soft acids and bases. *J. Am. Chem. Soc.*, **85**, 3533–3539.

Pearson, R.C. and Songstad, J. (1967) Application of the principle of hard and soft acids and bases to organic chemistry. *J. Am. Chem. Soc.*, **89**, 1827–1836.

Perinu, C., B. Arstad, A.M. Bouzga, K.J. Jens, (2014) The influence of solvating water molecules on the reactivity of monoethanolamine and related primary amines for CO_2 capture: a 15 N NMR study. *J. Phys. Chem. B* submitted.

Rahman, M.A., Maddox, R.N. and Mains, G.J. (1989) Reactions of carbonyl sulphide and methyl mercaptan with ethanolamines. *Ind. Eng. Chem. Res.*, **28**, 470–475.

Richardson, I.M.J. and O'Connell, J.P. (1975) Some generalizations about processes to absorb acid gases and mercaptans. *Ind. Eng. Chem. Process Des. Dev.*, **14**, 467–470.

Rivera-Tinoco, R. and Bouallou, C. (2007) Kinetic study of carbonyl sulphide (COS) absorption by methyldiethanolamine aqueous solutions from 415 mol/m^3 to 4250 mol/m^3 and 313 K to 353 K. *Ind. Eng. Chem. Res.*, **46**, 6430–6434.

Sada, E., Kumazawa, H. and Han, Z.Q. (1985) Kinetics of reaction between carbon dioxide and ethylenediamine in nonaqueous solvents. *Chem. Eng. J.*, **31**, 109–115.

Saha, A.K., Bandyopadhyay, S.S. and Biswas, A.K. (1995) Kinetics of absorption of CO_2 into aqueous solutions of 2-amino-2-methyl-1-propanol. *Chem. Eng. Sci.*, **50**, 3587–3598.

Sartori, G. and Savage, D.W. (1983) Sterically hindered amines for CO_2 removal from gases. *I&EC Fundam.*, **22**, 239–249.

Schwack, W. and Nyazi, S. (1993) Simultaneous UV-spectrophotometric determination of CS_2 and COS using the piperidine, pyrrolidine, ethylenediamine or morpholine reagent. *Fresenius J. Anal. Chem.*, **345**, 705–711.

Sharma, M.M. (1965) Kinetics of reactions of carbonyl sulphide and carbon dioxide with amines and catalysis by Brönsted bases of the hydrolysis of COS. *Trans. Faraday Soc.*, **61**, 681–688.

Teng, T.T. and Mather, A.E. (1990) Solubility of CO_2 in an AMP solution. *J. Chem. Eng. Data*, **35**, 410–411.

Tosh, J.S., Benson, H.E., Field, J.H., Anderson, R.B. (1960) Equilibrium Pressures of Hydrogen Sulphide and Carbon Dioxide Over Solutions of Potassium Carbonate. Bureau of Mines Report Number 5622, Bureau of Mines.

Umbreit, G.R. (1961) Spectroscopic determination of secondary amines. *Anal. Chem.*, **33**, 1572–1573.

Vaidya, P.D. and Kenig, E.Y. (2009) Kinetics of carbonyl sulphide reaction with alkanolamines: a review. *Chem. Eng. J.*, **148**, 207–211.

Versteeg, G.F., van Dijk, L.A.J. and van Swaaij, W.P.M. (1996) On the kinetics between CO_2 and alkanolamines both in aqueous and non-aqueous solutions. An overview. *Chem. Eng. Commun.*, **144**, 113–158.

Versteeg, G.F. and van Swaaij, W.P.M. (1988a) On the kinetics between CO_2 and alkanolamines in both aqueous and non-aqueous solutions – I. Tertiary amines. *Chem. Eng. Sci.*, **43**, 573–585.

Versteeg, G.F. and van Swaaij, W.P.M. (1988b) On the kinetics between CO_2 and alkanolamines in both aqueous and non-aqueous solutions I, primary and secondary amines. *Chem. Eng. Sci.*, **43**, 587–591.

Wizig, H.W. (1985) An innovative approach for removing CO_2 and sulphur compounds from a gas stream. Laurance Reid Gas Conditioning Conference.

Wölfer, W. (1982) Helpful hints for physical solvent absorption. *Hydrocarbon Process., Int. Ed.*, **61** (1), 193–197.

Ying, J. and Eimer, D.A. (2013) Determination and measurements of mass transfer kinetics of CO_2 in concentrated aqueous monoethanolamine solutions in a stirred cell. *Ind. Eng. Chem. Res.*, **52**, 2548–2599.

5

Physical Chemistry Topics

5.1 Introduction

This chapter treats selected topics from physical chemistry that have particular relevance to gas treating. Some of the topics are not discussed in ordinary texts on physical chemistry. Other topics are on the speciality side, and often lightly treated. Finally there is material here that really belongs in introductory material, but is included because most practising engineers tend to forget their fundamentals, and in practice it was probably not taught all that thoroughly. An attempt has also been made to look into items of history in a small way. (My apologies if these fleeting references are found to be boring.)

Without a solvent, there is no solution. For this reason we shall start by discussing a few implications of the properties of solvents. Water is one such solvent.

Gas treating is very much about acid–base chemistry. For that reason, this will be the focus of extensive discussions.

5.2 Discussion of Solvents

It is appropriate to start a review of the chemistry with a general discussion of solvents and some non-aqueous solvent chemistry. Solution chemistry is very much influenced by the nature of the solvent. There are a few classification systems described by Popovych and Tomkins (1981). One due to Brønsted has often been referred to. Popovych and Tomkins (op cit.), however, discuss a more modern approach which is illustrated in Figure 5.1.

Water, alcohols and glycols are all classified as amphiprotic neutral solvents. These may exhibit both acidic and basic properties. The amines used to bind CO_2 and H_2S are classified as amphiprotic protophilic. These have essentially basic properties, but will also act as acids towards compounds that are more basic than themselves. The absorbent solutions used are mixtures of amines and a neutral solvent. These mixed solvents ought logically to have basic character.

Gas Treating: Absorption Theory and Practice, First Edition. Dag Eimer.
© 2014 John Wiley & Sons, Ltd. Published 2014 by John Wiley & Sons, Ltd.

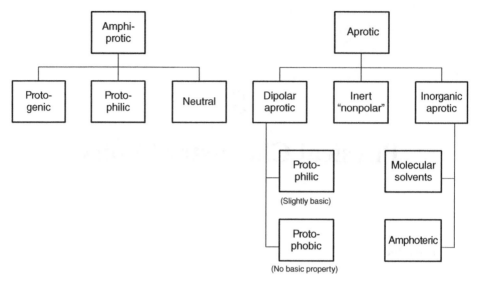

Figure 5.1 *Classification of solvents based on the work of Popovych and Tomkins (1981). Reproduced with permission from Eimer (1994).*

Brønsted subdivided the amphiprotic neutral solvents into two groups, one for high dielectric constants and one for low. There is no firm distinction between these subgroups and a firm dividing value of the dielectric constant is not given. It varies between authors from 15 to 30 (Popovych and Tomkins op cit.). Triethylene glycol (TEG) has a dielectric constant of 26 at room temperature (25°C). This places TEG right in the middle of the two amphiprotic neutral subgroups. Hence, Brønsted's subdivision of this group is not useful when considering TEG. The classification laid out in Figure 5.1 is better.

Water has a high dielectric constant. There are, however, solvents with much higher dielectric constants than water. The dielectric constant is an important parameter in the classification of solvents, one of Brønsted's three original parameters. Reaction rate constants are often correlated with dielectric constants.

The solvent's ability to ionise will give an indication of its role in reactions with ionic features. The autoprotolysis constant, pK_{ap}, describes the degree of self-ionisation of solvents. Some relation between pK_{ap} and dielectric constant may be expected. The simple plot of pK_{ap} versus the dielectric constant in Figure 5.2 shows a trend.

Autoprotolysis describes a solvent's ability to dissociate and produce protons. According to IUPAC's *Commission on Electroanalytical Chemistry* (Rondinini *et al.*, 1987) it may, for a generic solvent HSol, be defined as:

$$2\,HSol \leftrightarrow H_2Sol^+ + Sol^- \tag{5.1}$$

When there are mixed solvents, cross-coupling to make mixed ions is possible and must be accounted for.

In the case of the solvent being water, this dissociation becomes:

$$H_2O \leftrightarrow H^+ + OH^- \tag{5.2}$$

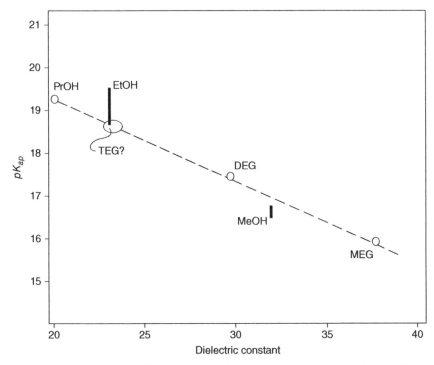

Figure 5.2 *A comparison of* pK_{ap} *to the dielectric constant of selected solvents. The uncertainty in the* pK_{ap} *value for TEG is big, hence the question mark. Reproduced with permission from Eimer (1994).*

The equilibrium constant of this reaction is referred to as the autoprotolysis constant:

$$K_{ap} = \frac{[H^+]\,[OH^-]}{[H_2O]} \tag{5.3}$$

In the case of water it is customary to define a constant K_w such that:

$$K_w = K_{ap}\,[H_2O] = [H^+]\,[OH^-] \tag{5.4}$$

Methanol, that is also used as a solvent in an absorption process (Amisol), can also autoprotolyse:

$$MeOH = MeO^- + H^+ \tag{5.5}$$

Its autoprotolysis constant is smaller than that of water by more than two orders of magnitude. Comparing glycols to methanol, it is seen that there is a lot more autoprotolysis taking place in monoethylene glycol (MEG) compared to TEG. The implication is that if, for example, an amine is added to the glycol solution to help catch H_2S, there is a lot more reactivity in MEG than TEG that is the one usually used for water removal from gas. If there is a significant amount of water in the solution, this will dominate the acid–base chemistry anyway. Even 1% water would give a profound effect.

To get the full picture of the acidity of the actual compounds dissolved in the solvent, their protonation constants, or pK_p-values, in the particular solvent must be obtained. If these solutes are present in significant quantities, they will also be part of the solvent and influence its properties. Using an amine as example, a protonation reaction may be written as:

$$R_1R_2NH_2^+ = R_1R_2NH + H^+ \qquad (5.6)$$

The equilibrium constant for this reaction may written as:

$$K_p = \frac{[R_1R_2NH]\,[H^+]}{[R_1R_2NH_2^+]} \qquad (5.7)$$

The pK_p (also referred to as the acidity constant, pK_a) value is simply:

$$pK_p = -log_{10}\left(K_p\right) \qquad (5.8)$$

There is no solution unless there is a solvent. This is obvious but may easily be overlooked as most teaching is based on aqueous systems where the solvent is taken for granted. Water is indeed our most important solvent. It is timely, however, to point out that the presence of significant amounts of alkanolamines in gas treating absorbents will influence the chemistry that takes place. These alkanolamines are also solvents. As their concentrations increase, they will more and more influence the solution's chemistry. A part of this chemistry is the solution's ability to dissolve the compounds present. Also activity coefficients will be influenced by the concentration of the amine.

There are commercial absorbents that use solvents other than water. The best known is probably Shell's Sulfinol absorbents that use 'Sulfolan' in addition to water. (Sulfolan is a trade name for a special organic solvent, 2,3,4,5-tetrahydrothiophene-1,1-dioxide, also referred to as tetramethylene sulfone). There are versions with both diisopropanolamine (DIPA) and methyldiethanolamine (MDEA) as the alkanolamines. Historically the alkanolamine processes used both water and glycol (particularly diethylene glycol (DEG)) as solvents. There are a number of physical solvents available including Selexol, Sepasolv, Rectisol (chilled methanol) and the Fluor solvent (propylene carbonate).

If we were to add an alkanolamine to the glycol solution in the natural gas water removal unit in order to remove H_2S say, care must be taken if this is to be a success. Looking at the values of the solvent autoprotolysis constants in Figure 5.2, it is clear that TEG would be a much poorer alternative than MEG as solvent. However, if there were a little water present this would most likely dominate the acid–base solvent involvement. There is, however, very little water left in the dry glycol entering the top of the water removal tower, but there is typically 5% (wt) water in the rich glycol leaving the absorber bottom.

5.3   Acid–Base Considerations

5.3.1   Arrhenius, Brønsted and Lewis

In the 1880s Svante Arrhenius (Swedish chemist) developed his theory of ionisation of aqueous solutions. He allocated acidic properties to the presence of H^+ ions and basic/alkaline properties to the presence of OH^- ions. This may be referred to as the water-ions definition, often thought of as the classic definition of acid and bases. However, in current work the later definitions due to Brønsted and Lowry and the even wider definition due to Lewis are those used. This was discussed in Chapter 4, but is fundamental to the ensuing discussion to the extent that a reminder is in order. (The name of Arrhenius lives on, however, as connected to the exponential relationship between reaction kinetic constants and absolute temperature.)

5.3.2   Weak and Strong Acids and Bases

When an acid or a base is fully dissociated in water, it is said to be 'strong'. If it is only partially dissociated, it is said to be weak. The CO_2, H_2S and alkanolamines normally met in gas treating are weak. Their dissociation is thus governed by relevant equilibrium relationships. Acidity constants for common alkanolamines are listed in Table 5.1. The gas compounds CO_2 and H_2S will not, of course, start to dissociate until they are dissolved in, for example water as they are normally encountered above their critical points.

The definition of weak and strong acids is also used for electrolytes in general. If a salt is fully dissociated, it is said to be a strong electrolyte. Likewise it is a weak electrolyte if it is not fully dissociated.

5.3.3   pH

The definition of pH today is: $\mathbf{pH} = -\mathbf{log}\ a_{H+}$ where a_{H+} is the H^+ ion activity (log with basis 10). This change from the first pH definition normally encountered, namely that $\mathbf{pH} = -\mathbf{log}\ c_{H+}$, that is a concentration based definition, is easy enough to handle. Some claim, however, that there is no such thing as the activity of an isolated ion. Hence, it could be said that pH is a controversial concept when taken to the extreme.

The use of pH as an acidity scale has proved to be a useful tool in acid–base chemistry. The measurement of pH seems simple and straightforward in principle but the reality is often very different. There are several types of problems. The electrode must be calibrated. The sample may be 'difficult' and cause contamination and/or drift. There may even be the question of finding a compatible electrode. Many practical aspects are discussed by Bates (1973). A very useful discussion is also given by Serjeant (1984/1991).

Table 5.1   pK_a *values of common alkanolamines.*

Amine	pK_a (25°C)[a]	Normal boiling point (°C)[b]	Molecular weight
MEA	9.45	170	61.08
DEA	8.88	269	105.14
TEA	7.77	360	149.19
MDEA	8.52	247	119.1
DIPA	9.00	250	133.19
AMP	~9.82	165	89.14
DGA	9.5	221	105.14

[a]Data from Dow, except DGA from Huntsman.
[b]Data from Huntsman and *Chemical Book*.

When measurements are made in aqueous solutions, there are plenty of available calibration standards. With respect to calibrating pH electrodes IUPAC (Covington, Bates and Durst, 1985) defines three types of standards:

- **RVS – Reference value standard**: This is the first among equals for calibrating. In any solution it is based on the emf (electromotoric force) value of 0.05 M KHPh in that solution. (KHPh = Potassium hydrogen phthalate).
- **PS – Primary standard**: IUPAC has defined another seven buffers in specified concentrations as PSs. These shall be considered to have equal standing as the RVS, and they cover the whole aqueous pH scale.
- **OS – Operational standard**: These standards shall be traceable back to the RVS or PSs and are the standards commercially available for calibrating any pH electrode used in the lab or otherwise. The distinction is that these OSs are made and checked against the RVS or a PS.

It is recommended to calibrate the pH electrode using OSs with values on both sides of the expected values to be measured. This is in recognition of the fact that electrodes for routine use, usually a combination electrode, have their peculiarities. Separate electrodes should be used if precision measurements are wanted. Measurements in very pure water with low conductivity may be treacherous. Electrodes need a minimum of conductivity in the solution to function. The use of a combination electrode to determine the pH of very pure water with a very low conductivity has in one case been known to give 9 as the answer. It caused a bit of worry until the problem, which was a measurement problem due to the very low conductivity, was cleared up.

5.3.4   Strength of Acids and Bases

So-called strong acids that include HCl, HBr, HI, HNO_3 and $HClO_4$ are fully dissociated in water, at least up to concentrations of 0.1 M. Weak acids are not strong, and are typified by incomplete dissociation in water. How much they dissociate may be described using a generic acid HA according to:

$$HA = H^+ + A^-$$ (5.9)

The equilibrium value for this reaction is:

$$K_a = \frac{[H^+][A^-]}{[HA]}$$ (5.10)

where K_a is the equilibrium constant. This is often referred to as the acid dissociation constant. (See also pK_p). The higher it is, the stronger the acid. It is for convenience often tabulated as pK_a values, and they are defined as:

$$pK_a = -log_{10}(K_a)$$ (5.11)

Due to this definition pK_a increases when K_a decreases. Hence, a lower pK_a means a stronger acid.

It could equally be said that A^- is base that reacts with acid H^+, and that the equilibrium constant for this reverse reaction is:

$$K_b = \frac{[HA]}{[H^+][A^-]}$$ (5.12)

It is obvious that $K_b \cdot K_a = 1$.

Usually these constants are all reported as pK_a values, even for bases like the alkanol-amines. The higher the pK_a value, the stronger they will be as a base and the weaker they will be as acids. Note that acid of an alkanolamine is the protonated species and the acid dissociation is considered as:

$$R_1R_2NH_2^+ = R_1R_2NH + H^+ \qquad (5.13)$$

The equilibrium reaction of this reaction is:

$$K_a = \frac{[R_1R_2NH]\,[H^+]}{[R_1R_2NH_2^+]} \qquad (5.14)$$

The pK_a of the amine defines its basic character, and its protonation reaction is often referred to. The higher the pK_a, the stronger is the basic character of the amine.

5.3.5   Titration

It is known from introductory chemistry courses that when a strong acid is titrated with a strong base, or vice versa, the pH at the equivalence point is 7. An example is an HCl solution titrated with NaOH. Here the salt formed, NaCl, is a strong salt that fully dissociates, and the solution is neutral. Hence the pH becomes 7 at the equivalence point. This is a situation that is easy to handle numerically.

When either acid or base is weak this is not quite as straightforward. If HCl is added to a solution of example given monoethanolamine (MEA), the development of pH as (pure) HCl is added is shown in Figure 5.3. It may be noted that the equivalence point occurs at a pH a little above 5, it is actually 5.28 not accounting for activities. This explains why this titration is done using, for example, methyl red as an indicator. The relatively flat part above pH 8 is due to the buffer between MEA and MEAH$^+$. After the equivalence point the pH curve again flattens out but continues to go down due to the addition of extra acid.

The relationship in Figure 5.3 is relatively straightforward to calculate when the equilibrium associated with the amine protonation is taken into account. The behaviour of pH in this titration curve is due to the buffering properties invoked by the presence of

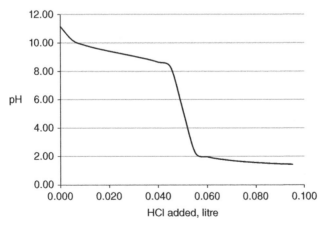

Figure 5.3   *A titration curve for 50 ml MEA HCl based on basic estimates. (No activities, concentration based.)*

the weak base MEA. The curve in Figure 5.3 is based on relating the concentration of H^+ ions, and thus pH, to the equilibrium of the reaction:

$$MEA + H_2O \leftrightarrow MEAH^+ + OH^- \tag{5.15}$$

Its equilibrium relation is:

$$K_b = \frac{[MEAH^+][OH^-]}{[MEA]} \tag{5.16}$$

Note that the equilibrium relation is written without accounting for the concentration of H_2O in line with normal practice. The value of the equilibrium constant K_b, which is the base constant of MEA is easily found since:

$$pK_b = pK_w - pK_a \tag{5.17}$$

This means that $pK_b = 14 - 9.6 = 4.4$ and $K_b = -antilog(pK_b) = 0.0000398$.

Chemically speaking there is also the reaction with HCl being used for titration, and where a salt is produced:

$$HCl + MEA = MEAH^+Cl^- \tag{5.18}$$

To establish the titration curve in Figure 5.3 it is necessary to evaluate the pH at varying amounts of HCl added. The procedure should start at the end with MEA in excess, actually it would start with the pH estimated for the original MEA solution sample. Here, it shall be assumed that a 0.050 l (50 ml) sample of 0.12 M MEA is to be titrated with 0.12 M HCl. There is no real significance in the two solutions having the same strength.

The pH at the start is calculated via the equilibrium relation 5.16 to determine OH^- concentration:

Initial concentration of MEA: 0.12 M
x moles of MEA converted to OH^- and $MEAH^+$
Concentration of MEA after this is $0.12 - x$.

From Equation 5.16

$$[OH^-] = \frac{K_b[MEA]}{[MEAH^+]} = \frac{(0.0000398)(0.12 - x)}{x} = x \tag{5.19}$$

$$x^2 + 0.0000398x - (0.0000398)(0.12) = 0 \tag{5.19a}$$

$$x = 0.00217 \tag{5.19b}$$

Using the relation of the water dissociation product:

$$[H^+] = \frac{K_w}{[OH^-]} = \frac{10^{-14}}{0.00217} = 4.618 \times 10^{-12} \tag{5.20}$$

$$pH = -\log_{10}(4.618 \times 10^{-12}) = 11.34 \tag{5.21}$$

As HCl is added, this estimate becomes more complicated. In this situation MEAH$^+$ is formed both from Equations 5.15 and 5.18. One sample calculation is as follows:

The start point is 0.050 l of MEA sample solution of strength 0.12 M.
This means that there are $(0.050\,l)(0.12\,M/l) = 0.0060$ mol MEA in the sample at the start.

If HCl is added such that MEA is in excess, we may do as follows:

Start is n moles of MEA $= 0.0060$ mol
Add p moles of HCl; 10 ml of strength 0.12 M: $p = 0.0012$ mol
Excess MEA is then $n - p$ moles of MEA
Finally, from Equation 5.15 x moles of MEAH$^+$ and OH$^-$ is formed.

The volume of sample plus HCl solution added is

$$V = 0.050 + 0.010 = 0.060\,l$$

Putting this information into the expression 5.16

$$K_b = \frac{\dfrac{p+x}{V}\dfrac{x}{V}}{\dfrac{(n-p)-x}{V}} = \frac{(p+x)(x)}{V(n-p-x)} \tag{5.22}$$

From this, a second order equation in x is found:

$$x^2 + \left(K_b V + p\right)x - K_b V(n-p) = 0 \tag{5.22a}$$

and

$$x = 0.5\left\{-\left(K_b V + p\right) + \sqrt{\left(K_b V + p\right)^2 + 4K_b V(n-p)}\right\} \tag{5.22b}$$

Note that one root has been chosen as x cannot be negative.
Putting in for the various values set out previously, we have that:

$$K_b V + p = \{(0.0000398)(0.060) + 0.0012)\} = 0.00120024 \tag{5.23}$$

$$x = 0.5\left\{-0.00120024 + \sqrt{(0.00120024)^2 - (4)(0.0000398)(0.060)(0.0060 - 0.0012)}\right\} \tag{5.22c}$$

$$x = 9.48 \times 10^{-6} \tag{5.22d}$$

$$[OH^-] = \frac{x}{V} = \frac{9.48 \times 10^{-6}}{0.06} = 1.58 \times 10^{-4} \tag{5.24}$$

From the water relation, Equation 5.4:

$$\left[H^+\right] = \frac{10^{-14}}{[OH^-]} = \frac{10^{-14}}{1.58 \times 10^{-4}} = 6.33 \times 10^{-11} \tag{5.25}$$

From the definition of pH:

$$pH = -\log_{10} 6.33 \times 10^{-11} = 10.20 \qquad (5.26)$$

5.3.6   Buffer Action in the NaOH or KOH Based CO_2 Absorbents

It is said to be buffer action in a solution when an acid or base can be added without drastically changing its pH. This kind of behaviour arises when either a weak acid or base is involved or when both the acid and base are weak. The reason is that the salt formed is partially dissociated and have common ions with the acid and base. In terms of gas treating there is a buffer effect since both CO_2 and H_2S are weak acids in aqueous solutions. An interesting buffer is the one where potassium or sodium carbonate is used to absorb CO_2 or H_2S. Figure 5.4 shows how the pH changes when CO_2 is added to a sodium hydroxide solution. The reactions involved are:

$$CO_2 + 2\,NaOH = Na_2CO_3 + H_2O \qquad (5.27)$$

Then

$$CO_2 + Na_2CO_3 + H_2O = 2\,NaHCO_3 \qquad (5.28)$$

Each reaction is mirrored in the two pH plateaus appearing in the figure. It is not until all the alkali is exhausted that the pH drops sharply as carbonic acid build up.

An actual gas treating process is based on potassium hydroxide. (The Benfield process was invented in the 1950s). Since both NaOH and KOH are strong bases, the buffer behaviours of their salts are similar. The buffer behaviour is illustrated in Figure 5.4 where the pH has been estimated based on adding CO_2 to 1 l of an aqueous solution of strength 5 M NaOH. Out of interest this exercise is started with pure NaOH that is gradually converted to Na_2CO_3 while the pH changes from 14.7 to 12.4. At that stage the solution is

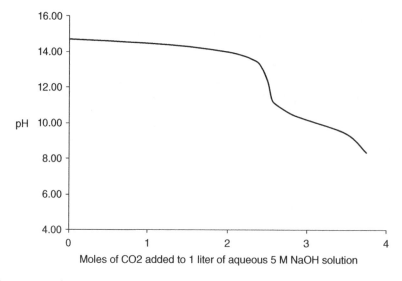

Figure 5.4   *The development of pH of an aqueous NaOH solution as CO_2 is added.*

2.5 M in Na_2CO_3. From a gas treating point of view, the interesting pH range is from 12.4 to 8.4 where the Na_2CO_3 has been converted to $NaHCO_3$ and is now a 5 M solution in this bicarbonate. In this range the slope of the pH curve is relatively flat, and it is the buffer action that leads to this. The numbers would be the same for a KOH solution. It is of course questionable to calculate pH based on concentrations at these solution strengths, but is done to illustrate the principle of this buffer.

The buffer behaviour of this solution may be accounted for by estimates as for the titration system discussed previously. In this case there is a weak acid reacting with a strong base. The relations are, to an extent, turned upside down.

The procedure for calculating the pH values shown in Figure 5.4 is simple enough once the technique is mastered. It is explained next.

As long as there is excess NaOH in the solution, the estimate of pH is very simple. The excess concentration is calculated, and this is also the concentration of OH^- ions. If 2 M of the 5 M solution was spent on conversion to Na_2CO_3, there would be $5 - 2 = 3$ M NaOH left, and the:

$$[OH^-] = 3\,M \tag{5.29}$$

$$[H^+] = \frac{K_w}{[OH^-]} = \frac{10^{-14}}{3.0} = 3.33 \times 10^{-15} \tag{5.30}$$

And:

$$pH = -\log_{10}\left(3.33 \times 10^{-15}\right) = 14.5 \tag{5.31}$$

Once all the NaOH has been converted, it becomes a little more complicated since both carbonate and bicarbonate will be present and the equilibrium between them accounted for. In this case there are five unknown concentrations:

$$[H^+] = x; [OH^-] = y; [CO_2] = z; [HCO_3^-] = v; [CO_3^=] = w \tag{5.32}$$

The concentration of Na^+ is given by the initial concentration of NaOH, it is 5 M and we use symbol S for this concentration.

There are three equilibrium conditions that give three equations, there is the charge balance, and it is known how much carbon has been added to the solution in the form of CO_2 so this gives a mass balance with respect to carbon (the carbon added is specified as $(C)(V)$ for now, upper case V is the liquid volume). These equations are (written in terms of the letters for convenience):

$$K_I = \frac{x \cdot v}{z} \quad (\text{1st dissociation of } H_2CO_3) \tag{5.33}$$

$$K_{II} = \frac{xw}{v} \quad (\text{2nd dissociation of } H_2CO_3) \tag{5.34}$$

$$K_w = x \cdot y \ (\text{Water dissociation}) \tag{5.35}$$

$$S + x = y + v + 2w \ (\text{Charge balance}) \tag{5.36}$$

$$C = z + v + w \ (\text{Carbon balance}) \tag{5.37}$$

It is not difficult to solve these equations for the five unknowns, but it involves trial and error. Straightforward really but cumbersome. An alternative is the use of computer software. There is another way, the chemist's way.

It is argued that to avoid an accumulation of H^+, the following equations must go forward by the same amounts:

$$HCO_3^- \rightarrow H^+ + CO_3^=$$
(5.38)

$$HCO_3^- + H^+ \rightarrow H_2O + CO_2$$
(5.39)

We can next assume that the overall reaction, by adding these equations, is

$$2HCO_3^- \rightarrow CO_3^= + H_2O + CO_2$$
(5.40)

Its equilibrium relation is:

$$K = \frac{[CO_2]\,[CO_3^=]}{[HCO_3^-]^2}$$
(5.41)

This constant K is unknown as it stands. However, this relation may be manipulated in the following way:

$$K = \frac{[CO_2]\,[CO_3^=]}{[HCO_3^-]^2}\frac{[H^+]}{[H^+]} = \frac{[H^+]\,[CO_3^=]}{[HCO_3^-]}\frac{[CO_2]}{[H^+]\,[HCO_3^-]} = \frac{K_{II}}{K_I}$$
(5.42)

All we did was to multiply the relation by 'one' and rearrange. The two equilibrium constants on the right are known. We can now proceed to estimate the various concentrations in the solutions as in the titration case previously. The estimates are now easy enough to handle with a spreadsheet.

5.4   The Amine–CO_2 Buffer System

Buffer behaviour will also take place if mixtures of amines are used as absorbents. The system will be become more complicated to estimate since there are more parallel equilibria to be taken into account. In the concentrated solutions used in practice activities must also be accounted for. This is why there is no really simple model for estimating absorption equilibria. Such models are discussed in Chapter 16. However, let us take a look at this buffer system from a physical chemistry point of view. Both the alkanolamine and the CO_2 are weak electrolytes. For the purpose of this discussion we shall choose to use MDEA as our model amine as we do not need to consider any carbamate in this case. If carbamate was included there would more unknowns to consider.

Let us start with a 5 M MDEA solution with no CO_2 loading. What is the pH?

The pK_a of MDEA (R_3N for the chemistry discussions) is 8.52, which means that its:

$$pK_b = pK_w - pK_a = 14 - 8.52 = 5.48$$
(5.43)

Hence, if x M R_3N is converted to R_3NH^+ and OH^-, then:

$$10^{-5.48} = 3.31 \times 10^{-6} = \frac{x.x}{5 - x}$$
(5.44)

The solution to this second order equation is straightforward and yields $x = 0.00407$. This means that $[R_3N] = 5 - x = 4.996$, $[OH^-] = [R_3NH^+] = 0.00407$.

Furthermore:

$$[H^+] = \frac{10^{-14}}{[OH^-]} = \frac{10^{-14}}{0.00407} = 2.46 \times 10^{-12}; \text{and pH} = 11.6. \tag{5.45}$$

There is certainly a buffer effect regarding pH as CO_2 is added to this solution. However, there is a further complication to the numerics in that the salt formed will hydrolyse. That is to say there are the following reactions to contend with in an equilibrium situation:

$$R_3NH^+ \rightarrow R_3NH + H^+ \tag{5.46}$$

and

$$HCO_3^- + H_2O \rightarrow H_2CO_3 \tag{5.47}$$

as well as:

$$H^+ + OH^- \rightarrow H_2O \tag{5.48}$$

When these are added, the net reaction of hydrolysis is:

$$R_3NH^+ + HCO_3^- \rightarrow R_3NH + H_2CO_3 \tag{5.49}$$

With the system at equilibrium, the relation is:

$$K_{HYD} = \frac{[R_3N]\,[H_2CO_3]}{[R_3NH^+]\,[HCO_3^-]} = \frac{[R_3N]\,[H_2CO_3]}{[R_3NH^+]\,[HCO_3^-]}\,\frac{[H^+]\,[OH^-]}{[H^+]\,[OH^-]}$$

$$= \frac{[H^+]\,[OH^-]}{\underbrace{\dfrac{[R_3NH^+]\,[OH^-]}{[R_3N]}}\,\underbrace{\dfrac{[HCO_3^-]\,[H^+]}{[H_2CO_3]}}} = \frac{K_W}{K_B K_I} \tag{5.50}$$

For MDEA, $pK_b = 5.48$, that is $K_b = 3.31 \times 10^{-6}$. Hence:

$$K_{HYD} = \frac{10^{-14}}{(3.31 \times 10^{-6})\,(4.16 \times 10^{-7})} = 0.00726 \tag{5.51}$$

At loading 1.0 mol CO_2/mol MDEA, all the MDEA has been spent, and the solution will be 'neutralised'. What is the pH of the solution at this condition? – The answer is not 7.

At this point the 5 M MDEA solution has in principle become a solution of 5 M bicarbonate balanced against R_3NH^+. Because of Reactions 5.46–5.48 and the deduced Equation 5.49, an amount of concentration change x of R_3NH^+ will be hydrolysed. As a consequence there is the same change in HCO_3^- and of course concentration x of R_3N and H_2CO_3 will materialise. Using Equation 5.50 (first step) it follows that:

$$K_{HYD} = 0.00726 = \frac{(x)\,(x)}{(5-x)\,(5-x)} \tag{5.52}$$

The solution to this is easy enough, and the answer is:

$$x = 0.393 \tag{5.53}$$

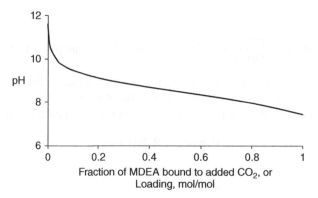

Figure 5.5   *The development of solution pH when CO_2 is added to a 5 M MDEA solution, and hydrolysis of the produced salt of weak acid and weak base is accounted for.*

It now follows that:

$$\left[R_3NH^+\right] = 5-0.393 = 4.61 \text{ M} = [HCO3^-]\,; \text{ and } \left[R_3NH\right] = \left[H_2CO_3\right] = 0.39 \text{ M}$$
$$(5.54)$$

For completeness in the reaction equations H_2CO_3 has been used here but equally we could have used CO_2 since this is the dominant form in the solution. It does not matter for the numerics. Next the $[H^+]$ may be found by using equation:

$$K_a = \frac{\left[R_3N\right]\left[H^+\right]}{\left[R_3NH^+\right]}; \left[H^+\right] = \frac{K_a\left[R_3NH^+\right]}{\left[R_3N\right]} = \frac{\left(3.02 \times 10^{-9}\right)(4.61)}{(0.39)} = 3.57 \times 10^{-8} \quad (5.55)$$

$$pH = -\log_{10}\left(3.57 \times 10^{-8}\right) = 7.45 \qquad (5.55a)$$

It is observed that the 'neutralisation point' for this system is into the region normally considered to be basic. This is a consequence of the hydrolysis of the salt formed when the weak base and the weak acid come in contact. In Figure 5.5 the pH of the whole range of CO_2 additions to a 5 M MDEA solution has been estimated by the same technique as described earlier.

How good is such a model? Such pH measurements were performed Lidal (1992), and his values are compared with the present estimation technique in Figure 5.6. The agreement is good at low loadings. At a loading of 0.6, the present technique estimates pH values about 0.3 pH units lower. It shows that making estimates of this kind is absolutely relevant in a situation where experiments are being planned, or any other circumstances where an estimate is needed. A pH measurement is quickly done, but preparing and verifying the solutions are time consuming tasks.

5.5   Gas Solubilities, Henry's and Raoult's Laws

The laws attributed to Henry and Raoult are not laws of nature like, for example Newton's law of gravitation. They are limiting laws applicable to vapour–liquid and vapour–gas systems as shall become clear. The laws will be explained and clarified by examples relevant

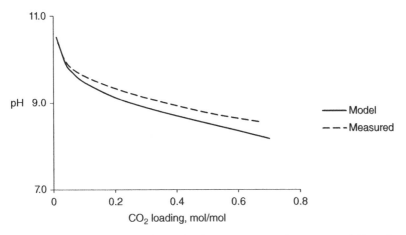

Figure 5.6 *Estimates of pH in CO_2 loaded solutions compared to measured. Data from Lidal and Erga (1991). The absorbent is aqueous MDEA.*

to gas treating. These limiting laws are commonly used as approximations for dilute and nearly pure systems. (William Henry was a British chemist 1774–1836, and the law dates from 1803. Francois-Marie Raoult was French 1830–1901, and the law dates from 1882.)

5.5.1 Henry's Law

Henry's law applies to a system when one or more components are present in dilute concentrations in the liquid phase. Figure 5.7 shows the relationship between the dilute component's partial pressure in the gas, or vapour, phase and its concentration in the liquid phase. The Henry's law may, for component i, formally be expressed as

$$H_i = \lim \left(\frac{p_i}{x_i} \right) \quad \text{as } x_i \to 0 \qquad (5.56)$$

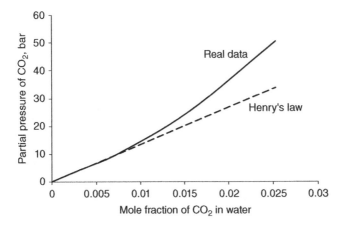

Figure 5.7 *Solubilities of CO_2 in water at 18°C. The 'real data' are from Perry (1984), and the Henry's law data are from Wilhelm et al. (1977).*

Or as commonly written (with infinite dilution being implicit):

$$p_i = H_i x_1 \tag{5.56a}$$

This is particularly relevant to gases at low pressures being physically dissolved in a solvent. H is the Henry's coefficient, p the partial pressure and x the mole fraction in the liquid.

There are more ways than one to represent the Henry's law relationship.

It may be done in terms of the partial pressure in the gas phase and the mole fraction as used previously, or the liquid concentration as moles per unit volume may be used. For convenience these relations, which are trivial, are developed here:

$$p_i = H_{Ci} C_i^L \tag{5.57}$$

Since

$$C_i^L = x_i C_{total}^L \tag{5.58}$$

It follows that

$$H_{Ci} = \frac{H_i}{C_{total}^L} \tag{5.59}$$

Another form found in the gas treating literature is by using concentrations in both phases whereby the Henry's coefficient becomes dimensionless:

$$C_i^G = m_i C_i^L \tag{5.60}$$

From the ideal gas law

$$C_i^G = \frac{p_i}{RT} \tag{5.61}$$

This, and in view of Equations 5.58–5.61:

$$\frac{p_i}{RT} = m_i x_i C_{total}^L \tag{5.62}$$

$$\frac{p_i}{RT} = RT m_i x_i C_{total}^L \tag{5.63}$$

and

$$m_i = \frac{H_i}{RT C_{total}^L} \tag{5.64}$$

Typical values for relevant gases are given in Table 5.2 and further data for CO_2 and H_2S in industrial solvents in Table 5.3. The form of the law used is that used in Equation 5.56(a).

To illustrate the limiting nature of Henry's law the partial pressure–liquid concentration relationship between CO_2 and water is depicted in Figure 5.7. It is clear that the relationship is non-linear and that Henry's law is thus only valid as the presence of CO_2 approaches 0. However, it will be a good approximation for partial pressures up to say 10 bar at least for the temperature shown. It is an absolute must to recognise the limitations of approximations used. They should always be checked.

Table 5.2 *Solubilities of selected gases in water.*

Temperature (°C)	Henry's coefficients (bar)		
	25	40	60
Helium	145 217	145 191	139 251
Nitrogen	86 462	102 492	114 781
Oxygen	44 134	54 145	63 711
Carbon dioxide	1 659	2 371	3 386
Hydrogen sulfide	548	746	1 006
Methane	40 443	51 261	62 582
Ethane	30 310	42 541	56 508
Propane	37 498	54 955	74 784
N_2O	2 313	3 425	5 159
SO_2	40	63	107

Wilhelm *et al.* (1977).

Table 5.3 *Solubilities, Henry's coefficients (bar), for H_2S and CO_2 in selected physical absorbents.*

	Average MW	25°C		40°C		60°C	
		CO_2	H_2S	CO_2	H_2S	CO_2	H_2S
Water (see Table 5.3)	18	1659	548	–	–	–	–
Propylene carbonate[a]	102	85.3	23.4	114	32.1	154	44.2
Selexol[a]	280	35.7	4.4	46.7	6.41	65.5	10.1
Sepasolv[b]	316	18	2.5	27	4	41	7
Sulfolane[c]	120	80	19	130	22	150	35

[a]Xu *et al.* (1992).
[b]Wölfer (1982) and density from Härtel (1985).
[c]Murrieta-Guevara *et al.* (1988).

5.5.2 Gas Solubilities

Gases are in general not very soluble in liquids as long as there is no chemical reaction involved. How much that may be dissolved will increase with the gas pressure over the solution and decrease when the temperature is increased. This should be clear from the previous section. There are a number of ways that gas solubility may be expressed. It is useful to know of them because they are all used in various encyclopedias of physical data.

Henry's law and its various possible forms have been described earlier. Adding the choice of pressure units to those means that there are many ways in which data may be reported. Attention must be paid to units.

The Bunsen coefficient is another. This is defined the volume of gas, as if it was at 0°C and 1 atm, dissolved per unit volume of solvent at the system temperature when the partial pressure of the gas is 1 atm.

An alternative variation on this concept is the Ostwald coefficient. This is defined as the volume of gas at the system temperature and its partial pressure dissolved in a unit volume of the solvent at the system temperature.

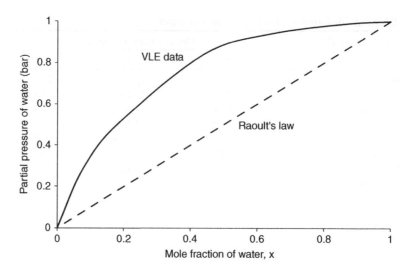

Figure 5.8 *Vapour–liquid equilibria for water-MEA. Data from Kohl and Nielsen (1997).*

Unfortunately the source of data will sometimes leave you to guess. In such cases it is useful to know the probable alternatives. Obviously the use of such data must be with care and preferably considering the effects of alternative guesses. Remember that not all reported data are to the standard of the best learned journals.

5.5.3   Raoult's Law

Raoult's law deals with the other end of concentrations, namely nearly pure liquids. It states that the partial pressure of a component is proportional to its concentration:

$$p_i = x_i p_i^0 \tag{5.65}$$

where p_i^0 is the vapour pressure of the pure component. The system water-MEA is used for illustration in Figure 5.8. (It is good practice to list the components in the order of ascending boiling points when discussing volatilities. This makes it easy to review the situation with respect to volatilities and which component boils off first.) It is clear from the figure that Raoult's law is not a good approximation for the system H_2O-MEA. The relative volatility is too high.

It is seen that the water-MEA separation is an easy one to carry out due to the big volatility difference. This is of great importance when recovering the alkanolamine in the top of the desorption column and its condenser and in the water wash in the top of the absorption column. Boiling points of common alkanolamines are listed in Table 5.1.

Although Raoult's law is of little use for water-MEA, it is clear that the limiting relationship due to Henry would be useful. The Henry's law line has not been indicated in the figure, but a visual inspection of Figure 5.8 shows that the Vapour-Liquid Equilibria (VLE) curve could be approximated by a straight line at least up to mole fraction 0.05 at either end.

5.6   Solubilities of Solids

It is well known that solids usually have limited solubility in liquids. If they are present in concentrations above a certain limit, precipitation will occur unless supersaturation is experienced. The latter is a metastable situation. For slightly soluble salts that are strong electrolytes, the solubility may be described by a solubility product. When this is available, it may also be used to predict the solubilities when there are two salts with common ions.

The most interesting salts in gas treating are perhaps the carbonate and bicarbonate of sodium and potassium. In this case the solubility situation is more complex due to these salts being involved with hydrolysis:

$$CO_3^{2-} + H_2O = HCO_3^- + OH^- \tag{5.66}$$

A similar situation occurs with the sulfides. In this case simple solubility products will not suffice as simultaneous equilibria must be considered.

Real solubility data must be determined to chart when precipitation may occur.

5.7   N$_2$O Analogy

The similarity between the CO_2 and N_2O molecules were observed by Hirschfelder, Curtiss and Bird (1954) and Amdur *et al.* (1952) according to Clarke (1964). The structural formulae are shown in Figure 5.9. It was first exploited for work in the context of CO_2 absorption into alkanolamines by Clarke. The fact that these molecules were both linear and had a central positive charge meant that they might be expected to behave similarly in solvents.

Clarke argued that the diffusivities measured for CO_2 and N_2O in water were within 2–3% of each other which were judged to be within experimental uncertainty. Hence he proposed to use the values measured for N_2O diffusivity in alkanolamine solutions in lieu of CO_2 data that could for obvious reasons not be measured.

When it came to solubilities the picture was seen as less clear as solubilities are much more sensitive to the properties of the solute than the diffusivities. However, it was argued that the ratio of solubilities between N_2O and CO_2 would remain constant between solutions.

The relation between the solubilities of CO_2 and N_2O was also explored by Laddha, Diaz and Danckwerts (1981). They found the solubilities of these gases were very similar in water, glycol, propanol, glycerol, pentanediol and diethyleneglycol. The solubility ratios varied from 0.95 to 1.05.

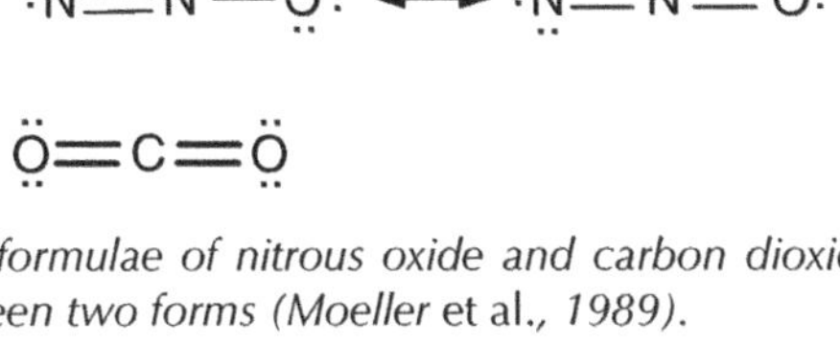

Figure 5.9 *Structural formulae of nitrous oxide and carbon dioxide. That of nitrous oxide shows resonance between two forms (Moeller et al., 1989).*

The so-called N_2O analogy has become generally accepted as the approach to use. Recent articles on the subject concentrate on the data scatter or lack of data rather than on the underlying principle. The so-called N_2O analogies may be written:

$$H_{CO_2, solvent} = H_{N_2O, solvent} \left[\frac{H_{CO_2, water}}{H_{N_2O, water}} \right] \tag{5.67}$$

and:

$$D_{CO_2, solvent} = D_{N_2O, solvent} \left[\frac{D_{CO_2, water}}{D_{N_2O, water}} \right] \tag{5.68}$$

Here H represents the solubilities in the form of the Henry coefficient and D represents the diffusion coefficient. The subscripts refer to each gas in either water or the chemical solvent.

The proof of the pudding is really that this technique has been successfully applied to represent absorption rate data. It must, however, be kept in mind that this analogy is mainly empirically based. There is no real fundamental theoretical foundation at the bottom of this. It is used because it has proved to be useful. Obviously the N_2O data thus used becomes part of the models for CO_2 reaction and absorption rates. Consistent combinations of models must be used. Various data representations proposed over time deviate at least within experimental error. There are no solubility and diffusivity data and models that are generally accepted as superior.

5.8   Partial Molar Properties and Representation

Partial molar properties are something engineers will be faced with now and again. Most people working with gas treating will certainly be acquainted with partial pressures. For ideal mixtures the partial molar property is simply explained with partial molar volumes as an example. Here the total volume, V, of the mixture will be

$$V = \sum_i x_i V_i \tag{5.69}$$

where x is mole fraction and V_i is the partial molar volume. For real mixtures this is not correct, however, and the partial molar property is defined as:

$$X_i = \left[\frac{\partial X}{\partial n_i} \right]_{T,P,n_{j \neq i}} \tag{5.70}$$

The molar property is thus the amount needed to cause a change dX in the total property when an incremental amount dn_j of j is added to the mixture.

The occasions are many when there is a need to represent properties of liquids as a function of the liquid composition. Redlich and Kister introduced a nice way of doing this a long time ago (1948). The method is applicable to any property really, and it has been used widely. They make use of a so-called excess property that is defined (using the generic property X) by:

$$X_{mix}^E = X_{mix} - \sum_i x_i X_i \tag{5.71}$$

And this excess property for a binary solution may then be correlated by the expression:

$$X_{mix}^{E} = x_1 x_2 \sum_{i}^{n} A_i \left(1 - 2x_2\right)^{i} \tag{5.72}$$

The expression in the summation operator is a polynomial where '*n*' defines the order. Summation is from $i = 0-\text{n}$. The product in front of the summation operator ensures that the excess property is zero for both the pure components. It has, for example proven to be a good way of correlating physical properties (Han *et al.*, 2012).

5.9 Hydration and Hydrolysis

Mixing the terms hydration and hydrolysis should be avoided. Like absorption and adsorption they are easily confused. Precision should always be strived for.

Hydration may imply a number of processes. It may be the reaction of a compound with the OH^- ion, for example:

$$CO_2 + OH^- = HCO_3^- \tag{5.73}$$

The term also covers the adding of so-called crystalline water to crystals, it could also be the solvation process in water where a compound is surrounded by water molecules. Retention of water by any tissue material is also a hydration process, and logically the absorption of water in desiccants like glycols. Absorption of water in glycols from natural gas is often referred to as dehydration rather than drying to avoid confusion with removal of heavier hydrocarbons. Talking about drying the gas to lower its dew point could refer to both the water and the hydrocarbon dew point.

Hydrolysis is a chemical process in which a water molecule is added to a substance resulting in the split of that substance into two parts (Wikipedia, http://en.wikipedia.org/wiki/Hydrolysis, 2011). When CO_2 is dissolved in water:

$$CO_2 + H_2O = H^+ + HCO_3^- \tag{5.74}$$

Here, H_2O is added to the CO_2 molecule to make H_2CO_3 which subsequently dissociate. This is a hydrolysis process.

5.10 Solvation

The IUPAC definition of solvation is:

> *Any stabilising interaction of a solute (or solute moiety) and the solvent or a similar interaction of solvent with groups of an insoluble material (i.e. the ionic groups of an ion-exchange resin). Such interactions generally involve electrostatic forces and van der Waals forces, as well as chemically more specific effects such as hydrogen bond formation (IUPAC, 1997).*

Wikipedia describes solvation as the clustering of solvent molecules around a solute molecule.

In the solvation process ions are surrounded by a number of solvent molecules. It is also expected that an amine molecule in aqueous solution will be surrounded by water molecules and this is supported by the forces named in the definition. This clearly have some bearing on the chemistry of the solution when it becomes very concentrated in amine or the amine derived products. See also Figure 4.5.

This subject is little discussed in the gas treating literature but is something to make a note of.

References

Amdur, I., Irvine, J.W., Mason, E.A. and Ross, J. (1952) Diffusion coefficients of the systems CO_2-CO_2 and CO_2-N_2O. *J. Chem. Phys.*, **20**, 436.

Bates, R.G. (1973) *Determination of pH. Theory and Practice*, 2nd edn, John Wiley & Sons, Inc., New York.

Clarke, J.K.A. (1964) Kinetics of absorption of carbon dioxide in monoethanolamine solutions at short contact times. *Ind. Eng. Chem. Fundam.*, **3**, 239–245.

Covington, A.K., Bates, R.G. and Durst, R.A. (1985) Definition of pH scales, standard reference values, measurement of pH and related terminology. *Pure Appl. Chem.*, **57**, 531–542 (Recommendations 1984, IUPAC Commission Report).

Eimer, D. (1994) Simultaneous removal of water and hydrogen sulphide from natural gas. Dr Ing. thesis. NTH, Trondheim.

Han, J., Jin, J., Eimer, D.A. and Melaaen, M.C. (2012) Density of water (1) + monoethanolamine (2) + CO_2 (3) from (298.15 to 413.15) K and surface tension of water (1) + monoethanolamine (2) from (303.15 to 333.15) K. *J. Chem. Eng. Data*, **57**, 1095–1103.

Härtel, G.H. (1985) Low-volatility polar organic solvents for sulphur dioxide, hydrogen sulphide, and carbonyl sulphide. *J. Chem. Eng. Data*, **30**, 57–61.

Hirschfelder, J.O., Curtiss, C.F. and Bird, R.B. (1954) *Molecular Theory of Gases and Liquids*, John Wiley & Sons, Inc., New York.

IUPAC (1997) *Compendium of Chemical Terminology*, 2nd (the 'Gold Book'). Compiled by McNaught, A.D. and Wilkinson, A. edn, Blackwell Scientific Publications, Oxford. XML on-line corrected version: http://goldbook.iupac.org (2006) created by Nic, M., Jirat, J. and Kosata, B.; updates compiled by Jenkins, A. ISBN: 0-9678550-9-8. doi:10.1351/goldbook.

Kohl, A. and Nielsen, R. (1997) *Gas Purification*, 5th edn, Gulf Publishing.

Laddha, S.S., Diaz, J.M. and Danckwerts, P.V. (1981) The N_2O analogy: the solubilities of CO_2 and N_2O in aqueous solutions of organic compounds. *Chem. Eng. Sci.*, **36**, 228–229.

Lidal, H. (1992) Carbon dioxide removal in gas treating processes. Dr Ing. thesis. NTH, Trondheim, Norway.

Lidal, H., O. Erga, (1991) Equilibrium model for CO_2 absorption in an aqueous solution of a tertiary amine. Proceedings Gas Separation International, Austin, TX, April 22–24.

Moeller, T., Bailar, J.C., Kleinberg, J., Guss, C.O., Castellion, M.E. and Metz, C. (1989) *Chemistry with Inorganic Qualitative Analysis*, Harcourt Brace Javonovich, Inc., Orlando, FL.

Murrieta-Guevara, F., Romero-Martinez, A. and Trejo, A. (1988) Solubilities of carbon dioxide and hydrogen sulfide in propylene carbonate, N-methylpyrrolidone and Sulfolane. *Fluid Phase Equilib.*, **44**, 105–115.

Perry, R.H. and Green, D. (1984) *Perry's Chemical Engineers' Handbook*, 6th edn, McGraw-Hill.

Popovych, O. and Tomkins, R.P.T. (1981) *Nonaqueous Solution Chemistry*, John Wiley & Sons, Inc., New York (Krieger-reprint).

Redlich, O. and Kister, A.T. (1948) Algebraic representation of thermodynamic properties and the classification of solutions. *Ind. Eng. Chem.*, **40**, 345–348.

Rondinini, S., Longhi, P., Mussini, P.R. and Mussini, T. (1987 on behalf of the Commission on Electroanalytical Chemistry, published) Autoprotolysis constants in nonaqueous solvents and aqueous organic mixtures. *Pure Appl. Chem.*, **59**, 1693–1702.

Serjeant, E.P. (1984) *Potentiometry and Potentiometric Titrations*, Krieger Publishing Company, Malabar, FL (Reprint edition, 1991).

Wilhelm, E., Battino, R. and Wilcock, R.J. (1977) Low-pressure solubility of gases in liquid water. *Chem. Rev.*, **77**, 219–262.

Wölfer, W. (1982) Helpful hints for physical solvent absorption. *Hydrocarbon Process.*, **61** (11), 193–197.

Xu, Y., Schutte, R.P. and Hepler, L.G. (1992) Solubilities of carbon dioxide, hydrogen sulphide sulphur dioxide in physical solvents. *Can. J. Chem. Eng.*, **70**, 569–573.

6

Diffusion

This chapter is only an introduction to diffusion. The object of this chapter is to serve as a primer for further studies, and also to lay a foundation for ensuing discussions within mass transfer. This would be impossible to deal with unless there is a fundamental understanding of diffusion processes and their mechanisms. There is a bibliography at the end of the chapter. Cussler's book is particularly recommended for basic studies and as a supplement to the present text. An introductory comment to this chapter is that this field of chemical engineering has seen more people achieving fame than is usual in this discipline. It is probably due to the mathematical-physical nature of this subject.

6.1 Dilute Mixtures

Dilute mixtures are solutions where the solutes have only a negligible influence on the solution with respect to its properties subjected to analysis. The dilute mixtures limitation may at first hand seem to be a special case of little interest. Not so, an analysis of dilute mixtures gives a good insight into fundamental problems in diffusion without complicating the issues with extensive mathematics that could easily make the key points fuzzy. The approximations implicit in assuming dilute solutions are also surprisingly applicable to many problems where one might expect the assumption of dilute mixture inapplicable.

The early work of Thomas Graham (1805–1869) and later, Adolf Fick (1829–1901) revolved around dilute mixtures. Graham formed the idea that diffusion was inversely proportional to molecular mass. Fick surmised that diffusion flux was proportional to a concentration gradient, and that the associated proportionality constant has in time become known as the diffusion coefficient.

Gas Treating: Absorption Theory and Practice, First Edition. Dag Eimer.
© 2014 John Wiley & Sons, Ltd. Published 2014 by John Wiley & Sons, Ltd.

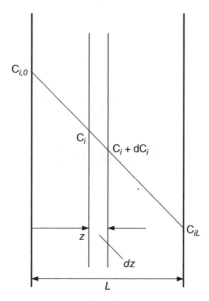

Figure 6.1   *Control element for analysing diffusion through a thin film in steady state.*

Fick's statement that has become known as Fick's first law of diffusion is

$$N_i = -D_i \frac{dC_i}{dz} \tag{6.1}$$

The minus sign comes about because the concentration slope is negative in the z-direction, and the flux is positive while the D_i is a physical transport property that logically should be positive.

It is desirable to develop an expression for the concentration profile for component 'i' through the 'film' where the component diffuses through or into. To do this the starting point is a combination of mass balance and rate equations.

Consider first a diffusion process through a film, and that the solution of diffusing species is so dilute that its movement does not influence the system. We shall only consider the resistance to mass transport in the film itself. This is also a good opportunity to introduce the 'control volume' that is fundamental to modelling of transport processes in chemical engineering. Figure 6.1 illustrates the system. According to Fick we should expect a linear concentration profile for the diffusing species, and that is reflected in the figure. A mass balance for component 'i' is made over the control element. Beware that its extent in the two other dimensions is '1'. As always the basis for the mass balance is:

$$\text{Accumulated} = \text{In} - \text{Out} \tag{6.2}$$

At this point in time we consider a steady state process, and hence there is no accumulation. The equation becomes:

$$0 = \left(-AD_i \frac{dC_i}{dz} \right) - \left(-AD_i \frac{d\left(C_i + \frac{dC_i}{dz} dz \right)}{dz} \right) \tag{6.3a}$$

and eventually:

$$D_i \frac{d^2 C_i}{dz^2} = 0 \tag{6.3b}$$

The boundary conditions for this differential equation are:

$$\text{At } z = 0, \ C_i = C_{i0} \tag{6.4}$$

$$\text{At } z = L, \ C_i = C_{iL} \tag{6.5}$$

Making use of the boundary conditions, the solution for the differential equation becomes:

$$C_i = \frac{z}{L} \left(C_{Li} - C_{0i} \right) + C_{0i} \tag{6.6}$$

This solution confirms the initial notion that the concentration profile is linear, but notice that this notion, or assumption, has not been used to arrive at the solution.

This solution is important in the sense that it is exploited for theoretical treatment of a number of mass transfer processes. The classic Whitman film theory for absorption mass transfer is one application, the membrane permeation process is another.

In the consideration earlier we looked at a steady state process through a finite film. We next want to remove these limitations. Figure 6.2 describes the situation. The fundamental description of a mass balance for component 'i' over the control volume is the same. Obviously the accumulation contribution is no longer zero. Another change is that one of the boundary conditions is altered.

$$Adz \frac{\partial C_i}{\partial t} = \left(-AD_i \frac{\partial C_i}{\partial z} \right) - \left(-AD_i \frac{\partial \left(C_i + \frac{\partial C_i}{\partial z} dz \right)}{\partial z} \right) \tag{6.7a}$$

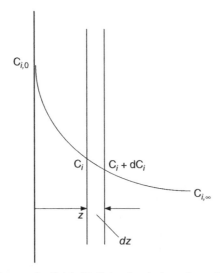

Figure 6.2   *A control element for fluid of infinite depth that where the concentration of 'i' will change over time.*

This may be rewritten as

$$Adz\frac{\partial C_i}{\partial t} = A\left(-D_i\frac{\partial C_i}{\partial z}\right) - A\left(-D_i\frac{\partial C_i}{\partial z} - D_i\frac{\partial^2 C_i}{\partial z^2}dz\right) \tag{6.7b}$$

$$\frac{\partial C_i}{\partial t} = D_i\frac{\partial^2 C_i}{\partial z^2} \tag{6.7c}$$

The Equation 6.7(c) is referred to as Fick's second law of diffusion.

Since there is no longer a finite film, the concentration of component 'i' now gradually approaches that in the bulk of the mixture it diffuses into as the distance z from the interface approaches infinity. The boundary conditions are:

$$\text{At } t = 0, \text{ at all } z, \ C_i = C_{i\infty} \tag{6.8}$$

$$\text{At } t > 0, \ z = 0, \ C_i = C_{i0} \tag{6.9}$$

$$Z \rightarrow \infty, \ C_i \rightarrow C_{i\infty} \tag{6.10}$$

The solution to Equation 6.7(c) is tricky, but an analytical solution was achieved by Boltzmann (an Austrian physicist, 1844–1906) who made a substitute variable to relate t and z in such a way that he was able to turn the equation into an ordinary differential equation that could be solved analytically. The new variable is defined as:

$$\varsigma = \frac{z}{\sqrt{2D_i t}} \tag{6.11}$$

The solution to the equation after use of the boundary conditions (redefined to fit in with the new parameter) is.

$$C_i = C_{i0} - \left(C_{i0} - C_{i\infty}\right) \cdot erf\left(\varsigma\right) \tag{6.12}$$

The function 'erf' is the error function. This result is exploited in the Higbie penetration theory and the various other surface renewal theories, out of which the one due to Danckwerts is the best known. These considerations are also used in mass transfer measurements using equipment like wetted wall, laminar jet, single sphere and more. A qualitative representation of the solution to Equation 6.7(c) with the boundary Equations 6.8–6.10 is shown in Figure 6.3.

6.2   Concentrated Mixtures

Diffusion causes convection. When the mixtures become concentrated, this can no longer be ignored. An example of a convective flux caused by diffusion is in the absorption of a reacting gas like CO_2 or H_2S into a chemical absorbent. Since these gases will be absorbed, there will be a concentration gradient to set up a diffusive flux of these gases towards the interface. However, there is a net flux of gas towards the interface from where at least the reacting species disappear into the liquid phase. There is also a significant amount of practically non-absorbing gas moving to the interface along with the reacting gas. This 'inert' gas will then be at a higher concentration at the interface than in the bulk of the gas. There will be a diffusion flux of this inert gas from the interface back into the bulk

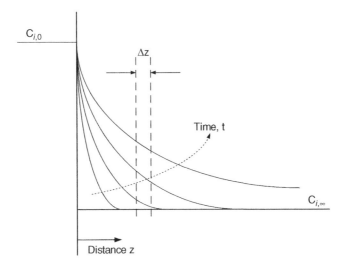

Figure 6.3   *A qualitative illustration of Fick's second law, that is, Equation 6.7(b).*

Table 6.1   *Recommendations for choice of base for Equation 6.13.*

Phase considered	Most likely base to avoid the convective term in Equation 6.13
Gas	Molar basis, volume basis
Liquid	Mass basis, volume basis
Membranes	Solvent basis

Cussler (2009).

of gas due to the concentration gradient set up. This diffusion flux will be relative to the convective flux set up by the reacting gas diffusing towards the interface. Mathematically the combined flux of a component 'j' may be described as:

$$N_j = -D_j \frac{dC_j}{dz} + C_j v^{convection} \tag{6.13}$$

Symbols are as before, but the $v^{convection}$ is some nominal velocity describing the convective flow. The mathematics describing these combined fluxes will be more or less complicated, depending on the reference frames chosen. To be more specific, convective flux can be expressed in terms of volume, mass, molar, solvent or Maxwell–Stefan flow. The trick is to choose the one out of these forms that will give the more benign mathematics. Cussler (2009) gives a good overview. Essentially speaking, his recommendations are summarised in Table 6.1.

A further illustration of this is given with reference to Figure 6.4. If the bulbs in the figure were filled with N_2 and He respectively, there would clearly be counter-diffusion of the two gases. Since N_2 is much heavier the bulb originally dedicated to He would be

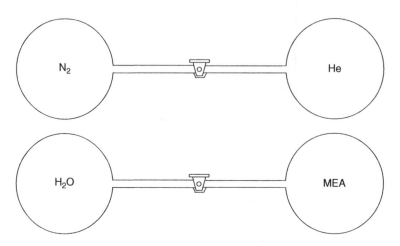

Figure 6.4   *Connected bulbs isolated by stopcock.*

become heavier, and vice versa. In this case it is easy to see that choosing mass (kg) as basis would involve a need to account for the movement of mass whereas the volumes remain unchanged, and so is the case for the number of moles. If the bulbs were filled with water and monoethanolamine (MEA), respectively, that is two liquids of similar densities but different molecular weights, it should be clear that a molar basis should be avoided whereas mass or volume would both be convenient with negligible convection.

6.3   Values of Diffusion Coefficients

Diffusion coefficients are often lacking for the system of interest. Even if some data are available, it is necessary to extrapolate to other concentrations, temperatures or pressures. There are available a number of estimation methods for both gases and liquids that can help us in our endeavours to obtain diffusion coefficients. In general they attempt to combine theory and empiricism. These methods are obviously not perfect. We still need to use them, and should be aware of their limitations and uncertainties involved.

It is not the purpose of the present text to present and review available methods. This has already been done by Poling, Prausnitz and O'Connell (2001). Their book reviews a number of methods and compares them for selected systems. For further discussion of this topic beyond what is done here, this book would be a good first stop. There is no panacea in this field. Judgement must be made when selecting which method to use in any one case. It may even be a case for having measurement made.

In the next two sub-sections the situation for systems used in absorption of CO_2 and H_2S is discussed. There is forever a shortage of data, and we need to know how to deal with that or do necessary improvisation if you like.

6.3.1 Gas Phase Values

A few measurements are available in the literature but there will be shortfalls. Often there is a value at some condition very different from the conditions of interest. Should the same value be assumed to be acceptable, or should a transposition of the data value be attempted? A decision must be made, so what are the alternatives? Here are a few possible avenues that may be open, formulated according to how they are presented by Poling, Prausnitz and O'Connell (2001). They introduce a form of the average molecular weight that is different from the sources of the correlations and this means that the constants are adjusted. The formulae are also adjusted to render the diffusion coefficients in metre square per second.

$$\text{Chapman-Enskog} : D_{12} = \frac{0.266 \times 10^{-6}T^{3/2}}{PM_{AVE}^{1/2}\sigma_{12}^2\Omega} \tag{6.14}$$

$$\text{Wilke-Lee} : D_{12} = \frac{\left[3.03 - \frac{0.98}{M_{AVE}^{1/2}}\right] \times 10^{-3}T^{3/2}}{PM_{AVE}^{1/2}\sigma_{12}^2\Omega} \tag{6.15}$$

$$\text{Fuller, Ensley and Giddings} : D_{12} = \frac{0.143 \times 10^{-6}T^{1.75}}{PM_{AVE}^{1/2}\left[(\Sigma adv)_1^{1/3} + (\Sigma adv)_2^{1/3}\right]^2} \tag{6.16}$$

The version of the Chapman–Enskog[1] equation given is based on a couple of simplifying choices. These equations will give estimates of the diffusion D (m^2/s) of component 1 in 2 or vice versa as these numbers are the same. They all show much the same temperature dependency although Fuller's equation predicts a little higher dependency. All shows an inverse relation to the total pressure P (bar). they are all inversely proportional to the average molecular weight, M_{AVE}, although Wilke and Lee (1955) has a minor (say 5%) adjustment of the estimated value through the numerator modification. Given the broad consensus between the three equations with respect the dependency of diffusion on temperature (T/K), pressure and molecular mass as represented by average molecular weight, it seems reasonable to use these relations when using data under conditions deviating from those of the data source.

The other parameters in these equations need to be introduced. Greek letter σ_{12} is a characteristic length involved with the two gases 1 and 2, and Ω is referred to as a collision integral. These parameters must be found from a table. A few relevant values are given in Table 6.2. Atomic diffusion volumes (Σadv) in the Fuller equation (1966, 1969) must also be found from tabulations, and a few have been included in Table 6.2. The collision

---

[1] It is almost impossible to trace this work back to an original publication although two are given (Chapman and Cowling, 1970; Enskog, 1917). Sydney Chapman from the UK and Daniel Enskog from Sweden arrived at the fundament of this equation independently before corresponding on the matter to take it further. Enskog received his doctorate on the matter at Uppsala in 1917, while Chapman already had an academic career doing research in the area.

Table 6.2 *Parameters for equations to estimate diffusivities in binary gas mixtures. (Data from Cussler, 2009; Fuller et al., 1969.)*

		ε/k	σ	$\Sigma(adv)$
Hydrogen	H_2	59.7	2.827	*6.12*
Nitrogen	N_2	71.4	3.798	*18.5*
Carbo monoxide	CO	91.7	3.690	*18.0*
Argon	Ar	93.3	3.542	*16.2*
Oxygen	O_2	106.7	3.467	*16.3*
Methane	CH_4	148.6	3.758	25.14
Ethane	C_2H_6	215.7	4.443	45.66
Propane	C_3H_8	237.1	5.118	66.18
Ammonia	NH_3	558.3	2.900	*20.7*
Carbonyl sulfide	COS	336.0	4.130	44.91
Carbonyl disulfide	CS_2	467	4.483	61.7
Carbon dioxide	CO_2	195.2	3.941	28.12
Hydrogen sulfide	H_2S	301.1	3.623	27.52
Nitrous oxide	N_2O	232.4	3.828	15.19
Methanol	MeOH	481.8	3.626	31.25
Ethanol	EtOH	362.6	4.530	51.77
Water	H_2O	809.1	2.641	*13.1*
Monoethanol amine	MEA	–	–	58.62

Atomic diffusion volume values given in italics are specific numbers for those molecules, not computed from addition of the atomic contributions. (Hirschfelder *et al.*, 1954; Fuller *et al.*, 1969).

integral used by Chapman–Enskog and Wilke–Lee is temperature dependent and must be calculated for each case:

$$\Omega = \frac{1.06036}{(T^*)^{0.1561}} + \frac{0.193}{\exp(0.47635T^*)} + \frac{1.03587}{\exp(1.52996T^*)} + \frac{1.76474}{\exp(3.89411T^*)} \quad (6.17)$$

Here $T^* = \dfrac{T}{\varepsilon/k}$ $\qquad\qquad\qquad\qquad\qquad\qquad\qquad\qquad\qquad$ (6.18)

$$\varepsilon = \sqrt{\varepsilon_1 \varepsilon_2} \qquad\qquad\qquad\qquad\qquad (6.19)$$

$$\sigma_{12} = 0.5 \left(\sigma_1 + \sigma_2\right) \qquad\qquad\qquad\qquad (6.20)$$

Be aware that the average molecular weight is calculated here according to:

$$M_{AVE} = 2\frac{M_1 M_2}{M_1 + M_2} \qquad\qquad\qquad (6.21)$$

If we need the diffusion coefficient for 5% CO_2 in N_2 at 40°C and 1.5 bar, it is straightforward to estimate this based on these formulae.

$M_{AVE} = 2[(28 \times 44)/(28 + 44)] = 34.2$; $\sigma = 0.5(3.941 + 3.798) = 3.870$
$\varepsilon/k = [(195.2)(71.4)]^{0.5} = 118.1$ $T^* = (40 + 273)/(118.1) = 2.651$
Putting this value for T^* into Equation 6.17 gives $\Omega = 0.983$
For the Fuller equation: $\left[(\Sigma adv)_1^{1/3} + (\Sigma adv)_2^{1/3}\right] = (28.12)^{1/3} + (18.5)^{1/3} = 5.686$.
Chapman–Enskog when putting in the values: $D_{12} = 0.114 \times 10^{-4}$ m²/s

Wilke–Lee $D_{12} = 0.123 \times 10^{-4}$ m²/s
Fuller, Ensley and Giddings $D_{12} = 0.107 \times 10^{-4}$ m²/s.

These values are close. Are they reasonable? One experimental value reported is for an unspecified mixture of CO_2 and N_2 at 1 bar and 25°C, 0.169×10^{-4} m²/s. If this is scaled for process temperature and pressure based on Chapman–Enskog, we get:

$$\left(0.169 \times 10^{-4} \right) (1/1.5)(313/298)^{1.5} = 0.121 \times 10^{-4} \text{m}^2/\text{s}$$

Amdur *et al.* (1952) report that the self-diffusion coefficient in CO_2 at 39.7°C and $P = 0.653$ bar is 0.13×10^{-4} m²/s when they refer their data to 1.013 bar. Estimates for the specification range from 0.12 to 0.13×10^{-4} m²/s.

If all our estimates agreed that well, we would be happy engineers. The good agreement is probably because the system is very much in the midst of the data bank used to develop these correlations. Estimates of diffusion coefficients of the amines used in gas treating in the gas phase would be more uncertain.

6.3.2   Liquid Phase Values

There are, in fact, many diffusivities reported in the literature for solutes in absorbents for the acid gases we are focusing on. However, there is almost always a shortage of data, and if there are data available, they may not be at the right conditions. Also for the liquid phase we need to have the ability to estimate diffusion coefficients. Here is a selection of correlations often referred to.

$$\text{Wilke-Chang (1955) } D = \frac{7.4 \times 10^{-12} T \sqrt{\varphi M_{SOLVENT}}}{\mu V_{SOLUTE}^{0.6}} \tag{6.22}$$

$$\text{Tyn-Calus (1975, a, b) } D = 8.93 \times 10^{-12} \left(\frac{V_{SOLUTE}}{V_{SOLVENT}^2} \right)^{1/6} \left(\frac{P_{A,SOLVENT}}{P_{A,SOLUTE}} \right)^{0.6} \left(\frac{T}{\mu} \right) \tag{6.23}$$

Where the Parachor $P_A = V\sigma^{1/4}$ \tag{6.24}

$$\text{Hayduk-Minhas (1982) } D = 1.25 \times 10^{-12} \left(V_{SOLUTE}^{-0.19} - 0.292 \right) T^{1.52} \mu_W^{\varepsilon^*} \tag{6.25}$$

where $\varepsilon^* = \left(10.2/V_{SOLUTE} \right) - 1.12$ \tag{6.26}

Beware that these are based on a good number of measured diffusion coefficients not particularly relevant to gas treating. We shall next demonstrate reasonable values for amines in aqueous solution. Water is a difficult molecule to deal with in this context because it differs a lot from the organic molecules we otherwise use. In the example in Section 6.7.3 this shall be further discussed in the context of H_2O in triethylene glycol (TEG) solutions.

In these three relations there are parameters appearing in all. It is seen that D is proportional to the absolute temperature T in two, but raised to the power of 1.52 in one, the Hayduk and Minhas (1982). This latter equation, which is for aqueous solutions only in the form given, also differs in how it relates D (m²/s) to the solvent's viscosity, μ (cP). In general the diffusivity is inversely proportional to the viscosity. The viscosity is of course

Table 6.3 *Diffusion coefficient for liquid systems from the literature.*

Solute	Solvent	Diffusion coefficient	Temperature (°C)	References
N_2O	Water	1.86	25	Eimer (1994)[a]
CO_2	Water	1.93	25	Eimer (1994)[a]
Methanol	Water	1.55	25	Derks et al. (2008)
CO_2	Ethanol	4.11	25	Snijder et al. (1995)
N_2O	Ethanol	4.26	25	Snijder et al. (1995)
MEA	0.04 M aq. MEA	1.12	30	Snijder et al. (1993)
DEA	0.01 M aq. DEA	0.84	30	Snijder et al. (1993)
MDEA	0.008 M aq. MDEA	0.79	30	Snijder et al. (1993)
DIPA	0.009 M aq. DIPA	0.71	30	Snijder et al. (1993)
DGA	Water	0.958	30	Chang et al (2005)
TEA	Water	0.796	30	Chang et al (2005)
AMP	Water	0.976	30	Chang et al. (2005)
2-PE	Water	0.826	30	Chang et al. (2005)
MEG	Water	1.241	25	Wang et al. (2009)
DEG	Water	1.039	25	Wang et al. (2009)
TEG	Water	0.815	25	Wang et al. (2009)

Multiply the diffusion coefficient given by 10^{-9} to obtain the value in m^2/s. The references have data at further conditions.
[a] These values are the average of 21 and 22 literature values adjusted for their different choice of gas solubility when interpreting their laboratory data.

also temperature dependent and decreases when the temperature increases. This means that also the two other equations have a higher temperature dependency than immediately seen.

As for the other parameters, there are a few special ones. However, V (l/kmol) is the molar volume of either the solute or solvent. ϕ is referred to as the association factor of the solvent and is dimensionless; it is 2.6 for water, 1.9 for methanol, 1.5 if ethanol. M is molecular weight. In Equation 6.24 σ is the surface tension of the solvent (dyn/cm).

More methods may be found in the literature. These will suffice for the purpose of making quick estimates when data are lacking. Such estimates will easily have error margins of ±20%. More accurate results can be expected when these equations are used to stretch existing data to other conditions.

A few literature values for diffusion coefficients of relevant compounds have been collected in Table 6.3. There are more values at other concentrations and temperatures in the references.

If we try to estimate the diffusion coefficient for MEA in water at infinite dilution, that is when the solvent has the properties of water, we can do as follows.

$\Phi = 2.6$ (for water) Molecular weights are 18.015 (H_2O) and 61.1 (MEA)

Viscosity of water at 40°C, $\mu = 0.651$ cP

Molar volumes are 18.16 l/kmol (water, steam tables), and $61.1(kmol/kg)(1\ kg/m^3) = 61.1$ l/kmol for MEA. (Density data from Han et al., 2012)

Surface tensions at 40°C are 69.6 and 46.7 dyn/cm for H_2O and MEA (Han et al., 2012)

Parachor ratio is then $P_{SOLVENT}/P_{SOLUTE} = (18.16/61.1)(69.6/46.7)^{1/4} = 0.328$

The size $\varepsilon^* = 10.2/61.1 - 1.12 = -0.953$ and $\mu_w^{\varepsilon^*} = 0.651^{-0.953} = 1.505$.

Using these data in the Equations 6.22, 6.23, and 6.25, we obtain:

Wilke–Chang: $D_{MEA} = 2.06 \times 10^{-9} \, m^2/s$
Tyn–Calus: $D_{MEA} = 1.66 \times 10^{-9} \, m^2/s$
Hayduk–Minhas: $D_{MEA} = 1.94 \times 10^{-9} \, m^2/s$.

According to Snijder *et al.* (1993), the diffusivity of MEA in water at 35 and 45°C was measured as 1.39×10^{-9} and $1.75 \times 10^{-9} \, m^2/s$, $1.59 \times 10^{-9} \, m^2/s$. We see that the estimate according to Tyn–Calus is the one nearest to the average of the two Snijder measurements. The values from Wilke–Chang and Hayduk–Minhas overpredicts by 30 and 22%, respectively. This is the situation we have to live with for our estimates when there is a lack of good experimental data.

6.4   Interacting Species

Anyone with some insight to chemistry, not the least to physical chemistry, would be quick to realise there have to be limitations to the aforementioned concepts of diffusion. In a liquid there are a number of effects present that could influence the rate of diffusion. It is important to gain a grasp of this field in order to know when to become precarious in our relations to diffusion.

Starting with strong electrolytes in solutions, it is clear that the nominal molecule will be present as fully dissociated into ions. A common example is NaCl, a neutral salt. Clearly the ions Na^+ and Cl^- cannot move independently. If they did, there would be local charge surpluses arising spontaneously, and even a separation process going on in breach of the second law of thermodynamics. It turns out that the two ions have different diffusivities in an aqueous solution but the faster one moves more in a 'zigzag' in order for the two to move at the same net speed in the direction specified. Care must be taken when obtaining or choosing values of diffusion coefficients to work with. Normally some weighted average to the ionic diffusion coefficients will give good results but the situation may force a deeper dive into the problem.

Conductivity of electrolyte solutions may be exploited to determine diffusion coefficients of molecules experimentally.

For weak electrolytes the picture becomes even more complicated because the solute, say MEA, will not be fully split into ions. The equilibrium relation between the concentrations of ions and molecule is defined by its pK_a value. In gas treating it does not stop there. The amine molecule will react with, for example CO_2 to produce carbamate and perhaps bicarbonate ions as well. For practical mass transfer considerations, it would be desirable to simplify the picture. It is, however, prudent to start with a proper understanding of the problem before simplifying approximations are made.

Weak electrolytes are only one form of associating solutes appearing in diffusion problems. Some surface active materials give rise to micelles. A further area to be aware of is the solution behaviour near phase splits (or the reverse) in liquids that may split into two liquid phases.

There is also solvation going on in a solution. This means that molecules or ions will form some form of cluster with a defined number of solvent molecules. When the solvent is water, the phenomenon is referred to as hydration. Clearly such solvation leads to a larger mass being involved with the diffusion process and this will slow it down.

6.5   Interaction with Surfaces

Catalysts and adsorbents are in general porous materials. (There are some surface catalysts used, however.) When dealing with these materials, care must be taken when analysing mass transfer. In addition to the resistance outside the particles themselves, the diffusion in the pores must be accounted for. If the pores are big, there may be viscous flow through them or bulk (molecular collisions driven) diffusion, but more likely the pore will be smaller, and the diffusion being controlled by molecule-surface collisions. This is known as *Knudsen diffusion*. For even smaller pores there may be surface diffusion or even capillary condensation.

This is a specialist area and most work is done in a limited number of specialist companies. It is, however, always a good idea to understand what is going on.

6.6   Multicomponent Situations

So far we have discussed diffusion more or less as if there was one component diffusing in another. In chemical engineering problems it is common to have multi-component mixtures. Since given binaries interact in different ways leading to different diffusion coefficients for different binary pairs, we would expect there to be analogous effects for ternary relations and so on. Irreversible thermodynamics has also been applied to this field.

There is indeed a lot written about this situation. Ideally we should take all these interactions into account. However, there is a general lack of data available which prevents us from making much impact in this direction. In view of this, multi-component diffusion may for the time being treated as something we ought to know about mainly for the sake of understanding. It is interesting to observe that there have been a few efforts at measuring ternary diffusion data in aqueous mixed amine systems, for example H_2O-DEA-MDEA (Lin, Ko and Li 2009).

A final word on diffusion: It is a transport phenomenon. As such it is not an equilibrium situation in which we can sit back and wait for to happen. To study it, we must gear up to do so while it takes place. It is a complicated process, and much of what has been written is empirically based although a theoretical base has often been the starting point. The wisdom to be extracted is to make use of knowledge gathered but apply healthy criticism. Never stop asking questions. Diffusional mass transfer is a field still being researched.

6.7   Examples

6.7.1   Gaseous CO_2-CH_4

Given a natural gas that is approximated as CO_2 in CH_4 with a pressure of 35 bar and a temperature of 40°C, we want to know the value of the diffusion coefficient.

We shall proceed by using the method suggested by Fuller *et al.* as set out previously in Equations 6.16–6.21:

$$\varepsilon/k = \sqrt{(195.2)(148.6)} = 170.3$$

$$T^* = \frac{273 + 40}{170.3} = 1.838$$

$$\Omega = \frac{1.06036}{(T^*)^{0.1561}} + \frac{0.193}{\exp(0.47635T^*)} + \frac{1.03587}{\exp(1.52996T^*)} + \frac{1.76474}{\exp(3.89411T^*)} = 1.108$$

$$\sigma_{12} = 0.5(3.941 + 3.758) = 3.846$$

$$M_{AVE} = 2\frac{(44)(16)}{44 + 16} = 23.5$$

$$D_{12} = \frac{(0.143 \times 10^{-6})(313^{1.75})}{(35)(23.5)^{1/2}\left[(28.12)_1^{1/3} + (25.14)_2^{1/3}\right]^2} = 5.5 \times 10^{-7}\ m^2/s$$

The values obtained based on the other techniques are 5.3×10^{-7} and $5.6 \times 10^{-7}\ m^2/s$ for the Chapman–Enskog and the Wilke–Lee methods, respectively.

Notice how the increased pressure leads to a much lower diffusion coefficient than what was found for CO_2 in N_2 at atmospheric pressure. There is a very strong pressure dependence for the diffusion coefficient.

6.7.2   Gaseous H_2O–CH_4

In gas drying absorption of H_2O in glycol the system H_2O–CH_4 is of interest. Let us this time demonstrate the Chapman–Enskog method Equations 6.14 and 6.17–6.21. The system temperature is 40°C and 35 bar as before.

$$\varepsilon/k = \sqrt{(809.1)(148.6)} = 346.7$$

$$T^* = \frac{273 + 40}{346.7} = 0.9027$$

$$\Omega = \frac{1.06036}{(T^*)^{0.1561}} + \frac{0.193}{\exp(0.47635T^*)} + \frac{1.03587}{\exp(1.52996T^*)}$$

$$+ \frac{1.76474}{\exp(3.89411T^*)} = 1.5158$$

$$\sigma_{12} = 0.5(2.641 + 3.758) = 3.200$$

$$M_{AVE} = 2\frac{(18)(16)}{18 + 16} = 16.94$$

$$\text{Chapman-Enskog}: D_{12} = \frac{0.266 \times 10^{-6}(313)^{3/2}}{(35)(16.94)^{1/2}(3.200)^2(1.5158)} = 6.6 \times 10^{-7}\ m^2/s$$

The values based on Wilke–Lee and Fuller *et al.*'s methods are 6.9×10^{-7} and $8.3 \times 10^{-7}\ m^2/s$ respectively.

6.7.3   Liquid Phase Diffusion of H_2O in TEG

To estimate the diffusion coefficient in TEG is actually not straightforward. Poling, Prausnitz and O'Connell (2001) discuss the various correlations listed previously. It appears that the Wilke–Chang correlation is not good when water is the solute. In the Tyn–Calus method water should be considered as a dimer. This means that its molar volume is doubled and this affects its Parachor value as well. Furthermore, this correlation is not recommended to be used for viscous solvents and the limit with respect to this is 20–30 cP. Pure TEG at 40°C has a viscosity of 25 cP. The variation of the Hayduk–Minhas' correlation previously presented is good for aqueous solutions, not for TEG. For this particular system their recommended correlation is:

$$D = \left(1.55 \times 10^{-12}\right) \frac{T^{1.29} P_{A,SOLVENT}^{0.5}}{\mu^{0.92} V_{SOLVENT}^{0.23} P_{A,SOLUTE}^{0.42}} \tag{6.27}$$

Relevant data at 40°C for these liquids are (Dow's TEG data (2007), water from 'steam tables')

	Molar volume (cm^3/mol)	Viscosity (cP)	Surface tension (dyn/cm)	Parachor	Parachor dimer
Water	18.2	–	43.5	52.3	104.7
TEG	135	25	69	346	–

In spite of the warning we try out the Wilke–Chang correlation out of interest:

$$D = \frac{7.4 \times 10^{-12} \, (313) \, \sqrt{(1.0)(150)}}{(25)(18.2)^{0.6}} = 2 \times 10^{-10} \text{ m}^2/\text{s}$$

We also try out the Tyn–Calus 'border line' correlation, but with H_2O dimer:

$$D = 8.93 \times 10^{-12} \left(\frac{2 \times 18.2}{135^2}\right)^{1/6} \left(\frac{346}{104.7}\right)^{0.6} \left(\frac{313}{25}\right) = 0.8 \times 10^{-10} \text{ m}^2/\text{s}$$

Finally we try the version of Hayduk–Minhas given in Equation 6.27.

$$D = \left(1.55 \times 10^{-12}\right) \frac{313^{1.29} 346^{0.5}}{25^{0.92} 135^{0.23} 52.3^{0.42}} = 1.5 \times 10^{-10} \text{ m}^2/\text{s}$$

We choose to use the latter value for later example estimates. The wisdom of this will be discussed at that point. At least we have an estimate for this diffusion coefficient, which is more likely than not better than a guess.

References

Amdur, I., Irvine, J.W. Jr., Mason, E.A. and Ross, J. (1952) Diffusion coefficients of the systems CO_2–CO_2 and CO_2–N_2O. *J. Chem. Phys.*, **20**, 436–443.

Chang, L.-C., Lin, T.-I. and Li, M.-H. (2005) Mutual diffusion coefficients of some aqueous alkanolamines solutions. *J. Chem. Eng. Data*, **50**, 77–84.

Chapman, S. and Cowling, T.G. (1970) *The Mathematical Theory of Non-Uniform Gases*, 3rd edn, Cambridge University Press (A 1991 version with foreword by Cercignani, C. is available).

Cussler, E. (2009) *Diffusion. Mass Transfer in Fluid Systems*, 3rd edn, Cambridge University Press, Cambridge.

Derks, P.W.J., Hamborg, E.S., Hogendoorn, J.A., Niederer, J.P.M. and Versteeg, G.F. (2008) Densities, viscosities, and liquid diffusivities in aqueous (piperazine + N-methyldiethanolamine) solutions. *J. Chem. Eng. Data*, **53**, 1179–1185.

Dow (2007) *Triehtylene Glycol*, Dow Chemical Co.

Eimer, D. (1994) *Simultaneous removal of water and hydrogen sulphide from natural gas*. Dr Ing. Thesis, NTH, Trondheim.

Enskog, D. (1917) Kinetische Theorie der Vorgange in massig verdunnten Gasen. Doctoral thesis. Uppsala Universitet, Uppsala. (Only quoted for completeness).

Fuller, E.N., Schettler, P.D. and Giddings, J.C. (1966) A new method for prediction of binary gas-phase diffusion coefficients. *Ind. Eng. Chem.*, **58** (5), 19–27.

Fuller, E.N., Ensley, K. and Giddings, J.C. (1969) Diffusion of halogenated hydrocarbons in helium. The effect of structure on collision cross sections. *J. Phys. Chem.*, **73**, 3679–3685.

Hamborg, E.S., Derks, P.W.J., Kerstens, S.R.A., Niederer, J.P.M. and Versteeg, G.F. (2008) Diffusion coefficients of N_2O in aqueous piperazine solutions using the Taylor dispersion technique from (293 to 333) K and (0.3 to 1.4) $mol.dm^{-3}$. *J. Chem. Eng. Data*, **53**, 1462–1466.

Han, J., Jin, J., Eimer, D.A. and Melaaen, M.C. (2012) Density of water (1) + monoethanolamine (2) + CO_2 (3) from (298.15 to 413.15) K and surface tension of water (1) + monoethanolamine (2) from (303.15 to 333.15) K. *J. Chem. Eng. Data*, **57**, 1095–1103.

Hayduk, W. and Minhas, B.S. (1982) Correlations for prediction of molecular diffusivities in liquids. *Can. J. Chem. Eng.*, **60**, 295–299.

Hirschfelder, J.O., Curtiss, C.F. and Bird, R.B. (1954) *Molecular Theory of Gases and Liquids*, John Wiley & Sons, Inc., New York.

Lin, P.-H., Ko, C.-C. and Li, M.-H. (2009) Ternary diffusion coefficients of diethanolamine and N-methyldiethanolamine in aqueous solutions containing diethanolamine and N-methyladiethanolamine. *Fluid Phase Equilib.*, **276**, 69–74.

Poling, B.E., Prausnitz, J.M. and O'Connell, J.P. (2001) *The Properties of Gases and Liquids*, 5th edn, McGraw-Hill.

Snijder, E.D., te Riele, M.J.M., Versteeg, G.F. and van Swaaij, W.P.M. (1993) Diffusion coefficients of several aqueous alkanolamine solutions. *J. Chem. Eng. Data*, **38**, 475–480.

Snijder, E.D., te Riele, M.J.M., Versteeg, G.F. and van Swaaij, W.P.M. (1995) Diffusion coefficients of CO, CO_2, N_2O, and N_2 in ethanol and toluene. *J. Chem. Eng. Data*, **40**, 37–39.

Tyn, M.T. and Calus, W.F. (1975a) Diffusion coefficients in dilute binary liquid mixtures. *J. Chem. Eng. Data*, **20**, 106–109.

Tyn, M.T. and Calus, W.F. (1975b) Temperature and concentration dependence of mutual diffusion coefficients of some binary liquid systems. *J. Chem. Eng. Data*, **20**, 310–316.

Wang, M.-H., Soriano, A.N., Caparanga, A.R. and Li, M.-H. (2009) Mutual diffusion coefficients of aqueous solutions of some glycols. *Fluid Phase Equilib.*, **285**, 44–49.

Wilke, C.R. and Chang, P. (1955) Correlation of diffusion coefficients in dilute solutions. *AIChE J.*, **1**, 264–270.

Wilke, C.R. and Lee, C.Y. (1955) Estimation of diffusion coefficients for gases and vapors. *Ind. Eng. Chem.*, **47**, 1253–1257.

Further Reading

Broad and Fundamental Discussion of Diffusion:

Sherwood, T.K., Pigford, R.L. and Wilke, C.R. (1975) *Mass Transfer*, McGraw-Hill.

Diffusion in a Transport Phenomena Perspective:

Bird, R.B., Stewart, W.E. and Lightfoot, E.N. (2002) *Transport Phenomena*, 2nd edn, John Wiley & Sons, Inc., New York.

Analytical Solutions to Specific Problems:

Carslaw, H.S. and Jager, J.C. (1947) *Conduction of Heat in Solids*, Clarendon Press, Oxford.

Crank, J. (1975) *The Mathematics of Diffusion*, 2nd edn, Clarendon Press, Oxford.

Associated Diffusion and Chemical Reaction:

Danckwerts, P.V. (1970) *Gas-Liquid Reactions*, McGraw-Hill.

7

Absorption Column Mass Transfer Analysis

7.1 Introduction

Absorption as a unit operation is rooted in various gases with different solubilities in the liquid used to wash out the desired gas components. It is a gas–liquid process. The liquid to be used is picked with care to provide to the best possible medium to effect the targeted separation.

The liquid used to bring about absorption is referred to as an '*absorbent*'. The gas absorbed is called the '*absorbate*'. The absorbent entering an absorption column (near its top) contains little absorbate and is often referred to as '*lean absorbent*' while the exiting absorbent with its higher load of absorbate is similarly referred to as the '*rich absorbent*'.

An absorption process is used to bring about a gas separation. In principle any amount of gas can be absorbed but the less to be absorbed, the easier it is to achieve the target. The liquid can be purely a non-reacting solvent, or it can be a combination with chemicals added to increase affinity through a chemical reaction, or reactions.

The application area for absorption is wide. It covers acid gas removal in gas treating where the amount of gas to be removed may vary from a few percent to 50% or more in extreme cases. For gas to pipelines the end specification for CO_2 is typically 2–4% while H_2S is removed to 2–4 ppm. A liquified natural gas (LNG) train would require CO_2 content to be lowered to 50 ppm. In synthesis gas trains like an ammonia plant, there will typically be 20% of CO_2 to be removed to as low a level as possible, with 100 ppm seen as an achievable target. These are tall columns like 20–30 m. Water removal from natural gas on the other hand is a relatively easy separation where the column could be as low as four trays or 3 m of packing. This kind of height may also be typical for many environmentally caused end-of-pipe clean-up operations.

The design challenge to be met by the chemical engineer is simple enough:

- Decide on an absorbent.
- Choose the column hardware.
- Determine the height of column needed.
- Find a diameter that will be good for mass transfer and pressure drop.

For simple tasks this can be done with pen and paper and perhaps a calculator. A bit of experience will also help. There are also design tools available to help out when the problem gets complex. All the well-known flowsheet simulators have column models included but these are of varying capability. Modern thinking on absorption column simulation goes along the line of using a rate based approach. Tray efficiencies are usually low. Even determining these would require some kinetic considerations if the numbers are to be meaningful unless experience numbers are available.

There are also operational challenges that must be considered at the design stage. The system to be used may have a propensity for foaming in which case the specific pressure drop (Pa/m) may need to be kept lower than otherwise optimal. The solution may cause solid deposits that must be allowed for both with respect to pressure drop and periodic cleaning. In gas treating with amines there is a need to filter the solution for maintenance but that is done externally to the column.

7.2   The Column

Most columns are designed for counter-current flow of gas and absorbent, but co-current flow is an alternative. For easy separations co-current flow is worth considering. Liquid must naturally always be fed from the top unless a liquid-continuous bubble column is planned for, but that is rare and we leave that aside as a special case.

The column's function is essentially to provide a gas–liquid contact area to facilitate the required separation. Contact area can be provided by sprays, by packings or by trays. The best choice depends on the situation. There are a few practicalities with respect to how this is done but that will be discussed in a later chapter. In the bottom of the column it is customary to provide a liquid sump that will provide a means of gas bubble disengagement, a seal against gas leaving with the rich liquid and also provision of a small buffer volume. An absorption column will also comprise a few seemingly 'inactive' parts in the sense that no, or negligible, separation takes part in these sections. In this chapter we shall stick to design of the active parts.

7.3   The Flux Equations

The flux equations have already been introduced in the rate of mass transfer chapter. A simple repeat is in order. The rate equations for absorption mass transfer may be written either in terms of the gas side phenomena:

$$N_i = k_G \left(C_i^{Gb} - C_i^{Gi} \right) \tag{7.1}$$

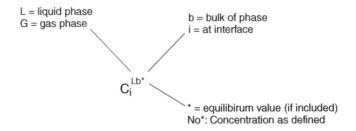

L = liquid phase
G = gas phase

b = bulk of phase
i = at interface

C_i^{Lb*}

* = equilibirum value (if included)
No*: Concentration as defined

Figure 7.1 *Notation used for concentrations.*

or alternatively in terms of the liquid side phenomena:

$$N_i = k_L^0 \left(C_i^{Li} - C_i^{Lb} \right) \tag{7.2}$$

There is a free choice. One is as good as the other, and they both express the same quantity, namely the rate of transfer of component 'i' from one phase to the other. Arguments could be raised for choosing either one. N_i is the flux of component 'i' (mol m^{-2} s^{-1}), k_G and k_L^0 the gas and liquid side mass transfer coefficients, respectively (m s^{-1}), C_i is the concentration of 'i' in situations as explained in Figure 7.1 (mol m^{-3}).

It is fully feasible to estimate the concentration values of 'i' at the interface, but it would involve a bit of trial and error. Using the interface values is not convenient, and for this reason we seek a way around. That may be done by introducing another concept, the overall mass transfer coefficients that will allow us to use the bulk values in the flux equations instead.

7.4 The Overall Mass Transfer Coefficients and the Interface

Let us start by reminding ourselves that the inverse of the mass transfer coefficient for a mass transfer region is a resistance to mass transfer. (The term 'region' is used to avoid the use of 'film' in order to avoid confusion with the Whitman two-film theory.) Figure 7.2 shows an overall picture of the situation of the various concentrations and the interface.

The resistances on the gas and liquid side are resistances in series. They may not be directly summed, however, since there is something happening on the interface itself. The argumentation goes as follows.

At the interface we assume that there is equilibrium between the concentrations of 'i' in the gas in the liquid. The partition coefficient that defines this equilibrium is defined as:

$$m_i = \frac{C_i^G}{C_i^L} \tag{7.3}$$

in terms of the concentrations exposed to each other.

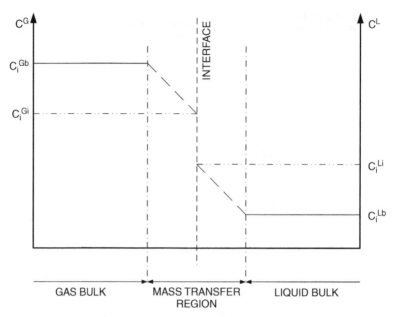

Figure 7.2   *The figure shows the overall picture of the various concentrations of the transferred component and the situation around the interface. The profiles in the transfer region are arbitrary.*

7.4.1   Overall Gas Side Mass Transfer Coefficient

We can now introduce the equilibrium relation 7.3 into the rate expression 7.1 which becomes:

$$N_i = k_G \left(C_i^{Gb} - m_i C_i^{Li} \right) \tag{7.1a}$$

From Equation 7.2 an expression is obtained for C_i^{Li}:

$$C_i^{Li} = \frac{1}{k_L^0} \left(N_i + k_L^0 C_i^{Lb} \right) \tag{7.2a}$$

Next we eliminate the interface value by substituting Equation 7.2(a) into Equation 7.1(a):

$$N_i = k_G \left\{ C_i^{Gb} - \frac{m_i}{k_L^0} \left(N_i + k_L^0 C_i^{Lb} \right) \right\} \tag{7.1b}$$

Solving for the flux N_i gives:

$$N_i = \frac{k_G}{\left(1 + \dfrac{m_i k_G}{k_L^0} \right)} \left\{ C_i^{Gb} - m_i \, C_i^{Lb} \right\} \tag{7.1c}$$

The last term in Equation 7.1(c) defines a gas concentration that is in equilibrium with the concentration in the bulk of the liquid. See Figure 7.2. This is defined by:

$$C_i^{Gb*} = m_i C_i^{Lb} \tag{7.3}$$

When Equation 7.3 is substituted into Equation 7.1(c), and Equation 7.1(c) is multiplied by $(1/k_G)/(1/k_G)$ and subsequently rearranged, we obtain a new expression for the flux of '*i*':

$$N_i = \frac{1}{\left(\dfrac{1}{k_G} + \dfrac{m_i}{k_L^0}\right)} \left\{ C_i^{Gb} - C_i^{Gb*} \right\} \tag{7.1d}$$

Inspecting Equation 7.1(d) we see that the driving force is expressed in terms of concentrations that are bulk properties of the gas and the liquid.

We next define a proportionality constant such that:

$$\frac{1}{K_G} = \frac{1}{k_G} + \frac{m_i}{k_L^0} \tag{7.4}$$

Equation 7.4 represents the sum of mass transfer resistances in series where the interface effect has been dealt with. Past research has ruled out any resistance to mass transfer by the interface itself. The new proportionality constant K_G we shall refer to as the overall gas side mass transfer coefficient. This is a constant to the extent that the partition coefficient and the two mass transfer region (film) coefficients are constants. Substituting Equation 7.4 into Equation 7.1(d), the rate equation may now be written:

$$N_i = K_G \left\{ C_i^{Gb} - C_i^{Gb*} \right\} \tag{7.1e}$$

7.4.2   Overall Liquid Side Mass Transfer Coefficient

The same exercise may be done by solving the other equation for the interface value and substituting as before.

This time we introduce the equilibrium expression into the rate expression 7.2 which becomes

$$N_i = k_L^0 \left(\frac{C_i^{Gi}}{m_i} - C_i^{Lb} \right) \tag{7.2b}$$

From Equation 7.1:

$$C_i^{Gi} = C_i^{Gb} - \frac{N_i}{k_G} \tag{7.1f}$$

Next we eliminate the interface value by substituting Equation 7.1(f) into Equation 7.2(b):

$$N_i = k_L^0 \left\{ \frac{C_i^{Gb}}{m_i} - \frac{N_i}{m_i k_G} - C_i^{Lb} \right\} \tag{7.2c}$$

The first term in Equation 7.2(c) defines a gas concentration that is in equilibrium with the concentration in the bulk of the liquid. See Figure 7.2. This is defined by:

$$C_i^{Lb*} = \frac{C_i^{Gb}}{m_i} \tag{7.3a}$$

Solving for the flux of 'i' N_i and using Equation 7.3(a) in Equation 7.2(c) gives:

$$N_i = \frac{k_L^0}{\left(1 + \dfrac{k_L^0}{m_i k_G}\right)} \left\{ C_i^{Lb*} - C_i^{Lb} \right\} \tag{7.2d}$$

When Equation 7.2(d) is multiplied by $(1/k_L^0)/(1/k_L^0)$ and subsequently rearranged, we obtain a new expression for the flux of 'i':

$$N_i = \frac{1}{\left(\dfrac{1}{k_L^0} + \dfrac{1}{m_i k_G}\right)} \left\{ C_i^{Lb*} - C_i^{Lb} \right\} \tag{7.2e}$$

Inspecting Equation 7.2(e) we see that the driving force is expressed in terms of concentrations that are bulk properties of the gas and the liquid. The first term is derived from the bulk concentration of i in the gas and the second term is simply the concentration of i in the liquid bulk.

We next define a proportionality constant such that:

$$\frac{1}{K_L} = \frac{1}{k_L^0} + \frac{1}{m_i k_G} \tag{7.5}$$

The new proportionality constant K_L we shall refer to as the overall liquid side mass transfer coefficient. This is a constant to the extent that the partition coefficient and the two mass transfer region (film) coefficients are constants. Substituting Equation 7.5 into Equation 7.2(e), the rate equation may now be written:

$$N_i = K_L \left\{ C_i^{Lb*} - C_i^{Lb} \right\} \tag{7.2f}$$

7.5   Control Volumes, Mass and Energy – Balances

Now for the intricate part of column design, how to estimate the height of active column needed. First, a few words on modelling in general. Models are built from a number of smaller 'part models'. An equilibrium model is one such part model and even that is divided into sub-parts. In flow situations it is customary to pick a so-called control volume, make all the appropriate balances (like mass and energy equations) over that volume. This will give one equation, or a set of equations, which may be solved. In complex chemical engineering problems this is a way of breaking the problem down into smaller bits that are manageable. There is a certain skill involved in defining the control elements and that can only be learned by experience. Managing and understanding control volumes are important. However, this is enough for an introduction, now we shall indulge in control volumes and equations.

7.5.1   The Relation between Gas and Liquid Concentrations

Let us first consider the control volume in Figure 7.3(a) defined by the column segment from z to dz. As usual a mass balance for component 'i' over the control volume is taken.

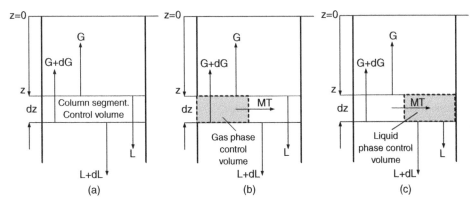

Figure 7.3   *(a–c) The control elements used to develop column models. The column cross–sectional area is A (m²). The total flows are indicated without subscripts and are generic with respect to units. 'z' is the distance from the top of the column.*

Our target is to obtain a relation between the gas phase and liquid phase concentrations of component '*i*' based on mass balance considerations. The approach is as always that:

$$\text{Accumulated} = \text{In} - \text{Out}$$

Since steady state is considered, this becomes:

$$0 = \left\{ \left[\left(G_V + dG_V \right) \left(C_i^{Gb} + dC_i^{Gb} \right) + L_V C_i^{Lb} \right] - \left[G_V C_i^{Gb} + \left(L_V + dL_V \right) \left(C_i^{Lb} + dC_i^{Lb} \right) \right] \right\}$$
(7.6)

This expression is correct but not easily useful. Its use would have to be as part of a set of equations to be solved numerically in a computer model. At this stage it is customary in text books to assume constant flow of gas and liquid. The Equation 7.6 then becomes:

$$0 = \left\{ \left[\left(G_V \right) \left(C_i^{Gb} + dC_i^{Gb} \right) + L_V C_i^{Lb} \right] - \left[G_V C_i^{Gb} + \left(L_V \right) \left(C_i^{Lb} + dC_i^{Lb} \right) \right] \right\}$$
(7.6a)

When this is ordered and rearranged, we obtain the following expression that can be integrated as follows:

$$\int_{C_i^{Lb0}}^{C_i^{Lb}} L_V dC_i^{Lb} = \int_{C_i^{Gb0}}^{C_i^{Gb}} G_V dC_i^{Gb}$$
(7.6b)

which yields:

$$C_i^{Gb} = \frac{L_V}{G_V} \left(C_i^{Lb} - C_i^{Lb0} \right) + C_i^{Gb0}$$
(7.6c)

The relationship in this form is straightforward to understand, but the limitations must be kept in mind. This equation could obviously be rearranged to yield the liquid concentrations as an explicit function of the gas concentrations:

$$C_i^{Lb} = \frac{G_V}{L_V} \left(C_i^{Gb} - C_i^{Gb0} \right) + C_i^{Lb0}$$
(7.6d)

The Equations 7.6(c,d) are referred to as the 'operating lines' in text books. Which one is used depends on whether we work liquid or gas-phase oriented.

7.5.2   Height of Column Based on Gas Side Analysis

The ground has now been prepared for analysing the need for column height. This may be done by taking a mass balance for component 'i' over the gas control volume shown in Figure 7.3(b). As before:

$$\text{Accumulated} = \text{In} - \text{Out}$$

Looking at the steady state condition and assuming constant gas flow, this balance becomes:

$$0 = \left[G_V \left(C_i^{Gb} + dC_i^{Gb} \right) \right] - \left[G_V C_i^{Gb} + N_i a A dz \right] \tag{7.7}$$

The mass transfer flux N_i is obtained from Equation 7.1(e). Substituting for this and eliminating the terms that cancel out:

$$0 = G_V dC_i^{Gb} - K_G \left\{ C_i^{Gb} - C_i^{Gb*} \right\} a A dz \tag{7.7a}$$

and by rearranging this equation:

$$dz = \frac{G_V}{K_G a A} \frac{dC_i^{Gb}}{C_i^{Gb} - C_i^{Gb*}} \tag{7.7b}$$

The equilibrium relation:

$$C_i^{Gb*} = m_i C_i^{Lb} \tag{7.3}$$

and the operating line:

$$C_i^{Lb} = \frac{G_V}{L_V} \left(C_i^{Gb} - C_i^{Gb0} \right) + C_i^{Lb0} \tag{7.6d}$$

give the extra information needed to allow Equation 7.7(b) to be integrated from the $z = 0$ to L with the corresponding values of the gas phase bulk concentration of 'i' as the other limits. It is easy to do this numerically using a computer.

7.5.3   Height of Column Based on Liquid Side Analysis

A similar analysis to that done for the gas side may be done for the liquid side. This may be done by taking a mass balance for component 'i' over the liquid control volume shown in Figure 7.3(c). As before:

$$\text{Accumulated} = \text{In} - \text{Out}$$

Looking at the steady state condition and assuming constant liquid flow, this balance is:

$$0 = \left\{ \left[L_V C_i^{Lb} \right] + N_i a A dz - \left[(L_V) \left(C_i^{Lb} + dC_i^{Lb} \right) \right] \right\} \tag{7.8}$$

The mass transfer flux N_i is obtained from Equation 7.2(f). Putting in for this and eliminating the terms that cancel out:

$$0 = K_L \left\{ C_i^{Lb*} - C_i^{Lb} \right\} a A dz - L_V dC_i^{Lb} \tag{7.8a}$$

And by rearrangement:

$$dz = \frac{L_V}{K_L aA} \frac{dC_i^{Lb}}{C_i^{Lb*} - C_i^{Lb}} \tag{7.8b}$$

The equilibrium relation:

$$C_i^{Lb*} = \frac{C_i^{Gb}}{m_i} \tag{7.3a}$$

and the operating line:

$$C_i^{Gb} = \frac{L_V}{G_V} \left(C_i^{Lb} - C_i^{Lb0} \right) + C_i^{Gb0} \tag{7.6c}$$

give the extra information needed to allow Equation 7.8(b) to be integrated from the $z=0$ to L with the corresponding values of the gas phase bulk concentration of 'i' as the other limits. It is easy to do this numerically using a computer.

7.6   Analytical Solution and Its Limitations

Going back to the gas side analysis there is an analytical solution for the case where the equilibrium relation is a straight line as well as the assumptions made earlier concerning constant gas and liquid flows. The equation is integrated from $z=0$ to L and the corresponding values for gas concentrations:

$$\int_0^L dz = \int_{C_i^{Gb0}}^{C_i^{GbL}} \frac{G_V}{K_G aA} \frac{dC_i^{Gb}}{C_i^{Gb} - C_i^{Gb*}} \tag{7.7c}$$

From Equations 7.3 and 7.6(d) it follows that:

$$C_i^{Gb*} = m_i \left\{ \frac{G_V}{L_V} \left(C_i^{Gb} - C_i^{Gb0} \right) + C_i^{Lb0} \right\} \tag{7.3b}$$

Introducing Equation 7.3(b) into Equation 7.7(c):

$$\int_0^L dz = \int_{C_i^{Gb0}}^{C_i^{GbL}} \frac{G_V}{K_G aA} \frac{dC_i^{Gb}}{C_i^{Gb} - \left[m_i \frac{G_V}{L_V} \left(C_i^{Gb} - C_i^{Gb0} \right) + m_i C_i^{Lb0} \right]} \tag{7.7d}$$

$$\int_0^L dz = \int_{C_i^{Gb0}}^{C_i^{GbL}} \frac{G_V}{K_G aA} \frac{dC_i^{Gb}}{\left(1 - m_i \frac{G_V}{L_V} \right) C_i^{Gb} - m_i \left[C_i^{Lb0} - \frac{G_V}{L_V} C_i^{Gb0} \right]} \tag{7.7e}$$

A solution to this may be found from a table of integrals and it is:

$$L = \left(\frac{G_V}{K_G aA} \right) \left(\frac{1}{1 - m_i \frac{G_V}{L_V}} \right) \ln \frac{\left[\left(1 - m_i \frac{G_V}{L_V} \right) C_i^{GbL} - m_i \left[C_i^{Lb0} - \frac{G_V}{L_V} C_i^{Gb0} \right] \right]}{\left[\left(1 - m_i \frac{G_V}{L_V} \right) C_i^{Gb0} - m_i \left[C_i^{Lb0} - \frac{G_V}{L_V} C_i^{Gb0} \right] \right]} \tag{7.7f}$$

This expression is easy enough to use but simplifications are feasible. After the integration terms may be eliminated in the denominator:

$$L = \left(\frac{G_V}{K_G aA}\right)\left(\frac{1}{1 - m_i\dfrac{G_V}{L_V}}\right)\ln\left[\frac{\left(1 - m_i\dfrac{G_V}{L_V}\right)C_i^{GbL} - m_i\left[C_i^{Lb0} - \dfrac{G_V}{L_V}C_i^{Gb0}\right]}{C_i^{Gb0} - m_iC_i^{Lb0}}\right] \tag{7.7g}$$

Next we may enter the values at the column bottom for the generic bulk values in Equation 7.6(c) such that this equation now becomes a mass balance with respect to '*i*' over the column, and rearrange this equation. It really now represents a new set of information and as such constitutes a new equation:

$$C_i^{Gb0} = C_i^{GbL} - \frac{L_V}{G_V}\left(C_i^{LbL} - C_i^{Lb0}\right) \tag{7.9}$$

Equation 7.9, which is a new relation, is substituted into the numerator of Equation 7.7(g):

$$L = \left(\frac{G_V}{K_G aA}\right)\left(\frac{1}{1 - m_i\dfrac{G_V}{L_V}}\right)$$

$$\times \ln\left[\frac{\left(1 - m_i\dfrac{G_V}{L_V}\right)C_i^{GbL} - m_i\left[C_i^{Lb0} - \dfrac{G_V}{L_V}\left[C_i^{GbL} - \dfrac{L_V}{G_V}\left(C_i^{LbL} - C_i^{Lb0}\right)\right]\right]}{C_i^{Gb0} - m_iC_i^{Lb0}}\right] \tag{7.7h}$$

After tidying up the numerator, this expression now becomes:

$$L = \left(\frac{G_V}{K_G aA}\right)\left(\frac{1}{1 - m_i\dfrac{G_V}{L_V}}\right)\ln\left[\frac{C_i^{GbL} - m_iC_i^{LbL}}{C_i^{Gb0} - m_iC_i^{Lb0}}\right] \tag{7.7i}$$

As for the gas side analysis carried out previously there is an analytical solution also for the liquid side analysis for the case where the equilibrium relation is a straight line as well as the assumptions made earlier concerning constant gas and liquid flows. The equation is integrated from $z = 0$ to L and the corresponding values for gas concentrations. The procedure is like that already shown previously.

The assumptions made in the development of this analytical solution are worth restating since they are important. They include:

- Constant flows of liquid and gas.
- A linear equilibrium relation for gas solubility. (The partition function):

  - Isothermal column would be implicit.

- Constant mass transfer coefficient.

- Constant contact area (a) and column cross-sectional area (A).
- Only one component needs to be analysed for absorption:

  - This has to do with interactions and the need to use two sets of mass balances.
  - Two, or more, components may also influence the absorption equilibrium of the other(s).

There are many instances where simplified solutions based on these approximations are in order. Early phase evaluations where accurate numbers are not required is one area where the assumptions behind the formulae may be stretched. At the end of the day, it is the engineer who does the analysis that will need to make a decision with respect to the accuracy needed and act accordingly. There is no fixed and absolute answer to what is a reasonable approximation! An engineer must use judgement!

7.7   The NTU–HTU Concept

Many textbooks make use of the concept of transfer units. NTU is the number of transfer units needed and this describes the difficulty of the separation at hand. HTU is the height of a transfer unit given in metres, and this is a property of the column in question. The height of the column becomes:

$$L = (HTU)(NTU)$$

From Equation 7.7(e) we have, after a fraction has been moved outside the integration sign, information that this column length is:

$$L = \frac{G_V}{K_G a A} \int_{C_i^{Gb0}}^{C_i^{GbL}} \frac{dC_i^{Gb}}{\left(1 - m_i \frac{G_V}{L_V}\right) C_i^{Gb} + m_i \left[C_i^{Lb0} - \frac{G_V}{L_V} C_i^{Gb0}\right]} \tag{7.7j}$$

Here the fraction in front of the integration sign is the HTU_G while the integration part is the NTU_G. The subscript G has been introduced since the relation in Equation 7.7(j) is related to the gas side analysis.

A similar consideration could be done for the liquid side to obtain HTU_L and NTU_L. In general:

$$HTU_G \neq HTU_L$$

and

$$NTU_G \neq NTU_L$$

However:

$$L = (HTU_G)(NTU_G) = (HTU_L)(NTU_L)$$

The column will be the same height no matter if the analysis is approached from the gas side or the liquid side.

There are also the terms height equivalent to a theoretical plate (HETP) and number of equivalents of a theoretical plate (NETP) that sometimes appear. A HETP is the height equivalent to a theoretical plate. This is given in metres and is measure of the apparatus' effectiveness, and is based on experience or measurement. The NETP is simply the number of theoretical stages (plates) estimated and is a measure of the difficulty of making the separation at hand. If the traditional McCabe–Thiele 'ladder-steps' approach is used instead of the previous rate based methodology, the number of steps is the NETPs. This is further described in traditional chemical engineering textbooks dealing with column design. An example is shown in Section 7.12 without going into detail.

7.8   Operating and Equilibrium Lines – A Graphical Representation

To understand the operating line it is useful to show it graphically along with the equilibrium curve. This is done in Figure 7.4(a). For absorption processes the operating line will always be above the equilibrium line. If it was not, there would be no driving force to give absorption. (For desorption processes the operating line would be below the equilibrium line.) The figure also shows that the driving force is quite small in the top of the column where the concentrations are lower. As these curves approach each other we say that the column is pinching. The pinch point is when these lines nearly touch. A pinch could be in the top, the bottom, and in the case of significant absorption heat effects, a pinch could also be somewhere in-between top and bottom.

In Figure 7.4(b) the integrand that is used to find the NTUs is plotted against the same concentration. The area under the curve represents the NTUs. It is easily seen that the more NTUs are needed per unit concentration change near to the pinching end of the column.

Making these kinds of plots is actually quite useful even if the whole estimation process is done with a computer tool. Viewing the situation graphically more easily reveals points like those discussed earlier.

It is a good idea to always plot the column profiles. Such plots can easily pinpoint sections that are little used because of pinching and they will also quickly tell if there are gains to

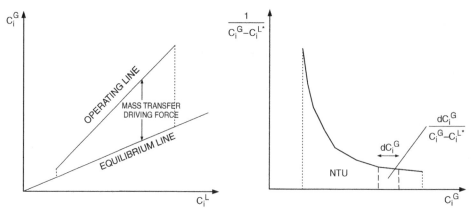

Figure 7.4   *(a) Graphical representation of operating and equilibrium lines. (b) Graphical representation of number of transfer units highlighting the increased need as a pinch point is approached.*

be made by making the column longer. This kind of consideration is important for column design. Engineers should always strive towards more cost effective units.

7.9   Other Concentration Units

The discussion in the preceding sections has been based on the use of molar concentrations of the species in both phases. This is the author's preference. The explanation is that dimensionless correlations for mass transfer coefficients provide these via the Sherwood number or equivalent and in this dimensionless number the mass transfer coefficient comes in the units $m\,s^{-1}$. Multiplying this with a concentration ($mol\,m^{-3}$) gives the flux. Sticking to this we can always compare mass transfer coefficients on a like basis, and this helps to develop a gut feel for the numbers. Such a reference frame is useful as a first check to see if the numbers obtained are reasonable. Furthermore, mass balances are easily dealt with based on volume flows and concentrations.

However, the literature uses a number of ways of expressing concentrations and thus driving forces for absorption. Before the event of computers it was useful to arrange equations and choose variables to keep flows constant through the column. 'Constant' is here defined by the equations to be used such as to make manual solutions less laborious. Let us relate the present scheme to the others one by one and see what the relations are.

Mole fraction is the first alternative. This may be used in both phases. It is convenient if the flow unit is $mol\,s^{-1}$. The relation is simply that the mole fraction of component '*i*':

$$x_i = \frac{C_i^L}{C_{tot}^L} \tag{7.10}$$

where the denominator is simply the molar density of the liquid phase in $mol\,m^{-3}$. This is the sum of all the molar concentrations for all species present. For the gas phase it is customary to use y for mole fraction, and hence:

$$y_i = \frac{C_i^G}{C_{tot}^G} \tag{7.11}$$

The flux equations for the liquid and gas phases, respectively, become:

$$N_i = K_L C_{tot}^{Lb} \left\{ x_i^{Lb*} - x_i^{Lb} \right\} \tag{7.12}$$

$$N_i = K_G C_{tot}^{Gb} \left\{ y_i^{Gb} - y_i^{Gb*} \right\} \tag{7.13}$$

The partial pressures are often used for the gas phase. When the ideal gas law applies, the relation to concentration is simply that:

$$C_i^G = \frac{p_i}{RT} \tag{7.14}$$

Using the appropriate value or units for the gas constant R will give the desired result.

The flux equation becomes:

$$N_i = \frac{K_G}{RT} \left\{ p_i^b - p_i^{b*} \right\}$$

(7.15)

Many defines the size $K_G/(RT)$ as a new mass transfer coefficient that is allocated another symbol. This will not be done here.

A fourth way of expressing concentrations is the mole ratio. This may be useful if the estimates are done by pen and paper, a method that should not be outrightly discarded. The trick introduced here, and it is a useful one, is to make the calculation based on flows of 'inert' gas or solvent. That is the gas or solvent components other that the one that is being absorbed. Upper case letters are usually used for this ratio. In this case:

$$X_i = \frac{C_i^L}{C_{tot}^L - C_i^L}$$

(7.16)

A similar expression may be made to find the gas phase ratio Y_i. In terms of mole fractions, the conversion is:

$$X_i = \frac{x_i}{1 - x_i}$$

(7.17)

The latter method is less common in textbooks, but it is well covered by Coulson and Richardson in their book *Chemical Engineering* (Coulson *et al.*,1978).

Obviously the operating lines, the component mass balances and any sub-function used like the partition function must also be expressed in corresponding units. Units may be chosen freely but the choices must be consistent. You are strongly recommended to make a choice and stick to it over time. Do not change your basis from case to case. It is too easy to make mistakes, and relating numbers from case to case becomes difficult. Now and again it is compelling to adapt the working environment, however.

7.10   Concentrated Mixtures and Simultaneous Absorption

Designs with concentrated mixtures and more than one component being absorbed, or even desorbed, represent the real world. Here there are no longer any parameters that remain constant along the column. For many systems there are also chemically active absorbents involved like alkanolamines. Such chemicals influence the mass transfer, and this is further complicated because the concentration of the 'free' chemical will vary strongly along the column height. The other aspect is that the absorption equilibrium relation is no longer simple to handle. This will be further discussed in later chapters.

Component mass balances for the control elements must be made as outlined earlier, but when more than one component is absorbed, mass balances must be made for all of them. Mass balances must also be made for all reacting species including those that are produced by these reactions.

Most likely there will be heat effects. Absorption is generally an exothermic process, but more so if chemical reactions between the absorbent and the absorbate are involved. It is possible to make one heat balance over the column control element (Figure 7.3a), but it is more stylish to make one for the gas and one for the liquid control volumes separately. The latter is principle more correct, but there is an amount of uncertainty involved

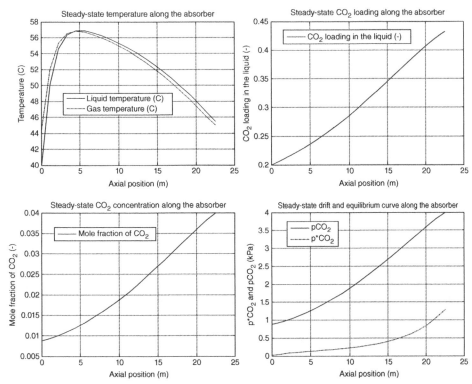

Figure 7.5 *(a–d) Column profiles produced by an absorption column simulator. The simulator is Statoil's SORBER and the profiles are related to case 4 in a paper by Svendsen and Eimer (2011). Upper left are temperature profiles for the gas and liquid phase, upper right is the profile of CO_2 loadings on the absorbent, lower left is the CO_2 concentrations in the gas and lower right depicts the operating and equilibrium lines for the column. (Reproduced with permission from Statoil © 2013.)*

in estimating the heat transfer coefficient between the two phases. If two heat balances are made, then separate temperature profiles will arise for the gas and liquid along the column. Studying those is interesting and illuminates the influence of feed stream temperatures. An example can be seen in Figure 7.5. The heat transfer coefficient between the phases can be estimated using the Chilton–Colburn analogy that relates heat transfer and mass transfer coefficients. The heat balance for the liquid phase control element is made as before:

$$\text{Accumulated} = \text{In} - \text{Out} + \text{Produced}$$

Here heat will be produced by both absorption and reactions, and all components and all reactions must be accounted for. The 'in and out terms' will include both energy entering and leaving due to mass transfer and sensible heat transfer driven by temperature differences in the gas and liquid. Steady state conditions shall be assumed. To avoid very long equations, the contributing elements shall be defined separately, given short names and then be combined.

We shall look at the liquid phase control element. Sensible heat in by absorption (IBA) with n components being absorbed and in by flow (IBF) from previous control element:

$$IBA + IBF = \sum_{i=1}^{n} N_i aAdz C_{p,i} \left(T^G - T^{\text{Ref}}\right) + L_V C_{Tot}^L C_{p,i} \left(T^L - T^{\text{Ref}}\right) \tag{7.18}$$

Sensible heat in by heat transfer (IBHT):

$$IBHT = h\left(aAdz\right)\left(T^G - T^L\right) \tag{7.19}$$

Sensible heat out by flow (OBF) to the next control element:

$$OBF = \left(L_V + dL_V\right) C_{Tot}^L C_{p,i} \left(T^L + dT^L - T^{\text{Ref}}\right) \tag{7.20}$$

If any of the components happen to desorb, that will be taken care of by N_i being negative in Equation 7.18.

Heat will be produced by absorption (PBA) and that heat will be released in the liquid phase. It is assumed that the heat of absorption of a component is independent of the presence of the others. For now it shall also be assumed that the heat of absorption is independent of how much that has already been absorbed. Hence PBA is:

$$PBA = \sum_{i=1}^{n} \left(N_i aAdz\right)\left(-\Delta H_{ABS}\right) \tag{7.21}$$

The final element in the heat balance is the heat effects of chemical reactions taking place in the liquid. It is, as an example, known that when enough CO_2 is absorbed in a solution of carbamate forming amines, the carbamate will start to convert to bicarbonate. For now it shall simply be said that the heat produced by reactions (PBR) in the liquid phase may be defined by:

$$PBR = \sum_{k=1}^{m} r_k \left(-\Delta H_{REAC}\right) \tag{7.22}$$

The reaction rate, r_k, is defined to yield a positive number. The reason for the minus signs in front of the ΔH is simply that by convention the reaction or absorption is exothermic when the value is negative. The minus sign ensures that this convention may be followed as well as obtaining a positive heat effect when the appropriate value is entered. The heat balance for the liquid phase may now be summarised by:

$$0 = [IBA + IBF + IBHT] - [OBF] + [PBA + PBR] \tag{7.23}$$

A similar heat balance may be made for the gas phase control element. In the gas phase there are usually no heat produced by exothermic processes, nor consumed by endothermic processes for that matter. However, in the absorption part of the nitric acid process there are chemical reactions going on in the gas phase, and this must then be taken care of. For gas treating processes this can thankfully be excluded.

In a complex absorption system it may no longer be assumed that mass transfer coefficients and physical properties are constant. The situation should be carefully considered. Fast chemical reactions will influence the mass transfer. This will be discussed in a later chapter. Another possibility is that a strong absorption of one component that sets up a convective flow towards the gas-liquid interface may influence the effective rate of diffusion of another marginal component. If deemed necessary sub-models must be made and accounted for as the solution of the equations progresses down the column.

A complex design problem has been described previously. An engineer should still ask if the problem may be simplified. This question should be answered in view of the circumstances. As an example, there is no need for a complex analysis if we just want to know the order of magnitude size of a column. Making a comprehensive analysis may just not be possible within the framework of the project at hand. The necessary software may not be available, and there may not be a budget to buy the service externally. Under those circumstances it becomes necessary to do the best analysis possible with all the simplifications that must be made.

To do a full analysis as outlined some form of computer code is in practice required. Those available in the commercial flowsheeting packages tend to be stage calculation oriented. There are rate based models available in the market but they tend to be expensive. An option is to program a model from scratch. That is not a cheap option, however, as the time to make the code must be accounted for. Whatever the choice made, economics and the model's usefulness to the organisation must be considered.

Figure 7.5 shows an example of what kind of information that may be obtained from a good model. This example is for the absorption of CO_2 from the exhaust of a combined cycle gas power plant into an aqueous solution of monoethanolamine.

7.11   Liquid or Gas Side Control? A Few Pointers

In dealing with absorption the question about gas or liquid side control of mass transfer crops up. It is really the rate of mass transfer that is the issue. A number of textbooks give some information on the matter but in general this seems to be a poorly understood issue with most chemical engineers. That is, most understand the concept but few can relate it to real systems.

We recall that the overall gas related mass transfer coefficient can be calculated from:

$$\frac{1}{K_G} = \frac{1}{k_G} + \frac{m_i}{k_L^0} \tag{7.4}$$

The first and the second term on the right-hand side represent the gas side and the liquid side resistance to mass transfer, respectively. To illustrate the situation figures are needed for the various terms. Cussler (2009) recommends two correlations in particular for k_G and k_L^0. Since the target is only a ball park understanding of the issue, it assumed a packed column with 2″ rings, gas velocity being 3 m/s, pressure 2 bar and the densities and viscosities being those of air and water at 20°C. On this basis it is estimated that:

$$k_G = 0.1 \text{ m/s and } k_L^0 = 0.0001 \text{ m/s}$$

In Table 7.1 the resistance to mass transfer for a few gases absorbed in water is summarised based on these considerations.

A general conclusion is that low solubilities drive the system towards liquid side resistance whereas high solubilities move the system towards gas side control. The Table 7.1

Table 7.1   *Solubilities and mass transfer resistances for selected gases into water at 20°C.*

Gas absorbed 'i'	Solubility m_i (–)	$1/k_G$ s/m	m_i/k_L^0 s/m	$1/K_G$ s/m	Resistance in liquid %
N_2	60	10	600 000	600 010	99.998
O_2	30	10	300 000	300 010	99.997
CO_2	1.1	10	10 800	10 810	99.9
H_2S	0.32	10	3 200	3 210	99.7
SO_2	0.023	10	230	240	96
NH_3	0.00064	10	6.4	16.4	39

Solubilities may be traced to Hougen *et al.* (1964) except for NH_3 where the data are from Perry, 6th edn., 1984).

may serve as a reference frame, but it is always a good idea to make specific checks for the column and system at hand. The earlier estimates are rough. (Using the values estimated in Section 11.7.2 will not change the overall conclusion.)

It is prudent to comment on the solubilities of SO_2 and NH_3. They both react with water to make sulfurous acid and ammonium hydroxide respectively. That may explain the high solubilities. How that will affect the mass transfer coefficients must be evaluated separately but that is deferred for now until mass transfer with chemical reaction has been discussed.

One system that has a genuinely high physical solubility for the absorbing component is water absorbed in glycol. Triethylene glycol (TEG) is routinely used in gas treating to remove water from natural gas to lower the gas' water dew point. Glycols are extremely hygroscopic if their water contents are low. Water absorption in TEG is a gas side controlled system.

A final word of warning is in order. The solubilities needed for these mass transfer resistance considerations are the physical solubilities. When faced with chemical solvents that have high affinities, for example CO_2, these chemical solubilities must not be used. The chemistry will certainly affect the absorption, but not quite in that way. Again this will be discussed in depth later.

7.12   The Equilibrium Stage Alternative Approach

Analysing the height requirement for a column by determining the number of equilibrium stages needed to reach the target is a classic method in chemical engineering. A graphical method to do this is due to McCabe and Thiele and features in virtually all textbooks dealing with distillation design (see, for example Coulson *et al.*, 1978). The method is simple and is illustrated in Figure 7.6. To obtain a straight operating line a constant ratio of gas to liquid flow is needed, but the graphical representation is not limited to straight lines like those used in Figure 7.6.

Modern day flowsheet simulators have column models that also include heat balances and can do this analysis numerically without the constraints of constant flow, and so on. This is convenient, but it is still useful to make the plot of stages shown to get an overview of the column. Such a plot will quickly reveal if the column has a pinch or if the separation would be much enhanced by a marginal increase in the column height.

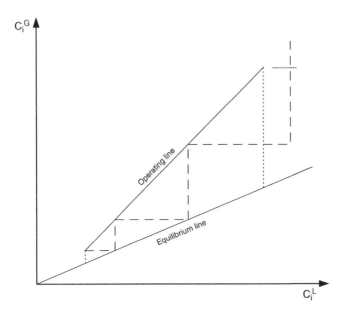

Figure 7.6   *The column height analysis based on the steps method due to McCabe–Thiele. Number of equilibrium stages shown is 3.7.*

This approach works well with distillation problems where stage efficiencies typically fall in the range 50–80%. At worst the real number of stages needed would be twice the theoretical number. In absorption with high driving forces, however, the stage efficiency could be 10% or less implying that the theoretical number of stages should be multiplied by 10 or more. Columns are still designed on this basis because there is a lot of design experience around based on extensive empirical column data. However, the trend is towards analysis based on the rate approach.

7.13   Co-absorption in a Defined Column

When removing CO_2 from natural gas, it will be of interest to estimate the co-absorption of the natural gas components, for example methane. This is easily done using Equation 7.7(i) as it stands by assuming a degree of methane saturation of the liquid leaving the absorber bottom, then estimating the length of column required to achieve this. If this length is shorter than the column at hand, then the degree of saturation will be higher. A new guess is made and so on. However, it is easy to re-arrange the Equation 7.7(i) to obtain a form explicit in the concentration of gas co-absorbed in the outgoing liquid:

$$C_i^{LbL} = \frac{1}{m_i} \left\{ C_i^{GbL} - \left(C_i^{Gb0} - m_i C_i^{Lb0} \right) \exp \left[\frac{K_G a A L}{G_V} \left(1 - m_i \frac{G_V}{L_V} \right) \right] \right\} \qquad (7.7k)$$

Or as the case might be it may be of interest to see what co-removal may be achieved when there is another component that is actually desirable to remove. An example might be traces of CS_2 in gas from a gasifier. In that case it would be convenient to have the equation in the form of:

$$C_i^{Gb0} = m_i C_i^{Lb0} + \frac{C_i^{GbL} - m_i C_i^{LbL}}{\exp\left[\dfrac{K_G aAL}{G_V}\left(1 - m_i \dfrac{G_V}{L_V}\right)\right]} \tag{7.7m}$$

In either case there is a bit of trial and error involved as the concentration of the co-absorbed species in the gas and liquid leaving cannot both be guessed independently. A mass balance for that species must also be used in combination with a convenient form of Equation 7.7.

It should be kept in mind that K_G or K_L may vary between species absorbed.

7.14   Numerical Examples

7.14.1   Ammonia Train CO_2 Removal with Sepasolv, NTUs

Recall the ammonia train CO_2 removal plant defined in Section 1.7.1.

The gas feed was 12 888 m^3/h at operating temperature and pressure.
The CO_2 in this feed constitute 2475 kmol/h and its mole fraction is 0.20.
It is desirable to bring the CO_2 level as low as possible since CO_2 is a catalyst poison in the ammonia synthesis. However, for the physical solvents to be analysed the target is set to 1000 ppm (always on a mole basis in gas) in the treated gas.
We shall assume that the process is operating at 40°C and 25 bar.

The first example is based on the use of the commercially available Sepasolv solvent. Data on this is obtained from Wölfer (1982).

The CO_2 solubility obtained from this source is 2.6 m^3/m^3 bar. There is no specification as to the state of the gas, but we shall assume it is at NTP (normal temperature and pressure). Under this assumption, the gas solubility may be recalculated to:

$$\frac{2.6\,\text{Nm}^3 \times 1000\,\text{mol/kmol}}{22.414\,\text{Nm}^3/\text{kmol}} = 116\,\text{mol/m}^3.\text{bar}$$

At 1 bar pressure the concentration of CO_2 in the liquid is then 116 mol/m^3.
The gas concentration in the gas at 1 bar and 40°C is:

$$\frac{(1\,\text{bar})\,(1000\,\text{mol/kmol})}{(0.083\,\text{m}^3\text{bar/kmol} \cdot \text{K})\,(313\,\text{K})} = 38.5\,\text{mol/m}^3$$

The dimensionless solubility we seek for the NTU estimate is then:

$$m = \frac{38.5}{116} = 0.331$$

The partial pressure of CO_2 at the column feed point is $(20/100) \times 25$ bar = 5 bar

The CO_2 saturation concentration in the absorbent at this point is:

$$(5\,\text{bar})\,\left(116\,\text{mol/m}^3 \cdot \text{bar}\right) = 580\,\text{mol/m}^3$$

Virtually all the CO_2 has to be removed. The minimum absorbent circulation rate is:

$$\frac{2475\,\text{kmol/h}}{0.580\,\text{kmol/m}^3} = 4269\,\text{m}^3/\text{h}$$

This needs to be increased, how much is our decision. In the end it is a question of economics. A higher circulation will facilitate a shorter column, but the energy for pumping increases. We shall try a 10% increase. Circulation rate is then:

$$(1.1)\,\left(4269\,\text{m}^3/\text{h}\right) = 4695\,\text{m}^3/\text{h}$$

There are two more decisions to make. One is the CO_2 concentration in the treated gas, and this we shall set to 1000 ppm (molar basis is always understood in a gas context). A lower value would have been nice, but experience suggests that it may not be realistic with a physical solvent. The second decision is the CO_2 content of the regenerated absorbent, and with respect to this we shall be optimistic and assume we can manage $1\,\text{mol/m}^3$. Using our standard notation, we can now calculate the CO_2 concentrations for the streams in and out of the column:

$$C_{CO_2}^{Lb0} = 1\,\text{mol/m}^3$$

$$C_{CO_2}^{Gb0} = \frac{(25\,\text{bar})\,\left(1000 \times 10^{-6}\right)\,(1000\,\text{mol/k mol})}{\left(0.083\,\text{m}^3\text{bar} \cdot \text{k mol}^{-1}\text{K}^{-1}\right)\,(313\,\text{K})} = 0.961\,\text{mol/m}^3$$

$$C_{CO_2}^{LbL} = \frac{580\,\text{mol/m}^3}{1.1} = 527\,\text{mol/m}^3$$

$$C_{CO_2}^{GbL} = \frac{(25\,\text{bar})\,(0.20)\,(1000\,\text{mol/k mol})}{\left(0.083\,\text{m}^3\,\text{bar} \cdot \text{k mol}^{-1}\,\text{K}^{-1}\right)\,(313\,\text{K})} = 192\,\text{mol/m}^3$$

The NTUs needed can now be estimated:

$$NTU_G = \left(\frac{1}{1 - m_i\dfrac{G_V}{L_V}}\right)\ln\left[\frac{C_i^{GbL} - m_i C_i^{LbL}}{C_i^{Gb0} - m_i C_i^{Lb0}}\right]$$

$$= \frac{1}{1 - 0.331\dfrac{12888}{4695}}\ln\left\{\frac{193 - 0.331 \times 527}{0.961 - 0.331 \times 1}\right\} = 37$$

This is the answer we were seeking. The NTU_G is high, but not unrealistic, and it reflects that this is a challenging separation to make. Estimation of the height of transfer units is deferred for now. However, if we increased the absorbent circulation rate from 110 to 120% of the minimum, the NTU_G would sink from 37 to 24. Inspecting the previous formula we quickly see that the denominator in the fraction in the log-function would become negative if more CO_2 was left in the regenerated absorbent. In a real design situation, the column top condition would be first priority for checking out in more detail. It could well be that the target of 1000 ppm of CO_2 in treated gas is too ambitious for this solvent.

7.14.2   Ammonia Train CO_2 Removal with Selexol, NTUs

For this problem we seek a direct comparison of an alternative absorbent. Selexol is another physical solvent offered for acid gas removal. It has to be realised that the comparison made here is based on limited literature data that is accepted on face value simply because we have no alternative.

According to Chen, Chen and Hung (2013) the solubility of CO_2 in Selexol at 40°C is given, based on validity of Henry's law, as 0.08 mol/(l bar). If we write Henry's law as:

$$p_{CO_2} = H_{C,CO_2} C_{CO_2}$$

Then the Henry's constant would be:

$$H = \frac{1}{0.08\,\text{mol} \cdot \text{L}^{-1} \cdot \text{bar}^{-1}} = 12.5\,\text{L} \cdot \text{bar/mol}$$

Combining Equations 5.55, 5.57 and 5.62, it follows that:

$$m_{CO_2} = \frac{H_{C,CO_2}}{RT} = \frac{12.5\,\text{m}^3 \cdot \text{bar} \cdot \text{k}\,\text{mol}^{-1}}{\left(0.083\,\text{m}^3 \cdot \text{bar} \cdot \text{k}\,\text{mol}^{-1}\,\text{K}\right)(313\,\text{K})} = 0.481$$

This is little higher than for Sepasolv, but a comparable value. It means that a higher absorbent circulation rate is needed. The concentrations of CO_2 in the streams in and out is set or estimated as before:

$$C_{CO_2}^{Lb0} = 1\,\text{mol/m}^3$$

$$C_{CO_2}^{Gb0} = \frac{(25\,\text{bar})\left(1000 \times 10^{-6}\right)(1000\,\text{mol/k}\,\text{mol})}{\left(0.083\,\text{m}^3\,\text{bar} \cdot \text{k}\,\text{mol}^{-1}\,\text{K}^{-1}\right)(313\,\text{K})} = 0.961\,\text{mol/m}^3$$

$$C_{CO_2}^{GbL} = \frac{(25\,\text{bar})(0.20)(1000\,\text{mol/k}\,\text{mol})}{\left(0.083\,\text{m}^3\,\text{bar} \cdot \text{k}\,\text{mol}^{-1}\,\text{K}^{-1}\right)(313\,\text{K})} = 192\,\text{mol/m}^3$$

$$C_{CO_2}^{LbL} = \frac{C_{CO_2}^{GbL}}{m_{CO_2}} = \frac{192\,\text{mol/m}^3}{0.481} = 400\,\text{mol/m}^3$$

$$C_{CO_2}^{LbL} = \frac{400\,\text{mol/m}^3}{1.1} = 364\,\text{mol/m}^3 = 0.364\,\text{k}\,\text{mol/m}^3$$

Circulation rate of absorbent is on this basis (110% of minimum flow):

$$L_V = \frac{2475\,\text{kmol} \cdot \text{h}^{-1}}{0.364\,\text{kmol} \cdot \text{m}^{-3}} = 6808\,\text{m}^3/\text{h}$$

Now:

$$NTU_G = \left(\frac{1}{1 - m_i \frac{G_V}{L_V}}\right) \ln \left[\frac{C_i^{GbL} - m_i C_i^{LbL}}{C_i^{Gb0} - m_i C_i^{Lb0}}\right]$$

$$= \frac{1}{1 - 0.481\dfrac{12888}{6808}} \ln \left\{\frac{192 - 0.481 \times 364}{0.961 - 0.331 \times 1}\right\} = 40$$

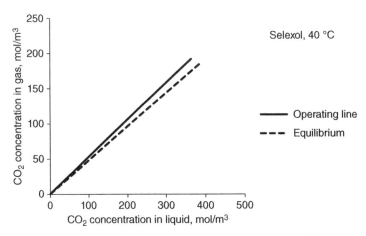

Figure 7.7   *Operating and equilibrium lines for the synthesis gas and absorption in Selexol case. The top of the column (lower concentrations) is the pinch point in this design.*

A taller column is needed given that the heights of transfer units are the same for both solvents. The liquid flow for Selexol is likely to lead to more gas–liquid contact area in the tower based on the use of packings but that is left for a later chapter (Figure 7.7).

7.14.3   Ammonia Train CO_2 Removal with Selexol, NTUs by Numerical Integration

An alternative way of estimating the NTUs for this process is to do numerical integration of Equation 7.7(c). For this purpose we split it as shown below. The NTUs are represented by the integral while the fraction now put in front is the HTU of this column. By taking it

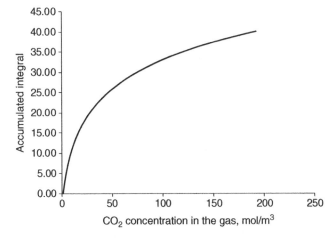

Figure 7.8   *Detail of numerical integration discussed previously. The sharp rise at the low concentrations illustrates the pinched condition at this end (the upper) of the column.*

Table 7.2 Details of the numerical integration of the relevant equation.

$C^{Gb}_{CO_2}$	$C^{Gb*}_{CO_2}$	$\dfrac{1}{C^{Gb}_{CO_2} - C^{Gb*}_{CO_2}}$	$dC^{Gb}_{CO_2}$	$\text{Integr} = \dfrac{dC^{Gb}_i}{C^{Gb}_i - C^{Gb*}_i}$	Sum of integrals
0.961	0.48	2.08	–	–	–
1	0.52	2.07	0.039	0.08	0.08
1.2	0.70	1.99	0.2	0.41	0.49
1.5	0.97	1.89	0.3	0.58	1.07
1.8	1.24	1.80	0.3	0.55	1.62
2	1.43	1.74	0.2	0.35	1.98
3	2.33	1.50	1	1.62	3.60
4	3.24	1.32	1	1.41	5.01
6	5.06	1.07	2	2.39	7.40
8	6.88	0.89	2	1.96	9.36
10	8.70	0.77	2	1.66	11.03
12	10.52	0.68	2	1.44	12.47
14	12.34	0.60	2	1.28	13.75
17	15.07	0.52	3	1.68	15.43
20	17.79	0.45	3	1.46	16.8
25	22.34	0.38	5	2.07	18.95
30	26.89	0.32	5	1.74	20.70
35	31.43	0.28	5	1.50	22.20
40	35.98	0.25	5	1.32	23.52
50	45.07	0.20	10	3.51	25.78
70	63.26	0.15	20	3.81	29.30
100	90.54	0.11	30	3.81	33.11
130	117.82	0.08	30	2.82	35.93
140	126.92	0.08	10	0.79	36.72
150	136.01	0.07	10	0.74	37.46
160	145.10	0.07	10	0.69	38.15
170	154.20	0.06	10	0.65	38.81
180	163.29	0.06	10	0.62	39.42
185	167.84	0.06	5	0.30	39.72
190	172.38	0.06	5	0.29	40.00
192.139	174.33	0.06	2.139053	0.12	40.12

outside the integral, we assume it is constant (Figure 7.8).

$$\int_0^L dz = \int_{C^{Gb0}_i}^{C^{GbL}_i} \frac{G_V}{K_G aA}\frac{dC^{Gb}_i}{C^{Gb}_i - C^{Gb*}_i} = \frac{G_V}{K_G aA}\int_{C^{Gb0}_i}^{C^{GbL}_i} \frac{dC^{Gb}_i}{C^{Gb}_i - C^{Gb*}_i}$$

It is seen that also this method leads to a need of 40 NTUs. This is not surprising, but the end result is sensitive to the size of steps in the integration. Since the numerical integration is done by a very primitive method, this is not surprising either (Table 7.2).

References

Chen, W.-H., Chen, S.-M. and Hung, C.-I. (2013) Carbon dioxide capture by single droplet using Selexol, Rectisol and water as absorbents: a theoretical approach. *Appl. Energy*, **111**, 731–741.

Coulson, J.M., Richardson, J.F., Backhurst, J.R. and Harker, J.H. (1978) *Chemical Engineering*, 3rd edn, vol. **2**, Pergamon Press.

Cussler, E. (2009) *Diffusion. Mass Transfer in Fluid Systems*, 3rd edn, Cambridge University Press, Cambridge.

Hougen, O.A., Watson, K.M. and Ragatz, R.A. (1964) *Chemical Process Principles*, 3rd edn, John Wiley & Sons, Inc., New York.

Perry, R.H. and Green, D. (1984) *Perry's Chemical Engineers' Handbook*, McGraw-Hill.

Svendsen, J.A. and Eimer, D.A. (2011) Case studies of CO2 capture columns based on fundamental modelling. *Energy Procedia*, **4**, 1419–1426.

Wölfer, W. (1982) Helpful hints for physical solvent absorption. *Hydrocarbon Process., Int. Ed.*, **61** (11), 193–197.

8

Column Hardware

The purpose of a column is to provide a means for contacting gas and liquid in such a way as to enable efficient mass transfer.

The object of this chapter is not to provide a complete and up-to-date design manual for column hardware. The aim is to help acquire an understanding for the variables that are important and how to look for solutions including fault finding for columns that do not do what they are intended to do. Most engineers will spend more time on that than designing new columns. Hopefully this material will provide a good platform to make the reader's need to navigate in this field easier. Improved data keep on being produced, and new and improved products are also coming. This part of chemical engineering looks very different today than a generation or two ago and is likely to change further.

8.1 Introduction

A column could simply be an empty shell with nozzles to spray the liquid as droplets that would fall in contact with the gas. The problem with this arrangement is that it tends to provide one equilibrium stage only as a maximum. For most problems in gas treating more stages are needed. Trays and packings have long histories of providing such contact.

Trays or packings? Both types of columns have been around for a long time, and there is no sign yet that one will oust the other. In the end it is a question of economics. The two types of column have different properties that will represent different advantages and disadvantages depending on the application. Tray columns will, for example, have higher pressure drops due to the layer of liquid on each tray that the gas has to bubble through. This will be an important factor for selection if there is next to no pressure drop available for use but is not likely to be very significant for a high pressure application.

Gas Treating: Absorption Theory and Practice, First Edition. Dag Eimer.
© 2014 John Wiley & Sons, Ltd. Published 2014 by John Wiley & Sons, Ltd.

The gas treating industry is conservative with respect to technology choices. This is not very different from the chemicals industry, but with present day prices of oil and gas it is more important to avoid operational problems than to save a little investment money when the plant is built. Foaming is for instance a recurring problem area in the hydrocarbon industry that could lead to downtime and lost revenue. The tendency is to repeat what is seen to have worked in the past rather than run the risk associated with trying something new. Innovation is more likely to come about in order to solve a problem than to make more or less marginal investment savings. For new applications the field is wide open though.

The preference between trays or packings may vary from one company to another and also between applications. This will be caused by those companies' operational experience and to an extent by what choice that was made in the first place when an application was first met.

In earlier times, say before 1980 perhaps, there tended to be vendors dealing with packings only and tray vendors. Today this situation has changed with take-overs and mergers restructuring the industry. Most companies of size will today offer both solutions.

8.2 Packings

Today, there are two main recognised varieties of packings being used. These are random packings that have been around since the 1920s at least, and structured packings that were first introduced in the 1960s as a speciality for difficult separations such as heavy water purification and vacuum distillation where pressure drop per stage and low liquid fluxes were challenges.

Today, structured packings have been developed into products that compete well with random packings for a large variety of applications.

Both types of packing have common needs in the form of liquid distributors, redistributors and supports.

A number of books have been written specifically about packings over the years. These provide a wealth of information and are summarised in a bibliography list in the reference section. The older ones may have mostly historic interest but they provide information on older products that still exist in various plants. It is important to be aware that data given for packings should be seen in the context of the liquid distributors and redistributors used. This information is not always available and adds to the uncertainty of the data available. Mass transfer data given is also subject to the mathematical analysis and sub-models used to identify them. In the early days good liquid distribution was not a big issue, and gas distribution not seen as a problem (Mackowiak, 2010). Packed columns in those days were usually less than a metre in diameter. Insight and understanding were sometimes also in short supply. There was a story about a column being packed with the packings still in their cloth sacks.

Packings are made to provide maximum mass transfer area per unit pressure drop needed to perform a separation. We generally think of the mass transfer area as the surface of the packing but that is not quite true. Firstly, not all packing area is wetted as some of it is 'shadowed' from the liquid. Secondly, some people claim the number of drip points provided is the important feature since they represent surface renewal, which is important in mass transfer. To an extent this may be due to arguments from sales engineers promoting their particular product. A third point is that liquid droplets may be dispersed in the gas phase and provide further contact area. Mackowiak talks about packing with walls and lattice packings. The latter resembling a cage more than a Raschig ring. Some indication of this is seen in Table 8.1. Clearly, drip points play a bigger role in a lattice packing.

Another type of development done on packings is to use membranes. In this case the membranes would be microporous without any separation on their own. Their benefits include a very high specific surface area, and since they separate gas and liquid physically, there should be no entrainment and flooding as we know it will not take place. However, there is clearly still an upper bound of gas and liquid loads. This technology was pursued from around the mid-1990s, but development seems to have petered out.

8.2.1   Types of Random Packings

Raschig rings represent the first generation of packings. These are tubes with length equal to their diameter. Sizes typically ranged from 10 to 100 mm, with 50 mm probably the most common size in the chemical industry. It was nearly impossible to wet both their inner and outer walls with liquid. Raschig rings were produced using ceramics, plastic and metal. A problem with ceramic material was attrition losses but they were popular for corrosive applications.

The Berl saddle was another early product but this was only made in ceramics. Intalox saddles came in both ceramics and plastic and are still on the market. These and Pall rings probably belong to what may be referred to as a second generation of packings. Particularly the plastic variety of the saddles and the Pall rings addressed the lack of access to all surfaces of the packing plus providing more drip points compared to the Raschig rings. Hy-Pak (a Norton Co trade name) was a further development of the Pall ring claiming to be more cost effective for a given performance. This is an interesting case as simple corrugations were used to make the ring stiffer allowing use of thinner metal thus reducing the amount of material and more tongues meant more available surface area, which was sacrificed to make the ring bigger, thus saving on the number of rings per unit volume. Hence, less metal, less machining and a lower price.

Newer packings, named third generation packings by Kister (1992), could perhaps be started with the Mini Rings made by Mass Transfer Ltd (a small independent company later taken over). They claimed much improved performance over Pall or Hy-Pak by making the height of the ring 1/3 of its diameter. The Norton Company was quick to introduce their Intalox Metal in the wake of this. This ring may be described as part saddle and part ring. Effectively its 'height' is order of 1/3 of its diameter like the Mini, but the Intalox Metal's height varies around its circumference. See Table 8.1.

Further competition was provided by Levapak and Nutter rings to name a couple of packings receiving (or acquiring) a bit of publicity. There are many more. Continental Europe seems to have a good number of additional packing types that are not normally come across in the Anglo–American culture. The German published books in the bibliography document this. Table 8.1 lists a few where names and pictures are given. Further information may be found in the bibliography cited if there is a need to pursue any of them. Producers' web sites are also good sources of information these days, there is no need to write letters to ask for leaflets.

Table 8.1   *Random packings. (This list is not exhaustive, there are many more.)*

Picture	Packing
	Raschig ring, ceramic (left), metal (right)
	Berl saddle, ceramic
	Intalox/Novalox/Torus
	Ceramic (left), plastic (right) Pall ring, ceramic (left), metal (right)
	Hy-Pak, metal. Notice doubling of tongues and the corrugation of sides compared to the Pall ring
	Tellerette, plastic
	Cascade mini ring, plastic left, metal right
	Intalox Metal Tower Packing (IMTP)
	Top-Pak ring, metal
	Nutter saddle, plastic left, metal right

8.2.2 Types of Structured Packings

Structured packings were invented around 1960. The driving force was the desire for high performance columns needed to distil heavy water, which is very demanding in terms of the number of separation stages needed. One application that springs to mind needed around 300 theoretical stages. A further problem area for packings at the time was vacuum distillation. This holds two distinct challenges, one is the need for a very low pressure drop and the other is the extremely low liquid fluxes ($m^3/m^2 \cdot h$) that are associated with this. Sulzer introduced two packings, called BX and CY, respectively, to deal with these challenges. These were made out of metal gauze folded and assembled in the characteristic way of structured packings.

For a long time structured packings were considered an expensive niche product. Eventually Koch and Nutter came up with alternatives that were designed with a broader range of applications in mind, and competition was provided. Sulzer in turn developed their Mellapak in the 1970s. This all led to structured packings finding a wider range of applications, more sales came about, and further interest in structured packings was created. Table 8.2 lists types of structured packings available today, or that have been on the market.

8.2.3 Fluid Flow Design for Packings

In packed columns gas and liquid share the same flow channels. In counter-current columns, the most common case, gas and liquid flow in opposite directions. When the gas flow becomes high enough, it prevents liquid from flowing downwards. The rate of liquid flow will influence when this situation occurs. This situation is referred to as *flooding*. It is this situation that determines the minimum diameter of a column for a given packing.

Table 8.2 Structured packings. (The list does not claim to be exhaustive.)

Koch–Glitsch	Flexipac
	Flexipac HC
	Flexipac S
	Intalox
	Goodloe
Raschig (Jaeger)	Raschig-Super-Pak
	Raschig-Pak
	MaxPak
Sulzer	Melapak
	Melapak Plus
	Gauze packings: BX, CY
Vff (Vereinigte Füllkörper Fabriken)	P250A
	P150B
	P125B
	P125C

Kister (1992) refers to publications where as many as 20 definitions of flooding are used. Most of these refer to incipient flooding while a few describes fully developed flooding. There is a certain amount of judgement involved in deciding when a column starts to flood although some may claim that a particular method that they use is based on objective measurements. That may be so, but what may seem clear when applied to one system may not be equally clear in another setting. Kister claims that flooding may be predicted with an accuracy of 10–15%. In view of this it may seem safe to design for maximum 85% of flood. Mackowiak (2010) provides a large data collection to test his flooding correlations and these show that the uncertainty is at least ±20%. Whenever data from similar applications are available it would be appropriate to base a design on these. The traditional flooding correlations were established based on the packings available at the time, and these may not be as good for modern packings. Mackowiak discusses this at some length and points out that flooding is likely to start differently in packing with defined walls compared to lattice packings. In the latter the liquid hold-up is to a much larger extent made up of droplet swarms compared to the films and rivulets on the walls of the other type. Moreover, he points out that visual observations made to determine flood points are inaccurate. In view of the spread of all historic data presented in his book, we could speculate that visual observations may have been a common technique.

It has been proposed (Strigle, 1987) that packed columns should be designed to a concept referred to as MOC (Maximum Operational Capacity). To understand this we need to refer to Figure 8.1 where the height equivalent of a theoretical plate (HETP) (or height of transfer unit: HTU) is shown as a function of gas flux. This type of diagram is well known from the distillation field. The MOC refers to the point where a further increase in gas flux would send the HETP steeply upwards. The concept is good but the form of the plot is not always as easy to interpret as that shown and the availability of data for MOC is not very good. Expensive experiments would most probably be needed to obtain the necessary data.

A third approach to capacity determination is to design for a pressure drop chosen by the designer. This pressure drop would be chosen to comply with published pressure drop

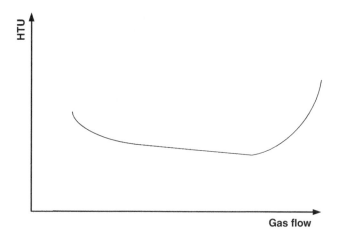

Figure 8.1   *Relationship between the gas flow and HTU for an absorption column.*

correlations. There are a number of such correlations or data banks available, see, for example, Kister (1992), but also information is provided for specific commercial products. The often referenced generalised pressure drop chart (GPDC) that has roots back to the Norton packing literature (1971) is, to a great extent, based on data for older, less efficient packings. Again care should be shown if applied to the latest products. This pressure drop chart is a development from the original flooding diagram published by Sherwood, Shipley and Holloway (1938).

If money was no object, it could be imagined that a column diameter was made extremely large to be absolutely sure that flooding would not occur and to add extra design margin. This is not a good idea as column operation would be jeopardised. There is a lower limit to gas flux in a column, but this limit is more diffuse than the maximum limit. A certain pressure drop is needed to ensure even gas flow distribution across the column, and this is needed to ensure proper contact between gas and liquid. The old Norton GPDC spans the pressure drop range 0.05–1.5 in. water column per feet packing (in/ft). (This is equivalent to 4.2–125 mm W.C./m). Kister mentions 0.1 in/ft (8.3 mm/m) as a minimum. (Flooding, by the way, starts somewhat above 2 in/ft or 170 mm/m.) A variety of this GPDC is produced in Figure 8.2. In this chart the flow parameter is defined as:

$$Flow\_parameter = \frac{L}{G}\sqrt{\frac{\rho_G}{\rho_L}} \tag{8.1}$$

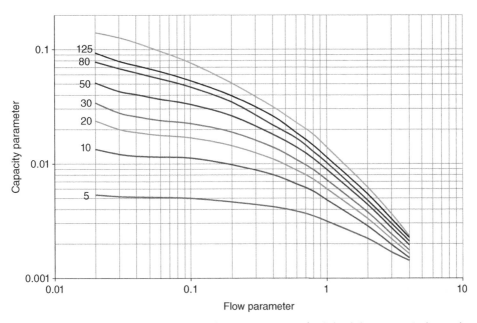

Figure 8.2 *Generalised pressure drop chart. Captions to the left of the curves indicate the pressure drop (mm H$_2$O column per m packing). The upper line is the flooding limit. (Note that the ordinate scale reflects that the original numeric factor included in the capacity parameter has been done away with.) (Based on data from Norton, (1977), Bulletin DC-11, "Design information for packed towers": Newer version of DC-10R.)*

The capacity parameter is defined by:

$$Capacity\_parameter = \frac{G^2 F \left(\frac{\mu_L}{\rho_L}\right)^{0.1}}{\rho_G \left(\rho_L - \rho_G\right)}$$ (8.2)

Physical properties are as per standard notation and are explained in the nomenclature. The gas flow G is a flux and has the units $kg \cdot m^{-2} \cdot s$. In the flow parameter this is of no relevance as long as the units are consequently used and render this size dimensionless. The parameter F is the interesting one to discuss. In this version of the diagram it is used as an empirical size and is specific to the packing. Actually, it is also influenced by liquid flow (Norton, 1977). One value is commonly given for a packing to help characterise it. They are found in packing leaflets. Originally in this GPDC, the packing factor F was calculated from:

$$F = {}^a\!/\!{}_{\varepsilon^3}$$ (8.3)

There are also limits to liquid flux. The upper bound has to do with liquid flooding the channels to the extent that gas would have to bubble through the liquid and there is a limit to the rate of rise of bubbles. The GPDC gives some guidance here since the curves take a dip when L/G essentially increases beyond a certain value. This value is influenced by the fluid densities. The lower bound has to do with the achievement of proper wetting of the packing. Data for this is much less easily available than gas capacity data. Coulson and Richardson's *Chemical Engineering*, volume 2 (Harker *et al.*, 2002) includes a chart for random packings that is probably based on older types. Kister (1992) refers to other works and suggests that the liquid load in random packings should be a minimum of $0.2-2$ US $gpm/ft^2 \cdot h$ ($0.5-5$ $m^3/m^2 \cdot h$ or $0.14-1.4$ $l/m^2 \cdot s$) depending on the material used. For structured gauze packing this could be as low as 0.05 (i.e. 0.12 $m^3/m^2 \cdot h$ or 0.03 $l/m^2 \cdot s$). Mackowiak (2010) gives a correlation for minimum liquid load. This is an awareness point to consider and if in doubt it is advisable to consult the packing vendor. A point to be aware of is that it is more challenging to achieve wetting when the surface tension of the liquid is higher rather than lower.

Mackowiak has developed an alternative method to the GPDC and presents this in his book. It is, however, more complex than the chart just discussed and is best suited to be combined with a bit of programming. His book is recommended for anyone with time available to make the initial effort to access the technique. The angle of attack is essentially to establish the liquid hold-up and treat the pressure drop as a packed bed with free space as reduced by the liquid. The shear forces between gas and liquid is also baked into the method.

It is of interest to know the hold-up of liquid in a packed column. One aspect is the weight it exerts on the packing support, another is the effect this may have on the process if, for example there is a chemical reaction going on in the bulk of the liquid not related to mass transfer rates. There are actually two types of hold-up, static and operational. The static hold-up is the liquid remaining on the packing after the column has been drained. This is small compared to the hold-up under operating conditions. During operation the hold-up will typically be several percent of the column volume. It is mainly decided by the liquid flow rate until the column load point is reached. Billet (1995) gives an expression for liquid hold-up (as fraction of packing volume) at the column's loading point. This is:

$$h_L = \left(\frac{12Fr_L}{Re_L} \right)^{1/3}$$ (8.4)

Here:

$$Fr_L = \frac{u_L^2 a}{g}$$ (8.5)

$$Re_L = \frac{u_L \rho_L}{a\mu}$$ (8.6)

The liquid velocity u_L is the liquid flux in m³/m² s, a is the packing surface density (m²/m³). The rest of the parameters are the standard physical data and the gravitational acceleration.

8.2.3.1   *Flooding, Pressure Drop Correlations, MOC*

Flooding was first correlated by Sherwood, Shipley and Holloway (1938) in the form of a diagram. The same diagram is still in use and may be found in most chemical engineering textbooks dealing with column design (e.g. Coulson: see Harker *et al.*, 2002). Although there may be as many as 20 definitions of flooding being used in the literature, the correlation is little affected by which one is used. If designing a column for 70% of flood as predicted by this correlation, it is reasonable to believe that the column will not flood (Kister, 1992; Strigle, 1987). Big companies have internal data banks that allow design closer to flooding with confidence. In such cases columns in the same application used previously is used as reference. Kister gives a simple correlation for prediction of pressure drop at incipient flooding:

$$\Delta P_{Fl} = 0.115F^{0.7}$$ (8.7)

F is the packing factor given by packing manufacturers.

The flooding diagram was developed by Lobo before Eckert introduced constant pressure drop curves based on the large data bank of the Norton Company. This diagram due to Eckert may be found in a Norton bulletin (1971), many commercial packing leaflets and textbooks. The use of these curves allows the designer to size a column such that it may operate within any given pressure drop restrictions (within reason!) in a process.

Due to the perceived uncertainties of defining flooding it was attempted to define a new concept to describe a column limitation. This was MOC. See Figure 8.1. In principle, this is a point where separation efficiency starts to deteriorate quickly while pressure drop starts to increase abruptly. It is a good concept but limited data available made it difficult to practise. Furthermore, it was not necessarily easy to determine the MOC point from experimental data in practice (Kister, 1992).

These days manufacturers have packing information available to be downloaded from their web pages. In these downloads pressure drops against fluid loads are given for specific packings. The data thus given are without any discussion of uncertainties but if they come from a trustworthy vendor they may be used. It is always interesting to make an evaluation to see how they compare with performance derived by general correlations. A double check is also a good guard against misinterpretation.

8.2.3.2   Special Considerations

When a system has a tendency to foam, it is necessary to design for a lower specific pressure drop. A column should, in principle, be able to operate without the addition of foam preventer even if it in practice is used once the column is in operation.

In gas treating some gas mixtures are retrograde, that means there is condensation when the pressure drops. Condensation may also occur when the absorbent enters colder than the gas feed. The condensation of hydrocarbons may cause foaming of an amine solution.

8.2.4   Operational Considerations

The comments on column design in the previous sections have been concerned with the maximum capacity of columns. Very often columns have to operate at reduced capacity due to demand being low for a period, or it may be that other sections of a plant suffer operational problems that lead to reduced overall plant capacity. For this reason it is also necessary to investigate how a given column design will perform at reduced load. How far down in load is it possible to go before separation efficiency suffers? This problem will become further complicated by the packing auxiliaries to be discussed next where it will become clear that pressure drops are important when liquid is to be distributed across the column.

8.3   Packing Auxiliaries

Putting packing auxiliaries into system, and getting acceptance from users that it is worth their while to spend serious money on the other column internals, has been a very important factor in the increased standing of packed towers over the last 30–50 years. Remembering back to the 1970s, it was hard to sell the need for spending as much money out of limited maintenance budgets on a liquid distributor as on the packings. (The author failed at least once).

A follow-up comment to this situation is that there must be many old columns, some are maybe still in operation, which are equipped with inadequate liquid distributors. This in turn must mean that there is a lot of measured mass transfer data affected by inadequate liquid distribution. These data are probably still used to make mass transfer correlations. It serves to remind us that chemical engineering is not just about punching numbers into formulae, but also about understanding challenges and making judgements on data sources.

Packing manufacturers provide a range of choices, and their products are generally well described in their company literature which may be downloaded from their web pages.

8.3.1   Liquid Distributors

Liquid distributors are extremely important for obtaining an optimum performance out of packings. Structured or random packings, you still need a good distribution. When liquid fluxes are low, the need for good liquid distribution is higher still. An example could be a glycol dehumidifier where the liquid flux is extremely low. Here the distributor could have drip points every 10 cm; that is, 100 per m^2.

There are three main types of liquid distributors:

- Ladder
- Pan
- Troughs

There will of course be many varieties within these main types.

Ensuring an even liquid distribution is in general achieved by the hydraulic liquid height being the same at each distribution point. That, at least, is the aim. There will be pressure loss in moving the liquid to each distribution point in the first place. Since these points are likely to at different distances from the column's liquid entry point, it should be clear that achieving a good liquid distribution is not without its challenges (Figures 8.3 to 8.5).

8.3.2 Liquid Redistributors

Liquid flowing down a column tends to drift to the column wall and will then stay there instead of finding its way back into the packing. For this reason the down-flowing liquid

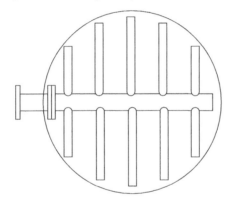

Figure 8.3 *Ladder type liquid distributor. The side arms have holes underneath where the liquid to be distributed can leave as jets.*

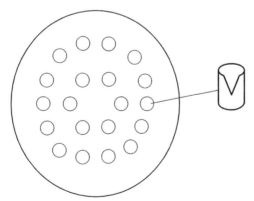

Figure 8.4 *Pan/riser/weir type liquid distributor. Gas rises through the pipes indicated while liquid runs down over the weir shown in the smaller, detailed figure.*

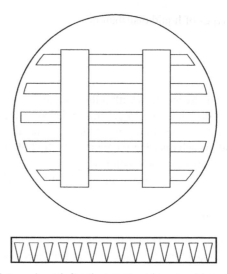

Figure 8.5 *Weir-trough type liquid distributor. Liquid is piped into the upper troughs. From there the liquid may flow into the lower troughs by holes in the bottom or by side weirs as shown in the lower sketch. The liquid from the lower troughs flow out through side weirs as shown in the lower sketch.*

is collected at intervals and redistributed to the packing. The distribution challenge is much the same as the original distributor but in addition there is the collection of liquid aspect.

In addition to the liquid handling, the redistributor may also be required to support the packing. This is because the weight of the packing would deform the packed beds if the bed of packing was allowed to be very deep. The rule of thumb is that packing depth should be no more than 10 m. (Norton Bulletin DC-10R (1971) gives 20 ft (6 m) as guidance with 30 ft (9 m) allowed 'under certain conditions'.)

8.3.3   Packing Support

It should be obvious that a packed bed must be supported to stay in place. Random packings are typically 50 mm in nominal size although this may range from 5 to 100 mm. A support must have openings small enough to retain the packings. This is not difficult as such but this support must be able to let the gas rise and the liquid drain. A modern metal packing has a void fraction in excess of 0.90. If the support grid was to match this, it would be too weak to keep a heavy load of packings from falling down. To avoid the support grid being the flow bottle neck and cause early flooding, it is common to design it as a corrugated grid as shown in Figure 8.6.

With the construction shown in the figure, it is possible to have an open area fraction higher than 1.00 if desirable! There will probably be a tendency for liquid to build up a little height in the 'downs' (or troughs) while the upward gas flow will dominate the 'highs'. This is well worth keeping in mind for whenever a support grid has to be improvised for a test column.

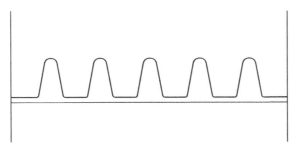

Figure 8.6 *Support grid/plate for random packing. The peaks shown are ridges crossing the column. Packings fill the 'valleys'. The ridge sides are slotted, or simply made of grids, to let the gas through sideways while some liquid may build up in the bottom of the 'valleys'.*

The vendors used to claim that the corrugation as such also helped to achieve a better packing randomness. No literature was ever found to back up this statement.

Structured packings come in the shape of 'building blocks'. They do not need such a fine grid for support.

8.3.4   Hold-Down Plate

Hold-down plates are sometimes used on top of a bed of random packings. The function is to prevent packing from being fluidised and/or disappear by entrainment in the gas. Plastic packings that are quite light are particularly susceptible to this. In fact it would take quite a hold-down plate to keep the packings in place if the column was operated under fluidising conditions. Flow surges arising from sudden pressure let-downs is another possibility that could cause entrainment of packings.

These plates are sometimes used. The need must be considered on a case-to-case basis. The plate itself is just a grid, most likely made of metal, and a ring around it to enable fixation to the column wall, or rather to a ring around its inside.

8.4   Tray Columns and Trays

Tray columns are sometimes referred to as plate columns (good to know if looking it up in a book's index). In a tray column the functional element for the separation process is a 'tray' where the liquid that flows downwards in the column flows laterally across the tray and the upwards flowing gas phase bubbles through the liquid layer formed on top of the tray. The process is illustrated in Figure 8.7. The liquid comes down to the tray through a liquid 'downcomer', and flows on to the tray under the downcomer skirt before it flows over a 'Steadying weir' before flowing on to the 'tray proper', the active tray area. Here, the gas will bubble through the liquid. When the liquid has crossed a tray, it flows over a weir and down into the next downcomer. It is this latter weir that is the main factor that determines the liquid height on the tray.

There are also trays with more than one downcomer. An example is shown in Figure 8.8. If the liquid load is high, this option may be necessary to use.

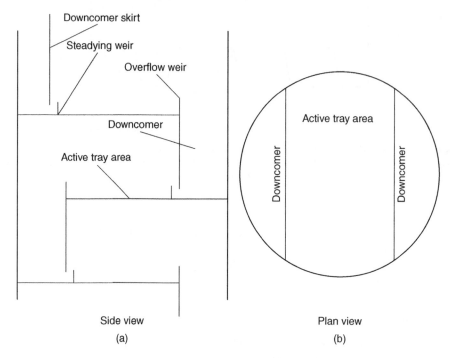

Figure 8.7 (a,b) The principle features of a tray with downcomers and weirs.

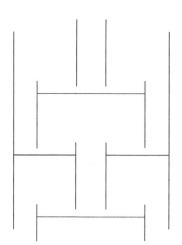

Figure 8.8 Tray column with dual downcomers.

8.4.1 Types of Trays

The active area of a tray has been designed in a number of different ways over the years. The difference has mainly been with respect to how the gas bubbles are dispensed into the liquid, but there are also tray designs where the downcomers are dispensed with. In this way the trays become cheaper to make and they make all the column cross sections active. The penalty is a poorer turn-down capability for the tray.

The main tray designs used today are:

- Sieve trays
- Valve trays
- Bubble cap trays (although present day use is mainly for special applications).

The active area of sieve trays is simply perforated with small holes that allow the gas to go through. If the gas velocity is high enough (giving enough pressure drop), the liquid will not leak down the hole. Valve trays have much bigger holes, but they have a device to cover it, but lifting enough to let the gas out sideways. The cover device is called a valve, and hence 'valve' trays (see Figure 8.9b). These trays have better turn-down capability than the sieve tray but are more expensive. Valve tray designs vary between vendors and their valve designs are under constant development to give the vendor a competitive edge.

Bubble cap trays are the longest serving design of the three and dominated the market until sieve trays gradually took over from the 1960s onwards. They are more expensive to make but still retain a market share. In some applications they have a special position due to 'proven' design and operation, and in some cases they are chosen due to their superior turn-down capability. The caps are submerged in the liquid layer, and the gas comes out of the vertical slots, see Figure 8.9(a).

8.4.2 Functional Parts of a Tray Column

The primary part of a tray column is of course the trays' 'active area' where the actual contact between gas and liquid is taking place. The design variables are liquid height and lateral liquid flow velocity, gas velocity and dispersion method, as well as the pitch of bubble dispersion positions and hole size. The aim is to play with these variables to achieve as much mass transfer as possible on the tray.

The downcomer is the other main feature of the tray column. It has a function beyond letting the liquid flow down to the tray below. The downward liquid velocity in the downcomer must be low enough to give bubbles following the liquid over the overflow weir a

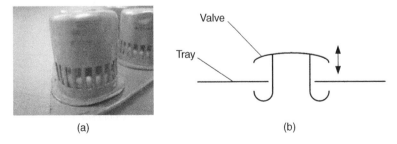

(a)   (b)

Figure 8.9 *Bubble caps (a) and valves (b).*

chance to disengage and rise back to the gas phase. If this is violated, there will be a very high volume fraction of bubbles in the downcomer, and a state of 'downcomer flooding' will arise. The liquid in the downcomer also provides a seal to prevent gas rising this way instead of going through the active tray area.

The downcomer skirt (also referred to as the downcomer apron) is the plate defining the downcomer section. It does not need to be vertical. As the liquid is cleared of bubbles as it moves downwards, the cross-sectional flow area may be reduced towards the bottom. Hence it is common for the skirt to slope outwards towards the column wall in the downwards direction. Its clearance to the plate below must be big enough to let the liquid through without too high a pressure drop. If this slit is too small there will be liquid backup in the downcomer. Too big a slit will not keep the liquid back and gas can bypass the active area by going up the liquid downcomer.

The 'steadying weir' is used to steady the liquid flow and make it flow evenly across the active section. Its distance from the skirt must be chosen to achieve this.

The 'overflow' weir is used to define a liquid height on top of the tray. However, the liquid height will be higher at the downcomer than at the overflow weir. This is because a height difference is needed as driving force to make the liquid flow across the tray.

8.4.3   Capacities and Limitations

As indicated in the previous sub-chapter trays must be operated within certain constraints. These constraints are basic to estimate for dimensioning a tray. Within these constraints the layout that gives the best efficiency and operational flexibility should be found. It is easy in principle, but difficult in reality due to the uncertainty of the various empirical design correlations available. Figure 8.10 explains the limitations of tray operations.

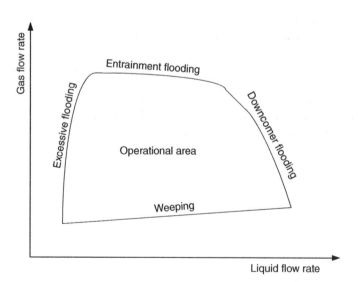

Figure 8.10   *Operational limitations for a tray. (Based on sieve tray performance diagram from, for example Kister (1993).)*

Kister (1992) gives a good account of design correlations that could be used for tray columns in general. In this text we are focusing on gas treating applications. The *GPSA Engineering Data Book* (GPSA, 1987) addresses gas treating in particular. It is worth noting that their quoted system factors demands an increased area by roughly 20% for amine columns and by 100% for glycol columns for water removal.

It is beyond the scope of the present text to review tray design methods in full. The intention is make the reader aware of the challenges that must be met.

An estimate of the maximum gas velocity in the column may be made using the Souders and Brown (1934) Equation 8.8:

$$C_{SB} = v_{G,flood}\sqrt{\frac{\rho_G}{\rho_L - \rho_G}} \qquad (8.8)$$

where C_{SB} is the Souders–Brown 'constant', $v_{G,flood}$ is the gas velocity at flood (or unacceptable entrainment) and ρ is the density of gas and liquid as indicated. The Souders–Brown constant is no longer looked upon as constant but is a function of tray spacing, liquid load, fractional hole area and hole diameter. Practical column designs are made based on a C_S-factor that is chosen based on available data. The C_{SB} is the C_S at flooding.

$$C_S = v_G\sqrt{\frac{\rho_G}{\rho_L - \rho_G}} \qquad (8.9)$$

The *GPSA Engineering Data Book* (1987) operates with a relation between C_{SB} and tray spacing where the latter varies from 25 to 90 cm ($10''-35''$). Higher tray spacing allows higher gas velocities. Allowable gas velocity tends to go down when the liquid load is increased but at low loads this may be the other way around. (Kister, 1992). Increasing the hole area will allow higher gas velocity. Kister (1992) states that the effect is more pronounced for fractions 0.05–0.10 than over 0.10. The C_{SB} will become lower when the hole diameter is increased. When designing, data must be found for the tray to be used.

An initial guidance to maximum downcomer velocities based on clear liquid is given by Kister (1992). According to this reference the maximum liquid velocity may vary from 0.06 to 0.09 m/s for tray spacing 45–75 cm; the higher the spacing, the higher the liquid velocity. These velocities are for amine and glycol systems met in gas processing.

In the literature cited and the bibliography there are alternative correlations offered, some of them from tray vendors, others from academic work in a broader sense. For a tray design it would be prudent to make the various estimates using more than one method to obtain a picture of the general uncertainty.

It should also be borne in mind that amine absorbers and desorbers are classified as foaming systems and should be derated accordingly. A glycol based absorber for water removal should be derated more than the amine system. To derate a column is simply to assign it a lower design capacity than that calculated from standard correlations.

8.4.4 Flow Regimes on Trays

According to Kister (1992) there are five main flow regimes on distillation trays:

1. *Bubble regime*: Distinct and individual bubbles rising through the liquid.
2. *Cellular foam*: This consists of bubbles separated by thin liquid walls.

3. *Froth* (also known as the mixed regime).
4. *Spray.*
5. *Emulsion.*

The bubble and cellular foam regimes are not likely to be found on industrial trays. If they appear on test trays, care must be taken in scale-up.

The froth regime is commonly found in industry. There is vigorous bubbling with bubbles of different sizes.

The spray regime takes over from the froth regime when the gas load is increased beyond a certain level. There is jetting of bubbles and in this regime the gas is the continuous phase. It is common in industry.

The emulsion regime is typically found when operating with high liquid loads and low vapour velocities. Gas bubbles are dispersed but are swept along the tray due to the high liquid flow velocity.

8.4.5   Tray Column Efficiencies

There are a number of ways to define the efficiency of a tray column.

- Overall column efficiency that refers to the ratio between actual and theoretical trays in a column.
- Murphree tray efficiency which is the efficiency expressed by approach to equilibrium at a tray.
- Point efficiency referring to the vapour composition at a specific point. There may be several such points on a tray.

The overall column efficiency is a questionable tool if the tray efficiency is in the region 5–25%, which is easily met in gas absorption due to the high driving forces encountered. In distillation work much higher efficiencies are normal. Care should be taken if the efficiency for a trace component is considered. In this case efficiencies may be abnormal if the component is swept to or from the interface by mass transport dominated by another and major component (O. Sandall, personal communication, 1994).

8.5   Spray Columns

Spray columns are quite simple. There is liquid distribution, most likely from spray nozzles in the top. Liquid is collected at the bottom. A maximum of one theoretical stage may be achieved with respect to mass transfer. Gas–liquid contact area depends on the nozzles used.

It is cheap to build and there are cases when this simple arrangement is good enough. Design information is hard to find. Mostly it will have to be based on the nozzle vendor's information on its spray characterisation. Beware of the turn-down characteristics of the spray nozzles.

8.6   Demisters

Information on demisting/droplet separation may be found in *Perry's Chemical Engineers' Handbook* (Perry and Green, 1997 or later) and there is even a book dedicated to this topic

(Bürkholz, 1989). There is also a wealth of information at hand in the literature that can be downloaded from the web sites of Koch, Sulzer and Vff (see References.) These three companies all provide demisters but there are also other providers.

Mists cover the size range of liquid particles from roughly 0.1 µm upwards. Sprays, that would include droplets, typically covers particles from 5000 µm diameter upwards. There is overlap and the borderlines between where the different names are used is far from clear. Hence, care must be exercised when evaluating information. There are also aerosols. These are important but aerosol problems are a little on marginal side for normal column demisting thinking.

In the past, say the 1970s, demisters were offered by specialist suppliers, a situation that was probably caused by specialist manufacturing skills. Today the situation has changed, and demisters will be offered by the packing and/or tray providers. The business has been restructured in this regard as well. Having said that, it is interesting to note that most of the eight demisting companies listed at the end of Bürkholz's book can still be found on the Internet, although with new owners in some cases.

Demisters, as they will be referred to here, work by one or more of the following mechanisms. The smallest particles will be caught by Brownian motion and this implies that the path to travel between 'capture hardware' must be short. The other way to catch droplets is by impingement, which could come about by 'capture hardware' being in the way of the travelling particle or by a variety of this called *inertia impingement* whereby the carrier gas is forced to alter course while the particles carry straight on and impinge.

It is useful to list standard droplet collecting techniques:

1. Aerosol filters.
2. Knitted wire mesh pads.
3. Vanes or chevrons.
4. Cyclones in various forms including swirl tubes.
5. Scrubbers (i.e. empty vessels where the droplets are supposed to fall out).

Numbers 2 and 3 are the most likely ones to be installed in a column and the discussion here will be restricted to those.

In general very high levels (like more than 99%) of mist or droplet removal is achievable. However, the removal efficiency will vary between droplet sizes. For aerosol removal special equipment is needed. An example is the so-called Brink filter developed by Monsanto around 1960 (see MECSGLOBAL.COM.ZA).

The vendors operate with a simple formula to allow judgement of capacity for various demisters. This is:

$$u_G = K\sqrt{\frac{\rho_L - \rho_G}{\rho_G}} \tag{8.10}$$

where u_G is superficial gas velocity and ρ represent the density of liquid and gas as indicated by subscript. The 'constant' K is specific for each type of demister, and is given, mostly in broad terms, by the vendor. The formula at least represents a check that a chemical engineer can do. It would serve to check what the first questions to the vendor might be.

Table 8.3 gives a few values for K-values to be used with Equation 8.10. ('Flexi' implies a Koch product and 'Mella' implies a Sulzer product). Any estimate is a first approach but in practice a direct contact with a vendor is needed. It is always a good idea to prepare well for this communication. At the end of the day, the responsibility for proper communication

Table 8.3 *A few representative K-values.*

Demister type	K (m/s)
Knitted wire mesh (basic)	0.08–0.11
Flexichevrons	0.05–0.35
Mellachevrons, simple	0.13–0.17
Mellachevrons, high capacity	0.15–0.45

of the problem lies with the chemical engineer making the enquiry. Guarantees are given relative to a specification after all.

8.6.1   Knitted Wire Mesh Pads

Knitted wire mesh pads have been around since at least 1970 or so. There was a company called KnitMesh making and selling them at the time. The trade mark KnitMesh is now claimed by Sulzer. They are built by knitting thin wire thread in a three dimensional pattern. Liquid would impinge on the wires and coalesce when drops find each other. The liquid in these bigger drops would, in the standard version, eventually drain back in to column. There are newer developments where liquid collection channels have been built into the pads. See, for example Sulzer's V-MISTER.

8.6.2   Vanes or Chevrons

Vanes are simply sheet metal that is bent at angles. The gas flow will be across the bends. Liquid droplets would impinge by inertia on the plates and drain. Figure 8.11 illustrates this arrangement. The figure also illustrates how liquid collection channels can be added to help prevent the liquid from re-entering the gas.

Vanes can be mounted horizontally for vertical flow of gas, or they can be mounted vertically for horizontal flow of gas. There are a number of proprietary designs on offer. The manufacturers have developed these designs to be more competitive for specific drop

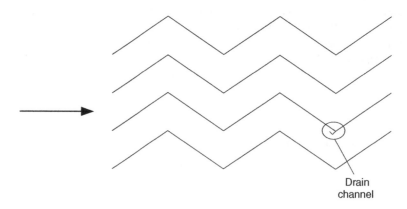

Drain
channel

Figure 8.11 *Vanes and chevrons.*

collection problems and they compete for this. It will be impossible to make the optimum choice from the vendor information given but it must always be the aim for a chemical engineer to be able to ask the right questions and ensure that good advice is given.

8.7 Examples

8.7.1 The Sepasolv Example from Chapter 7

It is interesting to take the example where the need for Number of Transfer Units (NTU) was estimated one step further and determine the diameter of column needed. We shall approach this using the GPDC in Figure 8.2. Variables and physical properties are summarised next:

Liquid flow, L_V: 4 695 m³/h
Gas flow, G_V: 12 888 m³/h.

The gas composition is needed to calculate the average molecular weight in order to estimate the gas density. It will be roughly as follows.

	MW	Mole fraction
H_2	2	0.580
N_2	28	0.198
CO	28	0.002
Ar	40	0.010
CH_4	16	0.010
CO_2	44	0.200
Mix	16.12	1.000

Liquid density, ρ_L: 1002 kg/m³
Gas density, ρ_V: 15.49 kg/m³ based on the gas being ideal, the previous composition, 40°C and 25 bar.
Liquid viscosity, μ_L: 0.004 kg/m·s (by extrapolation of data from Wölfer, 1982).

For the purpose of this estimate we shall choose 50 mm Pall rings in steel. They have a packing factor $F = 20$ (Norton, 1977).
The first action is to estimate the 'flow parameter':

$$Flow\_parameter = \frac{L}{G}\sqrt{\frac{\rho_G}{\rho_L}} = \frac{(4695 \, \text{m}^3/\text{h})(1002 \, \text{kg/m}^3)}{(12888 \, \text{m}^3/\text{h})(15.49 \, \text{kg/m}^3)}\sqrt{\frac{15.49 \, \text{kg/m}^3}{1002 \, \text{kg/m}^3}} = 2.93$$

We next go to the GPDC, use this value on the *x*-axis, and next choose a pressure drop to go with. In this case we notice that we are far to the right in the chart, and elect a pressure drop somewhere in the middle. The high liquid rate for this physical absorption causes the high flow parameter. Some discussion with a vendor would be in place but we carry on to obtain a fuller picture.

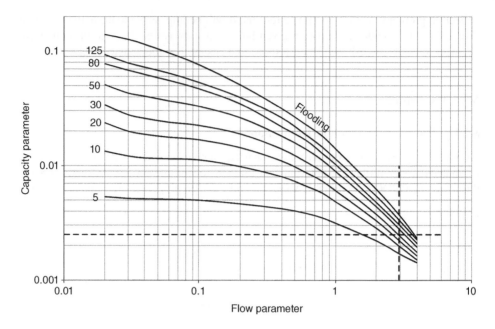

The chosen pressure drop is 30 mm W.C. per metre packing. From the figure it then follows that the capacity parameter is 0.0025.

$$G^2 = \frac{(Capacity\_parameter)\left(\rho_G\left(\rho_L - \rho_G\right)\right)}{F\left(\frac{\mu_L}{\rho_L}\right)^{0.1}} = \frac{(0.0025)(15.49(1002 - 15.49))}{20\left(\frac{0.004}{1002}\right)^{0.1}} = 6.62$$

$$G = \sqrt{6.62} = 2.57\,\text{kg/m}^2 \cdot \text{s}$$

Since we also know that $G' = (12\,888\,\text{m}^3/\text{h})(15.49\,\text{kg/m}^3)/3600 = 55.46\,\text{kg/s}$, the column cross-sectional area can now be estimated:

$$Cross - sec\,tional\_area = \frac{55.46\,\text{kg/s}}{2.57\,\text{kg/m}^2 \cdot \text{s}} = 21.55\,\text{m}^2$$

And the diameter is $D = \sqrt{21.55} = 5.24\,\text{m}$

8.7.2   The Selexol Example from Chapter 7

The parallel example where Selexol was looked at as an alternative differs from the above problem only in the amount of liquid circulation and of course the properties of the liquid. The deviating data are:

Liquid circulation, L_V: 6808 m^3/h
Liquid density, ρ_L: 1002 kg/m^3.

The flow parameter here becomes:

$$Flow\_parameter = \frac{L}{G}\sqrt{\frac{\rho_G}{\rho_L}} = \frac{(6808\,\text{m}^3/\text{h})\,(1035\,\text{kg/m}^3)}{(12\,888\,\text{m}^3/\text{h})\,(15.49\,\text{kg/m}^3)}\sqrt{\frac{15.49\,\text{kg/m}^3}{1035\,\text{kg/m}^3}} = 4.3$$

From the GPDC it is seen that this is on the very rightmost edge, or even the outside of the data range in the diagram. It is perfectly possible to carry on the estimate and obtain a diameter also for this case. The diameter obtained would be around 5.0 m. However, the large liquid flow may seem to be a challenge and in such a case specialist advice should be looked for.

In fact, even a water wash could be used in principle and has been built in the past, and in such a case even larger liquid to gas ratios would be encountered. There are specialist internals to be found for such cases.

A final word: The data found on Sepasolv and Selexol in the literature are not official solvent data from the licensors. Care should be exercised in interpreting the results.

8.7.3   Natural Gas Treating Example

Let us consider a relatively small natural gas treatment plant, one treating 30 MMSCFD. This takes place at 40°C and 35 bar.

Composition is described in this table:

Composition, mole fractions	Mole fraction	MW
-CH_4	0.865	16
-C_2H_6	0.050	30
-CO_2	0.080	44
-H_2S	0.005	32
Average MW calculated:		19.02

The specification will be to reduce the contents of CO_2 to 2% and the H_2S to 4 ppm. For now we want to consider mass transfer specifics. What size of mass transfer coefficients could be expected?

We start by converting the feed gas flow to metric value to concur with the availability of physical data for the system.

$$(10^6/\text{MM})\,(30\,\text{MMSCFD})\,(1\,\text{D}/24\,\text{h})\,/\,(836.62\,\text{SCF/kmol})$$

$$=1494\,\text{kmol/h} = 0.4150\,\text{kmol/s}$$

The actual volumetric flow is $= (0.4150)\,(23.645)\,(1.013/35)\,(313/288) = 0.3087\,\text{m}^3/\text{s}$

The first thing to do is to make an estimate of the column diameter needed since the linear velocities based on an empty column will be needed for mass transfer estimates. However, before that we need to estimate the absorbent circulation rate that is needed. One mole CO_2 will bind 2 mol monoethanolamine (MEA), while 1 mol H_2S will bind 1 mol MEA. We choose to work with a 30% wt aqueous MEA (5 M) solution for ease of data availability.

CO_2 to be captured is approximately: $(0.08 - 0.02)(0.4150) = 0.02490\,kmol/s$.
H_2S to be captures is approximately: $(0.005 - 0.000004)(0.4150) = 0.002075\,kmol/s$.

We shall assume a differential loading of MEA equal to 0.25 mol/mol.
If the flow of absorbent is L_v (m^3/s), it follows that:

$$L_v\,(5\,kmol/m^3)\,(0.25\,mol/mol) = [(2)\,(0.02490) + (0.002075)]\,kmol\ MEA\ needed/s$$

$$L_v = 0.04150\,m^3/s$$

Physical data needed are:

Liquid density: $1056\,kg/m^3$ (Han *et al.*, 2012)
Liquid viscosity: $0.0014\,kg/m{\cdot}s$ (Amundsen, Øi and Eimer, 2009)
Surface tension: $0.062\,N/m$ (Han *et al.*, 2012)
Gas density: $P/(RT) = 33.64\,kg/m^3$ assuming ideal gas.

Liquid mass flow is $L_w = (0.04150\,m^3/s)\,(1056\,kg/m^3) = 43.83\,kg/s$.
Gas mass flow is $G_w = (0.3087\,m^3/s)\,(33.64\,kg/m^3) = 10.38\,kg/s$.

The flow parameter, is $\dfrac{43.83}{10.38}\sqrt{\dfrac{33.64}{1056}} = 0.753$.

Choosing a pressure drop of 20 mm water column (WC) per metre packing, the capacity parameter becomes 0.008 based on Figure 8.2. We can now estimate the gas mass flux.

$$G_{wf}^2 = \frac{(Capacity\_parameter)\,\left(\rho_G\,(\rho_L - \rho_G)\right)}{F\left(\dfrac{\mu_L}{\rho_L}\right)^{0.1}} = \frac{(0.008)\,(33.64\,(1056 - 33.64))}{20\left(\dfrac{0.0014}{1056}\right)^{0.1}} = 53.24$$

$$G_{wf} = \sqrt{53.24} = 7.30\,kg/m^2 \cdot s$$

The column cross-sectional area is $\dfrac{G_w}{G_{wf}} = \dfrac{10.38}{7.30} = 1.423\,m^2$

The diameter is $\sqrt{\dfrac{4x1.423}{\pi}} = 1.346\,m$

8.7.4   Example, Flue Gas from CCGT

This is based on Section 1.7.4. The feed flow may be summarised as

Flue gas summary	Flow kmol/h	Mole fraction
Nitrogen	64 628	0.753
Oxygen	11 044	0.129
Water	6 866	0.080
CO_2	3 273	0.038
Total	85 811	1.000

Its average molecular weight is 28.3. The actual volumetric flow of gas is $G_v = (85\,811)$ $(22.414)\,(313/273)\,(1.013/1.05) = 21\,27\,474\,m^3/h = 591.0\,m^3/s$

We assume that the temperature will be 40°C and that the pressure will be 1.05 bar. Liquid properties will be the same as in the previous example. We shall assume ideal gas. Physical data needed are:

Liquid density: 1120 kg/m³ (Han *et al.*, 2012)
Liquid viscosity: 0.0014 kg/m·s (Amundsen, Øi and Eimer, 2009)
Surface tension: 0.062 N/m (Han *et al.*, 2012)
Gas density: 1.142 kg/m³.

The gas flow $G_w = (591.0)(1.142) = 675.0$ kg/s.

We assume that the lean and rich loadings of absorbent are 0.20 and 0.45 mol/mol, respectively. The volumetric liquid flow is L_v which may be found from the CO_2 balance over the absorption tower:

$$L_v (5)(0.45 - 0.20) = (0.90)(3273) = L_v = 2357 \, \text{m}^3/\text{h} = 0.6546 \, \text{m}^3/\text{s}$$

The liquid mass flow is $L_w = (1120)(0.6546) = 733$ kg/s

The flow parameter is $\dfrac{733}{675}\sqrt{\dfrac{1.142}{1120}} = 0.035$

We want a minimum of pressure drop and go for 5 mm WC per metre. The capacity parameter from Figure 8.2 is accordingly 0.005.

$$G_{wf}^2 = \frac{(Capacity\_parameter)\left(\rho_G\left(\rho_L - \rho_G\right)\right)}{F\left(\dfrac{\mu_L}{\rho_L}\right)^{0.1}} = \frac{(0.005)(1.142(1120 - 1.142))}{20\left(\dfrac{0.0014}{1120}\right)^{0.1}} = 1.201$$

$$G_{wf} = \sqrt{1.201} = 1.096 \, \text{kg/m}^2 \cdot \text{s}$$

The column cross-sectional area is $\dfrac{G_w}{G_{wf}} = \dfrac{675}{1.096} = 616 \, \text{m}^2$

The diameter is:

$$\sqrt{\frac{4 \times 616}{\pi}} = 28.0 \, \text{m}$$

and

$$L_{vf} = \left(0.6546 \, \text{m}^3/\text{s}\right) / \left(616 \, \text{m}^2\right) = 0.001 \, \text{m}^3/\text{m}^2 \cdot \text{s (or m/s)}$$

While $G_{vf} = ((591.0 \, \text{m}^3/\text{s})/(616 \, \text{m}^2) = 0.959 \, \text{m}^3/\text{m}^2.\text{s}$ (or m/s)

This diameter is rather large and probably bigger than previously built columns. This figure is also much higher than the result of a short cut approach suggested by Chapel, Mariz and Ernest (1999). In their work they suggest that the column diameter can be calculated from the formula:

$$(Coefficient)\sqrt{\frac{CO_2 \, (tonnes/d)}{\%CO_2}}$$

where the constant is 0.56 at 3% CO_2 and 0.62 at 13% CO_2 in the gas. By interpolation it is 0.565 at 4%. Tonnes per day refer to CO_2 removed. The amount of CO_2 removed is:

$$(0.9)(3273 \, \text{k mol/h})(44 \, \text{kg/k mol})(1 \, \text{tonne}/1000 \, \text{kg})(24 \, \text{h/d}) = 3111 \, \text{tonne/d}$$

and our CO_2 content is 3.8%:

$$diameter = 0.565\sqrt{\frac{3111}{3.8}} = 16.1\,m$$

If we repeat our initial procedure to calculate the column diameter while accepting a higher pressure drop like 30 mm WC per m, we could increase the capacity parameter to 0.028. Using this figure we find that the cross sectional area of the column becomes $260\,m^2$, and the diameter 18.2 m. This is a reasonable agreement with the short cut method. Columns of such a diameter may be built as one unit. The trade-off between pressure drop and investment in this case should be clear enough. There is obviously a huge impact. The penalty is increased fan power to drive the flue gas through without increasing the pressure at the gas turbine outlet. A 10 m high column would then have a pressure drop of 0.03 bar which must be seen in the context of assumed flue gas feed pressure of 1.05 bar. There are also more modern packing materials than the 2″ metal Pall rings specified for this calculation. They are adhered to here because of data availability and the fact that they are part of the data foundation for correlations to be discussed in Chapter 11. There are now special low pressure drop packings coming on the market with this application in mind.

References

Amundsen, T.G., Øi, L.E. and Eimer, D. (2009) Density and viscosity of Monoethanolamine + water + carbon dioxide from 25 to 80°C. *J. Chem. Eng. Data*, **54**, 3096–3100.

Billet, R. (1995) *Packed Towers (in Processing and Environmental Technology)*, VCH-Chemie Verlag, Weinheim.

Bürkholz, A. (1989) *Droplet Separation*, VCH Publishers, Weinheim.

Chapel, D.G., Mariz, C.L., Ernest, J. (1999) Recovery of CO_2 from flue gases: commercial trends. paper presented at the Canadian Society of Chemical Engineers Annual Meeting, Saskatoon, Saskatchewan, October 4–6.

GPSA (1987) *GPSA Engineering Data Book*, 10th edn, GPSA, Tulsa, OK.

Han, J., Jin, J., Eimer, D.A. and Melaaen, M.C. (2012) Density of Water (1) + Monoethanolamine (2) + CO_2 (3) from (298.15 to 413.15) K and Surface Tension of Water (1) + Monoethanolamine (2) from (303.15 to 333.15) K. *J. Chem. Eng. Data*, **57** (4), 1095–1103.

Harker, J.H., Backhurst, J.R. and Richardson, J.F. (2002) *Coulson and Richardson's Chemical Engineering*, 5th edn, vol. **2**, Butterworth-Heinemann.

Kister, H.Z. (1992) *Distillation – Design*, McGraw-Hill.

Mackowiak, J. (2010) *Fluid Dynamics of Packed Columns*, Springer.

Norton (1971) *Design Information for Packed Towers*, Bulletin DC-10R, Norton Company, Acron, OH.

Norton (1977) *Design Information for Packed Towers*. Bulletin DC-11, Norton Company, Akron, OH, (Newer version of DC-10R).

Perry, R.H. and Green, D.W. (1997) *Perry's Chemical Engineers' Handbook*, 7th edn, McGraw-Hill.

Sherwood, T.K., Shipley, G.H. and Holloway, F.A. (1938) Flooding velocities in packed columns. *Ind. Eng. Chem.*, **30**, 765–769.

Strigle, R.F. Jr., (1987) *Random Packings and Packed Towers. Design and Applications*, Gulf Publishing, Houston, TX.

Souders, M. and Brown, G.G. (1934) Design of fractionation columns. I. Entrainment and capacity. *Ind. Eng. Chem.*, **26**, 98–103.

Wölfer, W. (1982) Helpful hints for physical solvent absorption. *Hydrocarbon Process.*, **61**, 193–197.

Further Reading

Beck, R. (1969) *Ein neues Verfahren zur Berechnung von Füllkörpersäulen*, Vereinigte Füllkörper-Fabriken Gmbh + Co, Westerwald.

Kister, H.Z. (1990) *Distillation – Operation*, McGraw-Hill.

Koch-Glitsch L.P. (2007) Mist Elimination. Liquid-liquid coalescing (Downloaded 2012) Koch-Glitsch Bulletin MELLC-02, Rev 5, 2013. Online http://www.koch-glitsch.com/Document%20Library/ME_ProductCatalog.pdf (accessed 6 May, 2014).

Leva, M. (1953) *Tower Packings and Packed Tower Design*, 2nd edn, US Stoneware Company, Akron, OH.

Sulzer (2014) Gas/Liquid Separation Technology. 22.84.06.40 – II.14. Online http://www.sulzer.com/en/-/media/Documents/ProductsAndServices/Separation_Technology/Mist_Eliminators/Brochures/Gas_Liquid_Separation_Technology.pdf (accessed 6 May, 2014).

Vff (2014) Ihr Kolonnenausrüster und Berater, Your expert for tower packings, catalyst support material and column equipment. VFF 999 27 002. Ramsbach-Baumbach, Germany. Online http://www.vff.com/en/download (accessed 6 May, 2014).

9

Rotating Packed Beds

Rotating packed beds (RPBs) are the 'process intensification' (PI) version of conventional packed columns discussed in Chapter 8. The definition of PI, according to Colin Ramshaw, is a reduction of equipment size by 2 orders of magnitude. RPBs live up to this definition. An RPB is illustrated in Figure 9.1. The gas enters at the periphery and is forced through the packing to the open core from where the gas leaves. The liquid is distributed from a stationary device positioned in the core and travels through the packing aided by the gravitational forces set up by the rotation. There are seals to prevent the fluids from following other paths.

9.1 Introduction

The history of rotating gas–liquid contactors may be traced back to the 1930s (Reay, Ramshaw and Harvey 2008). RPBs, as they are known today, were first introduced in the early 1980s (Ramshaw, 1983). The RPB introduced by Ramshaw became known as Higee, and it caused an enormous interest. The only commercial installation at the time was in Traverse City (Michigan) where it was successfully used for stripping aviation fuel components from ground water.

In the late 1980s there were two test campaigns that were run to check the Higee's performance with respect to treating natural gas. Selective absorption of H_2S over CO_2 was particularly tested in one set-up where it was expected that the short residence time would favour the absorption of H_2S (Bucklin and Won, 1987; Bucklin, Buckingham and Smelser, 1988). This first test campaign was marred by experimental trouble not related to the Higee itself. Further tests were planned at the same site, but did not take place. Another test program was carried out to test absorption of CO_2 and also the dehydration process with absorption of water into glycol (Fowler, Gerdes and Nygaard, 1989). All RPB related aspects of these tests could be described as successful. However, introducing new

Gas Treating: Absorption Theory and Practice, First Edition. Dag Eimer.
© 2014 John Wiley & Sons, Ltd. Published 2014 by John Wiley & Sons, Ltd.

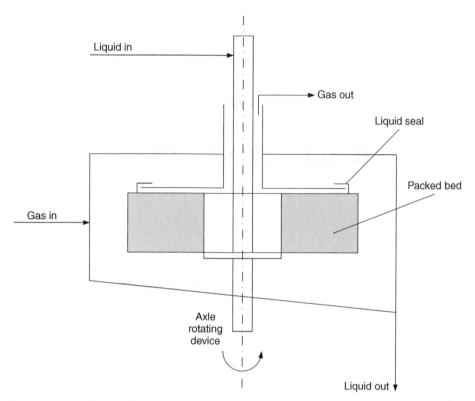

Figure 9.1   *A sketch of how a rotating packed is operated in counter-current gas–liquid contact. (Based on the description provided by Mallinson and Ramshaw (1981).)*

technology is heavy going and things ground to a halt. The Glitsch Company that had the rights eventually dropped their commercialisation work.

Meanwhile, or a little later, RPB were subjected to further research in China where a special research unit (High Gravity Engineering and Technology Centre (Higrav), at the University of Chemical Technology, Beijing) was set up. It appears that their Higee variation has been used on a large scale for de-aerating injection water in the Chinese oil and gas industry (Guo *et al.*, 2001). Other applications are also mentioned. The Higrav Institute is still very active in this field and many industrial size installations in actual plants are referred to (e.g. Chen, 2009; Qian *et al.*, 2010). A search for 'RPB' in a patent search engine quickly revealed more than a hundred Chinese and Taiwanese originating patents or patent applications.

There is also a commercial application in the chemical industry where it was found that RPB technology was an enabling factor when Dow Chemical Company succeeded in commercialising a new process for producing HOCl (Trent and Tirtowidjojo, 2001).

In the period from 1990 until today research work has been going on in a number of universities that include Newcastle-upon-Tyne where Colin Ramshaw became professor, and Chang Gung, Chung Yuan, National Taiwan and National Tsing Hua, all in Taiwan; and there is the Indian Institute of Technology in Kanpur. Last but not least there is the Higrav Institute at the University of Chemical Technology in Beijing.

Western industrial interest in RPB keeps appearing at intervals, and the technology is by no means dead although it would probably be difficult to find a supplier. The technology is open in so far that any patent applied for prior to 1995 would now be expired. A current interest from Alstom is described by Wolf and Aleksic (2012), a patent application that describes two RPBs on the same axle with co-current gas and liquid flow in the first followed by a counter-current flow bed. Another is by Strand, Fiveland and Eimer (2011) that deals with RPBs used for desorption. In the USA, recent patents include Dutra *et al.* (2012) and Park and Nelson (2007) assigned to Chevron and Cleveland Gas Systems respectively. The number drowns in all the Eastern patent applications.

General design correlations for RPBs have only recently been summarised but the literature is not extensive. It is an object of this chapter to attempt pointing to sources of information that may at least be used as starting points. Published data are, in general, based on small laboratory machines a few centimetres in diameter except for the two papers reporting pilot plant data from natural gas treating in New Mexico and Louisiana in the 1980s (Bucklin, Buckingham and Smelser 1988; Fowler, Gerdes and Nygaard, 1989).

9.2   Flooding and Pressure Drop

Flooding capacities of RPBs may be described by the traditional flooding diagrams due to Sherwood *et al.* (1938) for standard packed beds. The curve for stacked rings should be used. The situation is summarised by Reay, Ramshaw and Harvey (2008). A diagram is given in Figure 9.2. It shows that a RPB has an enormous capacity for gas and liquid, and this is the reason why units used in university laboratories are small. Even so they have been operated far away from flooding. In this figure the flow parameter and the capacity parameter are:

$$Flow\ parameter = \frac{L}{G}\sqrt{\frac{\rho_G}{\rho_L}} \tag{9.1}$$

$$Capacity\ parameter = \frac{v_G^2 a \rho_G}{g\varepsilon^3 \rho_L} \tag{9.2}$$

The gas pressure drop through a packing can be significant. The rotor will act as a fan trying to push the gas outwards while in counter-current operation the gas flow will be inwards. Clearly this will add to the gas pressure drop. The 'fan efficiency' will vary from rotor to rotor. No design rules have been established. Any value found for small units must be scrutinised in case there are other sources like packing support giving rise to pressure drop. Yang *et al.* (2010) have derived pressure drops for a given packing at speed using CFD (computational fluid dynamics). A more directly practical approach was made by Zheng *et al.* (2000). They analysed the unit section by section, including the peripheral gas volume, the rotating packing, the core in the middle of the packing and also specifying the gas exit channel. Their equations are complex and need more explanations than is reasonable to duplicate here. The reader is referred to the referenced paper. Pressure drop was also discussed by Agarwal *et al.* (2010), and they in turn refer to work done by Sandilya, Rao and Sharma (2001). Since rotors are arranged a little differently, it is left to the reader to go into the references for details.

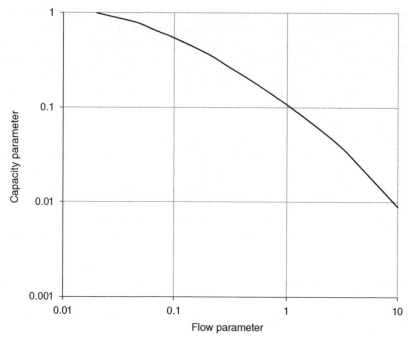

Figure 9.2 *The relationship between flooding and flows in an RPB. The parameters are defined by Equations 9.1 and 9.2. The figure is based on data from Reay, Ramshaw and Harvey (2008).*

9.3   Fluid Flow

Fluid flow in the rotor is an interesting subject. How is liquid distributed on to the packing? And to what extent will the liquid distribution be improved as it flows through the packing? Good initial distribution must be targeted. The rotor will tend to conserve an initial poor distribution. This was studied and concluded by Burns and Ramshaw (1996).

Another line of investigation has been taken by Yang *et al.* (2010) where they used CFD to analyse flows in the rotor. The problem is complex and their conclusion was only that the insight gained was useful in understanding the flow problems involved.

9.4   Mass Transfer Correlations

There are many factors that need to be taken into account when doing mass transfer design for an RPB. Each situation must be evaluated afresh. Table 9.1 lists a number of publications where mass transfer studies have been focused on or been discussed. The last two references describe older pilot plant work. The scatter is bigger than for laboratory units, but the rotors are at least closer to industrial scale. Chen has derived expressions for both the liquid and gas side mass transfer coefficients for RPBs (2006, 2011). The liquid side coefficient model

Table 9.1 *A limited literature overview of published research on RPBs (Newcastle University omitted).*

Rotating Packed Beds	Year Published	Cross- or Counter Flow	Research Theme
A limited literature overview			
Taiwan			
Cheng *et al.* (2013)	2013	Counter	Desorption of CO_2 from amine solutions
Lin and Chen	2011	Cross	Efficient CO_2 abs = f (rpm, G, L, abs type) using MEA, AMP, PZ
Chen	(2011)	Counter	VOC. MT coefficients, gas and liquid sides
Chen *et al.*	2008	Cross	VOC. MT coefficients, gas and liquid sides
Lin and Chen	2008	Cross	Overall gas side MT coefficient Kg = f(rpm, L, G); CO_2-NaOH
Lin *et al.*	2008	Cross	CO_2-NaOH. dP = f(rpm, L, G); Kg = f(rpm, L, G)
Lin and Liu	2007	Counter	CO_2-NaOH and O_2-des. dP = f (rpm, L, G); Kg and Kl = f (rpm, L, G)
Lin and Chen	2007	Cross	CO_2-NaOH. Abs efficiency = f (G, L, rpm)
Tan and Chen	2006	Counter	CO_2, MEA, AMP, PZ. Kg = f (G, L)
Cheng and Tan	2006	Counter	CO_2, MEA, PZ, AEEA. EFF = f (amines, T, L, G)
Chen *et al.*	2006	Counter	Packing sizes and shapes (random)
Chen, Lin and Liu	2005	Counter	Phys abs or des. Effect of varying diametre
Lin, Liu and Tan	2003	Counter	CO_2-removal. Kg = f (rpm, G, L, CO_2/abs fed)
Chen and Liu	2002	Counter	VOC absorption Kg = f(rpm, L, G)
India			
Rajan *et al.*	2011	Counter	Novel split RPB. Flood, surface area a, K; = f(L, G, rpm)
Rao, Bhowal and Goswami	2004	Counter	An appraisal. Lists RPB details of papers < 2000/2001
China			
Luo *et al.*	2012	Counter	CO_2-NaOH. Contact area determination
Qian *et al.*	2010	Counter	H_2S selective absorption w.r.t CO_2 using MDEA
Sun *et al.*	2009	Counter	Simultaneous absorption of CO_2 and NH_3
Yi *et al.*	2009	Counter	Modelling MT-coefficient based on Benfield-CO_2
USA/Glitsch supported pilots			
Fowler, Gerdes and Nygaard	1989	Counter	MT rates, H_2O-TEG, CO_2-DEA
Bucklin, Buckingham and Smelser	1988	Counter	Performance w.r.t selective H_2S absorption

Note:
abs: absolute
rpm: revolutions per minute
des: desorption
w.r.t: with respect to

established first was:

$$\frac{k_L a_e d_p}{D_L a} \left\{ 1 - 0.93\frac{V_o}{V_t} - 1.13\frac{V_i}{V_t} \right\} = 0.35Sc_L^{0.5}Re_L^{0.17}Gr_L^{0.3}We_L^{0.3}\left(\frac{a}{a_p'}\right)^{-0.5}\left(\frac{\sigma_c}{\sigma_w}\right)^{0.14} \tag{9.3}$$

Note that the correlation for k_L is really a correlation for $k_L a$.

In the newer work, k_G was derived from data for the overall coefficient K_G. Because the gas in the rotor could be seen as rotating with the rotor, the first attempt used old correlations from standard packed columns, perhaps a bit surprising. He starts with the correlations from Onda, Takeuchi and Okumoto (1968) for estimating both the gas side mass transfer coefficient and a correlation by Puranik and Vogelpohl (1974) for gas-liquid contact area made for standard packed columns. These correlations, as given by Chen, are as follows:

$$\frac{k_G(ad_p)^2}{D_G a} = 2Re_G^{0.7}Sc_G^{1/3} \tag{9.4}$$

$$\frac{a_e}{a} = 1.05Re_L^{0.047}We_L^{0.135}\left(\frac{\sigma}{\sigma_c}\right)^{-0.206} \tag{9.5}$$

Note that Chen's version of a_e as in Equation 9.5 has a different exponent on Re from the original (0.047 instead of 0.041). Both equations have also been reorganised here to render dimensionless entities.

It was, however, found to be unsatisfactory to use the correlation of k_G from Onda *et al.* A new one was developed based on the availability of 430 $K_G a$ data from the literature in combination with the $k_L a$ correlation the past work of Chen *et al.* in combination with the contact area model from Puranik and Vogelpohl that was still used in lack of a specific correlation for RPBs. The new correlation for the k_G is:

$$\frac{k_G a}{D_G a^2} \left\{ 1 - 0.9\frac{V_o}{V_t} \right\} = 0.023Re_G^{1.13}Re_L^{0.14}Gr_G^{0.31}We_L^{0.07}\left(\frac{a}{a_p'}\right)^{-0.5}\left(\frac{a}{a_p'}\right)^{1.4} \tag{9.6}$$

The dimensionless numbers in Equations 9.3–9.6 are defined as follows:

$$Fr = \frac{La_t}{\rho^2 a_c} \quad Gr_L = \left(\frac{d_p^3 a_c \rho^2}{\mu^2}\right)_L \quad Gr_G = \left(\frac{d_p^3 a_c \rho^2}{\mu^2}\right)_G$$

$$Re_L = \left(\frac{L}{a}\mu\right)_L \quad Re_G = \left(\frac{G}{a}\mu\right)_G \quad Sc_L = \left(\frac{\mu}{\rho}D\right)_L \quad Sc_G = \left(\frac{\mu}{\rho}D\right)_G$$

$$We_L = \left(\frac{L^2}{\rho}a\sigma\right)_L \tag{9.7}$$

L and G in Equation 9.7 are fluxes, kg/(m^2 s), ρ, μ, σ and D are density, viscosity, surface tension and diffusivity, respectively. The rest will be explained next.

More recently research work on gas-liquid contact area in an RPB was published by Luo *et al.* (2012). A contact area was developed based on data from their specific RPB made from wire mesh wound around a core. The absorption system used was the old classic CO_2 and a NaOH solution. Their relation is:

$$\frac{a_e}{a} = 66510 Re_L^{-1.41} Fr_L^{-0.12} We_L^{1.21} \Phi^{-0.74} \qquad (9.8)$$

In Equation 9.8 the Re and *We* numbers differ from those in Equations 9.3–9.7 in that the characteristic linear dimension is defined differently. Chen and co-workers use $1/a$ while Luo *et al.* use d_p.

These correlations cannot be used without a few explanations and definitions. First of all there are various specific areas used in the above correlations. Here a without any indexes refers to the nominal contact area between gas and liquid. Subscript e refers to the effective value for the packing and a_p' is the specific surface area of a packing of 2 mm diameter beads ($1800\,m^2/m^3$) that is the reference case of Chen and co-workers. Then there is d_p that is the 'spherical equivalent diameter' of the packing. This is defined as:

$$d_p = \frac{6(1 - \varepsilon)}{a\psi} \qquad (9.9)$$

where ψ is the sphericity of the packing, also known as the 'shape factor'. It is dimensionless and is defined as the ratio between the surface area of a sphere with the same volume as the packing and the surface area of the packing. (Sounds precise, but I have seen references to Raschig rings sphericity as both 0.3 and 0.56. Be careful.)

The volumes V referred to in Equations 9.3 and 9.6 are defined as V_o the volume of the rotor, V_i the volume in the core of the rotor and V_t as volume of the whole house defined by the rotor housing diameter and the height of the rotating bed. It reflects the fact that mass transfer is going on also in these volumes. Care has to be exercised when using such corrections for scale-up.

The factor Φ used in Equation 9.8 comes from the experiments where Luo and co-workers were using wire mesh as packing. It is defined as:

$$\Phi = \frac{c^2}{(d + c)^2} \qquad (9.10)$$

The size c is the width of the square opening of the mesh while d is the diameter of the steel fibres used to make the mesh.

A general pointer for size of RPBs needed is that in general the size of a height of a transfer unit (HTU) is in the order of a few centimetres. Hence many units may be included in one rotor.

It might be useful to remind ourselves about what the various dimensionless groups represent. This is done briefly in Table 9.2.

9.5   Application to Gas Treating

When Higee was first introduced all sorts of process plant applications were discussed. It was in the oil and gas industry, however, where the biggest interest was raised. This is

Table 9.2   *Dimensionless groups involved in the correlations.*

Group	Ratios
Froude	Inertia to gravitational forces
Grashof	Buoyancy to viscous forces
Reynolds	Inertia to viscous forces
Schmidt	Viscous to molecular diffusion rates
Weber	Inertia to surface tension forces

reflected in the two pilot plants operated in the USA. No one took the first step, however, and it was in China that the first oil and gas use was made by de-aerating injection water rather than treating gas. In the West it was Dow Chemical that put it to use in a chemical process (Trent and Tirtowidjojo, 2001). The Higrav Institute at Beijing University of Technology has carried out successful trials for full scale CO_2 capture in a refinery (Chen, 2009).

9.5.1   Absorption

The obvious application to study was selective absorption of H_2S in the presence of CO_2. This is due to the very short retention times found in an RPB. The pilot plant work reported by Bucklin, Buckingham and Smelser (1988) supports this idea, but their data were not satisfactory. Plans to improve the pilot plant were never executed. Qian *et al.* (2010) later picked up the thread in a small laboratory unit in Beijing.

Plain CO_2 removal was investigated in another pilot plant reported by Fowler *et al.* (1989) where they looked at absorption in diethanolamine (DEA) solutions. The same team also studied the absorption of water in triethylene glycol (TEG) for dehydrating the gas. Both investigations were successful and the unit worked well mechanically. Again, the work came to a stop.

Because of the pressure drop counter-current RPBs are not feasible in treating flue gas. However, a cross-flow configuration could be used but the unit would be very big and it has not been studied in depth. A variation based on co-centric rotating plates has been patented (Eimer, 1999).

9.5.2   Desorption

Desorption of CO_2 from monoethanolamine (MEA) solutions was attempted at the University of Newcastle-upon-Tyne around 2000 (Jassim *et al.*, 2007). It worked, but it was found to work better when the MEA concentration was increased. Ramshaw (1993) had already pointed out the possibility of running an RPB with a reboiler at the perimeter. It would be very good to combine the desorption column in the absorption–desorption process with the reboiler. There have been at least two efforts in this direction reported in the literature (Cheng, Chih Lai and Tan, 2013; Strand, Fiveland and Eimer, 2011). The first of these is an elaborate effort to include all units related to desorption in one unit while the second is a research effort to replace the desorption column itself with an RPB. A surprising side effect of their use of an RPB in this application is that they observe a reduction in the need for energy when using the highly conventional 30% (wt) MEA in aqueous solution.

9.6   Other Salient Points

Enhancement of mass transfer is particularly good for systems with liquid side mass transfer limitations. RPBs are very good for handling viscous liquids due to the high gravitational field set up. The same gravitational field also helps to abate foaming. Foam will collapse quicker when subjected to force fields.

The enhancement of mass transfer due to chemical reaction is a little lower than in normal packed columns due to the much higher mass transfer coefficient. This will become clear when these effects are discussed in Chapter 12.

RPBs are also better for mass transfer for gas side limited systems, but since the gas has a tendency to rotate with the packing the relative velocity is not as high as for the liquid. The main effect is due to the very high specific surface area allowed in an RPB.

9.7   Challenges Associated with Rotating Packed Beds

Design of seals is a matter that demands focused attention. Liquid seals will easily provide resistance to rotation and increase power needed if not designed intelligently.

The design must allow for easy access for maintenance. This is akin to any rotating equipment. In terms of that it is common to compare RPBs with centrifugal pumps or centrifuges for level of mechanical challenge and complexity. Since such a machine will be required to operate continuously, a robust mechanical design must be aimed for.

References

Agarwal, L., Parvani, V., Rao, D.P. and Kaistha, N. (2010) Process intensification in HiGee absorption and distillation: Design procedure and applications. *Ind. Eng. Chem. Res.*, **49**, 10046–10058.

Bucklin, R.W., P.A. Buckingham, S.C. Smelser, (1988) The Higee demonstration test of selective H2S removal with MDEA. Laurance Reid Gas Conditioning Conference, Norman, OK, March 7–9.

Bucklin, R.W., K.W. Won, (1987) Higee contactors for selective H2S removal and superdehydration. Laurance Reid Gas Conditioning Conference, Norman, OK, March 2–4.

Burns, J.R., C. Ramshaw, (1996) Process intensification: Visual study of liquid maldistribution in rotating packed beds. *Chem. Eng. Sci.*, **51**, 1347–1352, Comment and reply (1997): Chem. Eng. Sci., 52, 453 and 455.

Chen, J.-F. (2009) The recent developments in the HiGee technology. GPE-EPIC, Venice, Italy, June 14–17.

Chen, Y.-S. (2011) Correlations of mass transfer coefficients in a rotating packed bed. *Ind. Eng. Chem. Res.*, **50**, 1778–1785.

Chen, Y.-S., Hsu, Y.-C., Lin, C.-C., Tai, C.Y.-D. and Liu, H.-S. (2008) Volatile organic compounds absorption in a cross-flow rotating packed bed. *Environ. Sci. Technol.*, **42**, 2631–2636.

Chen, Y.-S., Li, F.-Y., Lin, C.-C., Tai, C.Y.-D. and Liu, H.-S. (2006) Packing characteristics for mass transfer in a rotating packed bed. *Ind. Eng. Chem. Res.*, **45**, 6845–6853.

Chen, Y.-S., Lin, C.-C. and Liu, H.-S. (2005) Mass transfer in a rotating packed bed with various radii of the bed. *Ind. Eng. Chem. Res.*, **44**, 7868–7875.

Chen, Y.-S. and Liu, H.-S. (2002) Absorption of VOCs in a rotating packed bed. *Ind. Eng. Chem. Res.*, **41**, 1583–1588.

Cheng, H.-H. and Tan, C.-S. (2006) Reduction of CO_2 concentration in a zinc/air battery by absorption in a rotating packed bed. *J. Power. Sources*, **162**, 1431–1436.

Cheng, H.-H., Chih Lai, C.-C. and Tan, C.-S. (2013a) Thermal regeneration of alkanolamine solutions in a rotating packed bed. Int J. *Greenhouse Gas Control*, **16**, 206–216.

Cheng, H.-H., Chih Lai, C.-C. and Tan, C.-S. (2013b) Thermal regeneration of alkanolamine solutions in a rotating packed bed. Int J. *Greenhouse Gas Control*, **16**, 206–216.

Dutra, E.M.M., L. Manuel, P. Krishniah, B. Brossard, (2012) Liquid distributor for a rotating packed bed. US Patent 2012198999.

Eimer, D. (1999) A method and a device for gas treatment. Norsk Patent 326844.

Fowler, R., K.F. Gerdes, H.F. Nygaard, (1989) A commercial-scale demonstration of Higee for bulk CO2 removal and gas dehydration. Offshore Technology Conference, Houston, TX, May 1–4.

Guo, F., Y. Zhao, J. Cui, K. Guo, J. Chen, C. Zheng, (2001) Effect of inner packing support on liquid controlled mass transfer process in rotating packed beds, 107–113, in *Proceedings from 4th International Conference on Process Intensification for the Chemical Industry, Brugge, Belgium, September 10–12*, M. Gough (ed.). BHR Group Ltd., Cranfield.

Jassim, M.S., Rochelle, G., Eimer, D. and Ramshaw, C. (2007) Carbon dioxide absorption and desorption in aqueous monoethanolamine solutions in a rotating packed bed. *Ind. Eng. Chem. Res.*, **46**, 2823–2833.

Lin, C.-C. and Chen, B.-C. (2007) Carbon dioxide absorption into NaOH solution in a cross-flow rotating packed bed. *J. Ind. Eng. Chem.*, **13**, 1083–1090.

Lin, C.-C. and Liu, W.-T. (2007) Mass transfer characteristics of a high-voidage rotating packed bed. *J. Ind. Eng. Chem.*, **13**, 71–78.

Lin, C.-C. and Chen, B.-C. (2008) Characteristics of cross-flow rotating packed beds. *J. Ind. Eng. Chem.*, **14**, 322–327.

Lin, C.-C. and Chen, Y.-W. (2011) Performance of a cross-flow rotating packed bed in removing carbon dioxide from gaseous streams by chemical absorption. *Int. J. Greenhouse Gas Control*, **5**, 668–675.

Lin, C.-C., Chen, B.-C., Chen, Y.-S. and Hsu, S.-K. (2008) Feasibility of a cross-flow rotating packed bed in removing carbon dioxide from gaseous streams. *Sep. Purif. Technol.*, **62**, 507–512.

Lin, C.-C., Liu, W.-T. and Tan, C.-S. (2003) Removal of carbon dioxide by absorption in a rotating packed bed. *Ind. Eng. Chem. Res.*, **42**, 2381–2386.

Luo, Y., Chu, G.-W., Zou, H., Zhao, Z., Dodukovic, M. and Chen, J.F. (2012) Gas-liquid effective area in a rotating packed bed. *Ind. Eng. Chem. Res.*, **51**, 16320–16325.

Mallinson, R.H., C. Ramshaw, (1981) Mass transfer apparatus and process. EP Patent 0 053 881 A1.

Onda, K., Takeuchi, H. and Okumoto, Y. (1968) Mass transfer coefficients between gas and liquid phases in a packed column. *J. Chem. Eng. Jpn.*, **1**, 56–62.

Park, J., G. Nelson, (2007) Method for degassing a liquid. US Patent 2007295662.

Puranik, S.S. and Vogelpohl, A. (1974) Effective interfacial area in irrigated packed columns. *Chem. Eng. Sci.*, **29**, 501–507.

Qian, Z., Xu, L.-B., Li, Z.-H., Li, H. and Guo, K. (2010) Selective absorption of H_2S from a gas mixture with CO_2 by aqueous N-methyldiethanolamine in a rotating packed bed. *Ind. Eng. Chem. Res.*, **49**, 6196–6203.

Rajan, S., Kumar, M., Ansari, M.J., Rao, D.P. and Kaistha, N. (2011) Limiting gas liquid flows and mass transfer in a novel rotating packed bed. *Ind. Eng. Chem. Res.*, **50**, 986–997.

Ramshaw, C. (1983) HIGEE DISTILLATION – An example of process intensification. *Chem. Eng.*, **389**, 13–14.

Ramshaw, C. (1993) The opportunities for exploiting centrifugal fields. *Heat Recovery Syst. CHP*, **13**, 493–513.

Rao, D.P., Bhowal, A. and Goswami, P.S. (2004) Process intensification in rotating packed beds (HIGEE): an appraisal. *Ind. Eng. Chem. Res.*, **43**, 1150–1162.

Reay, D., Ramshaw, C. and Harvey, A. (2008) *Process Intensification*, Butterworth-Heinemann.

Sandilya, P., Rao, D.P. and Sharma, A. (2001) Gas-phase mass transfer in a centrifugal contactor. *Ind. Eng. Chem. Res.*, **40**, 384–392.

Sherwood, T.K., Shipley, G.H. and Holloway, F.A. (1938) Flooding velocities in packed columns. *Ind. Eng. Chem.*, **30**, 765–769.

Strand, A., Fiveland, T., D.A. Eimer, (2011) Rotating desorber wheel. PCT WO 2011/005118.

Sun, B.-C., Wang, X.-M., Chen, J.-M., Chu, G.-W. and Chen, J.-F. (2009) Simultaneous absorption of CO_2 and NH_3 into water in a rotating packed bed. *Ind. Eng. Chem. Res.*, **48**, 11175–11180.

Tan, C.-S. and Chen, J.-E. (2006) Absorption of carbon dioxide with piperazine and its mixtures in a rotating packed bed. *Sep. Purif. Technol.*, **49**, 174–180.

Trent, D. and Tirtowidjojo, D. (2001) Commercial operation of a rotating packed bed (RPB) and other applications of RPB technology, in *Proceedings from 4th International Conference on Process Intensification for the Chemical Industry, Brugge, Belgium, September 10–12* (ed M. Gough), BHR Group Ltd., Cranfield, pp. 11–19.

Wolf, H., P. Aleksic, (2012) Rotating packed bed. European Patent Application EP 2 486 966.

Yang, W., Wang, Y., Chen, J. and Fei, W. (2010) Computational fluid dynamic simulation of fluid flow in a rotating packed bed. *Chem. Eng. J.*, **156**, 582–587.

Yi, F., Zou, H.-K., Chu, G.-W., Shao, L. and Chen, J.-F. (2009) Modeling and experimental studies on absorption of CO_2 by Benfield solution in rotating packed bed. *Chem. Eng. J.*, **145**, 377–384.

Zheng, C., Guo, K., Feng, Y. and Yang, C. (2000) Pressured drop of centripetal gas flow through rotating packed beds. *Ind. Eng. Chem. Res.*, **39**, 829–834.

10

Mass Transfer Models

In the previous chapter, models for mass transfer coefficients in columns were alluded to. In this chapter we shall discuss a few classic theories that are used to lay a fundament for understanding mass transfer in columns. Ideally, the mass transfer coefficient should be predictable from knowledge of the diffusion coefficient of the species transferred and the flow conditions. This is far from fulfilled by these classic theories.

Even so, it is useful to discuss these old theories. For instance, they serve to give a basis for the use of mathematics to predict the effect of chemical reactions on the mass transfer. Hopefully these models will also provide some deeper understanding of what is going on at the gas–liquid interface. Lastly a discussion of the fluid dynamics of the flows of gas and liquid will serve to give some insight into the complexity of the problem.

The oldest of these mass transfer theories originates from more than 100 years ago although it is usually referred to a younger paper published about 90 years ago, and the youngest about 50 years. It could be claimed that they have stood the test of time. They have their weaknesses but so far they are the best we have. They should be treated with the respect merited by age but used critically.

10.1 The Film Model

This is the mother of the mass transfer theories dating from 1923 (i.e. 90 years ago) and is sometime referred to as the Whitman (1923) film theory (it was, however, suggested already in 1904 by Nernst: Sherwood, Pigford and Wilke, 1975). According to this picture of mass transfer from gas to liquid (or vice versa) there is a thin stagnant film on each side of the gas–liquid interface. These films provide resistance to mass transfer; their thickness remains constant but of course differs between gas and liquid.

Gas Treating: Absorption Theory and Practice, First Edition. Dag Eimer.
© 2014 John Wiley & Sons, Ltd. Published 2014 by John Wiley & Sons, Ltd.

When this picture is used to arrive at the classic expression for mass transfer:

$$N_i = \frac{D_i}{l} \left(C_i^{Li} - C_i^{Lb} \right) \tag{10.1}$$

(as written for the liquid phase). l is the film thickness. It is implicitly assumed that the component 'i' is present in a dilute mixture such that its movement does not influence the mixture as such. This is the basis for the classic linear concentration profiles used to illustrate it. It is, furthermore, implicitly assumed as a steady state situation. The assumption of dilute solution could be dropped if there instead was a situation with equimolar counter-diffusion, which is often a good approximation in distillation mass transfer. It is further implicitly assumed that there is no temperature effect that would change the mixture properties significantly (in this context).

The argument for the presence of a stagnant film is mainly that there is turbulence or eddy diffusion in the bulk phases on each side of the interface. However, this eddy flow will not penetrate the interface and therefore tend to slow down and stop as the interface is approached. Also at the interface the relative velocities of gas and liquid will be zero. In a zone near the interface itself there is arguably no convective flow, and the only transport perpendicular to the interface must then be by diffusion (Figure 10.1).

Equation 10.1 is the result of taking a mass balance over the control volume and integrating the resulting differential equation from 0 to L. As developed in Chapter 6 the result becomes:

$$C_i = \frac{z}{L} \left(C_{Li} - C_{0i} \right) + C_{0i} \tag{6.6}$$

If this relation is combined with Fick's first law that defines the rate of diffusion through the film:

$$N_i = -D_i \frac{d}{dz} \left\{ \frac{z}{L} \left(C_{Li} - C_{0i} \right) + C_{0i} \right\} \tag{10.2}$$

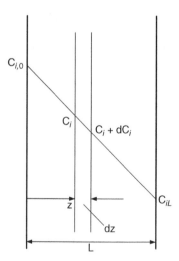

Figure 10.1   *The diffusion through a film with control element shown. The linear profile implies dilute mixture w.r.t. component 'i'.*

which becomes:

$$N_i = -D_i \left\{ \frac{1}{L} \left(C_{Li} - C_{0i} \right) + 0 \right\}$$ (10.3)

And this is the result stated in Equation 10.1. A similar exercise may be done for the gas phase. A key result is that the film theory postulates that the rate of mass transfer is directly proportional to the diffusion coefficient.

This model may be simple and crude but it has turned out to be useful for explaining the effect of chemical reaction on mass transfer rates. And it has survived as a teaching tool in chemical engineering for a long time.

10.2 Penetration Theory

It was reckoned that the film theory was too simple an explanation for practical gas–liquid contact in columns. The direct proportionality to the diffusion coefficient was not confirmed experimentally. On this background Higbie (1935) introduced the so-called *penetration theory*. In this model elements of fluid come to the surface and stay there for a fixed period while the component transferred diffuses into this stagnant liquid element. If the fluid element stayed for an infinite time, it would become saturated with the component, but the fixed period is very short. The period is assumed to be the same for all elements.

To analyse this situation where the component 'i' diffuses into the stagnant liquid element, we need to make use of Fick's second law developed earlier. It is recalled that the concentration profile of component 'i' could be described by:

$$C_i = C_{i0} - \left(C_{i0} - C_{i\infty} \right) .erf \left(\varsigma \right)$$ (6.12)

where the auxiliary variable ς is defined by:

$$\varsigma = \frac{z}{\sqrt{2D_i t}}$$ (6.11)

There are a number of ways to proceed from here, but Danckwerts (1970) uses a simple and elegant approach that is worth using. He argues that the rate of mass transfer per unit area through any plane parallel to the interface is:

$$\left(N_i \right)_z = -D_i \frac{dC_i}{dz}$$ (10.4)

And hence the rate of mass transfer per unit area at any time t is:

$$N_i = -D_i \left(\frac{dC_i}{dz} \right)_{z=0}$$ (10.5)

As was done in developing the film theory expression, the function describing C_i versus the length z, and here also time t, is substituted into Equation 10.5. The concentration function is available from Equations 6.12 and 6.11. Hence:

$$N_i = -D_i \left\{ \frac{d}{dz} \left[C_{i0} - \left(C_{i0} - C_{i\infty} \right) .erf \left(\varsigma \right) \right] \right\}_{z=0}$$ (10.6)

This next paragraph could be referred to as a *mathematical interlude*. It is worth going through to take the mystery out of the development. Firstly the definition of the error function where $\zeta = f(z, t)$ is:

$$erf(\varsigma) = \frac{2}{\sqrt{\pi}} \int_0^\varsigma e^{-\sigma^2} d\sigma \qquad (10.7)$$

Next:

$$\frac{d}{dz} erf(\varsigma) = \frac{2}{\sqrt{\pi}} e^{-\sigma^2} \frac{d\sigma}{dz} \qquad (10.7a)$$

$$\frac{d\sigma}{dz} = \frac{1}{2\sqrt{D_i t}} \qquad (10.7b)$$

$$\frac{d}{dz} erf(\varsigma) = \frac{2}{\sqrt{\pi}} e^{-\sigma^2} \frac{1}{2\sqrt{D_i t}} \qquad (10.7c)$$

$$N_i = D_i (C_{i0} - C_{i\infty}) \left\{ \frac{d}{dz} [erf(\varsigma)] \right\}_{z=0} \qquad (10.6a)$$

$$N_i = D_i (C_{i0} - C_{i\infty}) \left\{ \frac{2}{\sqrt{\pi}} e^{-\sigma^2} \frac{1}{2\sqrt{D_i t}} \right\}_{z=0} \qquad (10.6b)$$

At $z = 0$, $\sigma = 0$. Hence

$$N_i = D_i (C_{i0} - C_{i\infty}) \left\{ \frac{2}{\sqrt{\pi}} \frac{1}{2\sqrt{D_i t}} \right\}_{z=0} \qquad (10.6c)$$

And:

$$|N_i|_{z=0} = (C_{i0} - C_{i\infty}) \left\{ \frac{D_i}{\sqrt{\pi D_i t}} \right\}_{z=0} \qquad (10.6d)$$

$$|N_i|_{z=0} = \sqrt{\frac{D_i}{\pi t}} (C_{i0} - C_{i\infty}) \qquad (10.6e)$$

This is the end of the mathematical manipulation alluded to earlier.

The rate of absorption expressed in Equation 10.6e is the rate at any time t of the stagnant element's stay at the interface. To find the total amount of component 'i' absorbed during the element's stay, this expression must be integrated over the staying time θ. This is:

$$Q_i = \int_0^\theta |N_i|_{z=0} dt \qquad (10.8)$$

resulting in:

$$Q_i = 2\sqrt{\frac{D_i \theta}{\pi}} (C_{i0} - C_{i\infty}) \qquad (10.8a)$$

The average absorption rate in that period is:

$$N_i\big|_{average} = \frac{Q_i}{\theta}$$ (10.9)

and hence from Equation 10.8(a):

$$N_i\big|_{average} = 2\sqrt{\frac{D_i}{\pi\theta}} \left(C_{i0} - C_{i\infty}\right)$$ (10.9a)

Equation 10.9(a) is the expression arrived at by Higbie (1935) and has since been known as the *penetration theory*. It predicts that the rate of mass transfer is proportional to the square root of the diffusivity. This is nearer to the relation observed than the film theory prediction, but the penetration theory seems to underestimate the dependence on the diffusion coefficient. The penetration theory has also been used to predict the effect of chemical reaction on the rate of mass transfer with success.

By inspection of Equation 10.9a it is clear that:

$$k_L^0 = 2\sqrt{\frac{D_i}{\pi\theta}}$$ (10.10)

10.3 Surface Renewal Theory

Danckwerts (1951) suggested that fluid elements did not stay at the surface for *one* fixed period. The lifetime of an element at the interface would be random and follow a probability distribution with respect to this time period. Higbie's penetration model is also a surface renewal theory where the probability is 100% that the element stays for *one* defined length of time.

Danckwerts' approach was that the staying time of any element at the interface is independent of how long it has already stayed there. In short the fraction of the surface that has been there between θ and $\theta + d\theta$ is $se^{-s\theta}d\theta$ where s is the fraction of the surface that is renewed per unit time, a surface renewal rate. The instantaneous rate of absorption per unit area of surface that has been exposed for a time θ is known from the penetration theory work previously:

$$N_i = \sqrt{\frac{D_i}{\pi\theta}} \left(C_{i0} - C_{i\infty}\right)$$ (10.6f)

The average rate of absorption into the liquid accounting for all surface elements is (Danckwerts, 1970):

$$N_i\big|_{average} = s\int_0^\infty \sqrt{\frac{D_i}{\pi\theta}} \left(C_{i0} - C_{i\infty}\right)e^{-s\theta}d\theta$$ (10.11)

$$N_i\big|_{average} = s\sqrt{\frac{D_i}{\pi}} \left(C_{i0} - C_{i\infty}\right) \int_0^\infty \frac{1}{\sqrt{\theta}}e^{-s\theta}d\theta$$ (10.11a)

$$N_i\big|_{average} = \sqrt{D_i s} \left(C_{i0} - C_{i\infty}\right)$$ (10.11b)

Further details of the mathematical development are given in Danckwerts' original paper and in Coulson and Richardson's *Chemical Engineering*, volume 1 (1965). Comparing the Equations 10.11(b) and 10.9(a) it is easily seen that they predict the same relation between the mass transfer rate and the diffusion coefficient. The units of s and $1/\theta$ are also the same, but there is a difference as to how they have been derived.

10.4   Boundary Layer Theory

Boundary layer theory (BLT) is not normally taught in the same context as the preceding theories. This has to do with BLT taking a very different approach to mass transfer. In BLT the focus is on a fluid flowing over a flat plate forming a film from its edge where the film thickness is zero from where it grows due to the friction between the fluid and the wall that gives rise to a velocity profile in the film.

It is a lengthy development that leads to a relation between the dimensionless groups referred to as the Sherwood, Reynolds and Schmidt numbers. The eventual prediction is that the mass transfer coefficient is proportional to the diffusion coefficient raised to the power of 2/3. This relation is closer to observed results than any of the preceding theories.

BLT has not been used in explaining the effect of chemical reaction on the rate of mass transfer. It is elected not to discuss the BLT in any depth in this text. The reader is referred to the books by Cussler (2009), Sherwood, Pigford and Wilke (1975) and Bird, Stewart and Lightfoot (2002).

The relevance of a thin film flow developing on surfaces in an industrial absorption column may be debated.

10.5   Eddy Diffusion, 'Film-Penetration' and More

There are also other ways to achieve a 'power of 2/3' relation between mass transfer and the diffusion coefficient. An example is non-ideal flow in the column, for instance in the form of backmixing. This has to do with a more overall view of the column and the transports taking place within. More theoretically based approaches have also been made.

Lin, Moulton and Putnam (1953) introduced the concept of eddy diffusivity, ε, that involves the movement of whole elements of the type referred to by the penetration theories. This movement takes place in such a way that it boosts the mass transfer, and the transport equation may be written as:

$$N_i = \left(D_i + \varepsilon\right) \frac{dC_i}{dz} \tag{10.12}$$

Toor and Marchello (1958) and Marchello and Toor (1963) took a different approach to arrive at essentially the same form of equation. In their original work they pointed out that as the elements referred to in the penetration theories became old in the sense that penetration became significant, the rate of mass transfer would tend towards the situation described by the film theory. On this basis they pointed out that the film and penetration theories were not in conflict, but rather limiting cases of the mass transfer situation. There are some differences in the transition behaviour between the cases compared with the work of Lin, Moulton and Putnam but the limiting behaviour is essentially the same. In their later paper Marchello and Toor generalised their model, referred to as a 'film-penetration' model, by allowing for the 'elements' to influence the mass transfer rate also when not actually coming all the way to the surface. Also elements that take part in mixing of elements in the mass transfer zone would bear on the mass transfer rates. Their equations at the end of their development look like Equation 10.12, but their way of deriving the extra diffusivity ε is different.

These approaches can explain fractional power dependencies on diffusivities with respect to the rate of mass transfer. Sherwood, Pigford and Wilke (1975) discuss eddy diffusivities in mass transfer in more depth.

References

Bird, R.B., Stewart, W.E. and Lightfoot, E.N. (2002) *Transport Phenomena*, 2nd edn, John Wiley & Sons, Inc.

Coulson, J.M. and Richardson, J.F. (1965) *Chemical Engineering*, vol. **1**, revised 2nd edn, Pergamon Press.

Cussler, E.L. (2009) *Diffusion*, 3rd edn, Cambridge University Press, Cambridge.

Danckwerts, P.V. (1951) Significance of liquid film coefficients in gas absorption. *Ind. Eng. Chem.*, **43**, 1460–1467.

Danckwerts, P.V. (1970) *Gas-Liquid Reactions*, McGraw-Hill.

Higbie, R. (1935) The rate of absorption of a pure gas into a still liquid during short periods of exposure. *Trans. Am. Inst. Chem. Eng.*, **35**, 365–389.

Lin, C.S., Moulton, R.W. and Putnam, G.L. (1953) Mass transfer between solid wall and fluid streams. *Ind. Eng. Chem.*, **45**, 636–640.

Marchello, J.M. and Toor, H.L. (1963) A mixing model for transfer near a boundary. *Ind. Eng. Chem. Fundam.*, **2**, 8–12.

Sherwood, T.K., Pigford, R.L. and Wilke, C.R. (1975) *Mass Transfer*, McGraw-Hill.

Toor, H.L. and Marchello, J.M. (1958) Film-penetration model for mass and heat transfer. *AIChE J.*, **4**, 97–101.

Whitman, W.G. (1923) The two-film theory of absorption. *Chem. Met. Eng.*, **29**, 147–148.

11

Correlations for Mass Transfer Coefficients

11.1 Introduction

In Chapter 7 the way in which mass transfer coefficients come into the picture when analysing columns was explained, while in Chapter 10 the way in which mass transfer coefficients may be theoretically explained was discussed. Chapters 8 and 9 described various hardware options available when designing a column. The choice of hardware has, naturally, a profound effect on the rate of mass transfer that may be achieved.

There has been a long and continuous development of hardware where the criterion for optimisation in many ways has been 'more mass transfer for the same or lower pressure drop'. In view of this it would be expected that empirical correlations become dated due to improved equipment ousting the old. To an extent this is true, but surprisingly, some of the old correlations do remarkably well. Care should always be taken, however, when a correlation is used to describe equipment that was not around when the correlation was made.

Most, if not all, chemical engineering textbooks on mass transfer include correlations that may be used to predict mass transfer coefficients and/or the interfacial gas–liquid area. The reader should ensure that he/she fully understands how this material is to be used, which units should be used, unless dimensionless and not least which limitations should be observed when applying it. It also has to be remembered that general chemical engineering textbooks cannot be expected to be as authoritative on such matters as a specialist text.

11.2 Packings: Generic Considerations

There are a number of books available that specialise in packed columns. These include the works of Strigle (1987), Kister (1992), Billet (1995) and Mackowiak (2010). A vast

Gas Treating: Absorption Theory and Practice, First Edition. Dag Eimer.
© 2014 John Wiley & Sons, Ltd. Published 2014 by John Wiley & Sons, Ltd.

amount of company literature is also available from packing manufacturers. Recently mass transfer design of packed columns was summarised by Wang, Yuan and Yu (2005) and this work is still the best summary of mass transfer correlations as Mackowiak is more pressure drop/capacity oriented in his book.

It is not as simple as picking the latest correlation, inserting the relevant numbers and expecting an absolutely reliable mass transfer coefficient to be returned. Which correlation to use is open for debate. Strigle points out that the often resorted to correlation by Onda, Takeuchi and Okumoto (1968) is as inaccurate as $\pm 40\%$. The same author says that the newer correlation by Bolles and Fair (1979) is good for $\pm 25\%$. Clearly these correlations give estimates that are useful, but, for example the latter will need 25% extra height added to the column height estimate as a necessary safety margin. If this means that an extra liquid redistributor is needed, it represents an expensive insurance. Strigle also points out that the Bolles and Fair correlation includes a correction for column diameter that is claimed to be rooted in the performance of the test column they used, which introduces some bias. The correlations listed by Wang, Yuan and Yu are listed in Tables 11.1 and 11.2 by author and year.

Figure 11.1 is the idealised relation between height of a transfer unit (HTU) and the rate of gas flow. The various points referred to is not always as easy to identify.

Before doing column mass transfer calculations it is useful to consider the behaviour of packed columns and that may change over loading conditions that may be met. Figure 11.1 shows how HTU (proportional to $1/K_{OG}$) will vary as more or less gas flow is taken through at constant ratio L/G. It is seen that the mass transfer coefficient is significantly lower at low gas load. This had to do with the difficulty of providing good liquid distribution below a certain turn-down ratio. At point A the loading is high enough for the mass transfer to settle at a predictable value, and this situation continues past points B and C. Point C has traditionally been referred to as the loading point of the column. At point D the gas load is so high that it starts to interfere with the liquid like making it more turbulent and this leads to higher mass transfer coefficients. As the gas load is increased further towards point E, there is entrainment starting and this reduces the mass transfer effectiveness. The column is still stable to operate at E. Big companies with many columns have internal data banks that can be utilised to get the maximum out of columns.

The author of this text recalls a vacuum distillation column, used to recover an organic acid, which seemed to defy normal column wisdom. It was packed with ceramic packings and had a propensity for fouling and should really be cleaned out at regular intervals. However, the acid recovery improved as deposits gathered and the pressure drop went up. This went on beyond where the column should stop being operable. The reason for this behaviour may be attributed to poor liquid distribution whereby HTU versus gas load relation is more like that described in Figure 11.2 rather than in Figure 11.1.

11.3 Random Packings

The technological development of random packings has been discussed in Chapter 8. It should be obvious that the same development has also had an impact on mass transfer performance. A key consideration has been to achieve more mass transfer per unit pressure drop. A further object has been to shrink the sizes of columns needed. Packings are paid for by unit volume, and a more efficient packing could be priced higher (Figure 11.3).

Table 11.1   *Models for mass transfer in random packings published in the literature based on the work of Wang, Yuan, and Yu (2005).*

Due to whom?[a] (Year)	k_G	$k_L{}^0$	a	HETP HTU	Institution
Sherwood and Holloway (1940)	–	Y	–	–	MIT
van Krevelen and Hoftijzer (1948)	Y	Y	–	–	Staatsmijnen, Geleen, NL
Shulman and de Grouff (1952)		–	–	–	Clarkson College, NY
Morris and Jackson (1953)	–	–	–	Y	
Shulman, Ullrich, Proulx, Zimmerman, (1955)	Y	Y	Y	–	Clarkson C.
Cornell, Knapp, Close, and Fair (1960a,b)	–	–	–	Y	Monsanto
Bolles and Fair (1982)	–	–	–	Y	Monsanto
Onda, Takeuchi and Okumoto (1968)	Y	Y	Y	–	Nagoya University
Puranik and Vogelpohl (1974)	–	–	Y	–	University of Karlsruhe
Kolev (1976)	–	–	Y	–	Bulgarian Science Lab, Sofia
Bravo and Fair (1985)	Y	Y	Y	–	University of Texas, Austin
Zech and Mersmann (1979)	Y	Y	Y	–	TU München
Mangers and Ponter (1980)	–	Y	–	–	Swiss Fed Institution of Technology, Lausanne
Linek, Petricek, Benes and Braun (1984)	–	–	Y	–	Institution of Chemical Technology, CZ
Shi and Mersmann (1986)	Y	Y	–	–	TU München
Mersmann and Deixler (1986)	Y	Y	–	–	TU München
Rizutti and Brucato (1989)	–	–	Y	–	University of Palermo
Costa Novella, Escudro, Zamorano and Gomez (1992)	–	–	Y	–	
Billet and Schultes (1993)	Y	Y	Y	–	Ruhr University – Bochum
Wagner, Stichlmair and Fair (1997)	Y	Y	–	–	University of Texas, Austin
Nakajima, Maffia and Meirelles (2000)	–	–	Y	–	UNICAMP, Campinas, Brazil
Mackowiak (2011)	–	Y	Y	–	Envimac GmbH

[a]The reader must consult Wang *et al.*'s paper for further details of correlations.

Wang, Yuan and Yu (2005) has meticulously listed a large number of correlations. Their review can stand for itself. Here, only the models as represented by the authors, and a few relevant data are listed in Table 11.1.

A red thread in all these correlations is the concept that the maximum interfacial area in a packed column is the total dry surface of the packing. Pall rings and Raschig rings have exactly the same dry surface area. Clearly it will be very much harder to bring the wetting efficiency up in a bed of Raschig rings compared to a bed of Pall rings. Yet surface area

Table 11.2 *Models for mass transfer in structured packings published in the literature based on the work of Wang, Yuan and Yu (2005).*

Due to whom?[a] (Year)	k_G	$k_L{}^0$	a	HETP HTU	Institution
Bravo, Rocha and Fair (1985)	Y	Y	–	–	University of Texas, Austin
Shi and Mersmann (1985)	–	–	Y	–	TU München
Spiegel and Meier (1987)	–	–	Y	–	–
Nawrocki, Xu and Chuang (1991)	Y	Y	–	–	University of Alberta, Edmonton
Bravo and Fair (1990)	–	–	Y	–	University of Texas, Austin
Henriques de Brito, von Stockar, and Bomio (1992)	–	Y	–	–	Swiss Federal Institute of Technology, Lausanne
Henriques de Brito et al. (1994)	–	–	Y	–	Swiss Federal Institute of Technology, Lausanne
Hanley, Dunbobbin and Bennett (1994)	Y	Y	Y	–	APCI, Allentown
Brunazzi, Nardini, Paglianti and Potarca (1995)	–	–	Y	–	University of Pisa
Rocha, Bravo and Fair (1996)	Y	Y	Y	–	University of Texas, Austin
Brunazzi and Paglianti (1997)	Y	Y	–	–	University of Pisa
Gualito, Cerino, Cardenas and Rocha (1997)	–	–	Y	–	Institution of Technology Celaya, Mexico
Olujic, Kamerbeek and de Graauw (1999)	Y	Y	Y	–	Delft University of Technology
Shetty and Cerro (1997)	–	Y	–	–	University of Tulsa, OK
Siminiceanu, Friedl and Dragan (2002)	–	–	Y	–	–
Xu, Afacan and Chuang (2000)	Y	Y	Y	–	University of Alberta, Edmonton
Olujic (2002)	–	–	Y	–	Delft University of Technology

[a]The reader must consult Wang et al.'s paper for details of references.

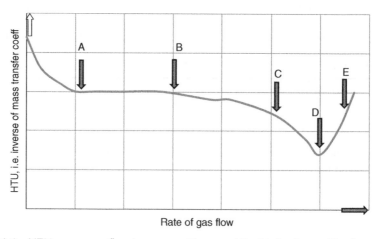

Figure 11.1 *HTU versus gas flow in a case with a good liquid distributor. The figure is based on a constant ratio of G/L throughout.*

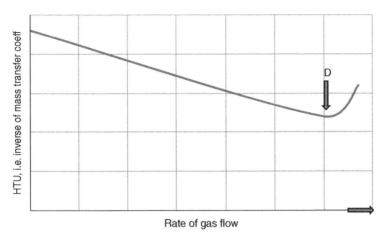

Figure 11.2 *HTU versus gas flow in a case with a poor liquid distributor. The figure is based on a constant ratio of G/L throughout.*

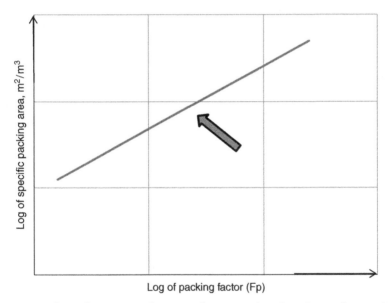

Figure 11.3 *A relation between packing specific area and packing factor. Current third generation packings have a relation that has moved in the direction of the big arrow from first and second generation. Capacity is higher when packing factor is low and efficiency is higher when specific area is high. Based on Kister, H.Z. (1992)* Distillation – Design. *McGraw-Hill.*

correlations are the same for both. This suggests that these correlations should be used with care. There is a lot of Raschig Ring based data behind the correlations. This was after all the dominating packing of its day.

The reader is advised to try a number of correlations when making column estimates. This will guard against using correlations under circumstances where they should not really be used. Secondly, it should be clear that mass transfer design has a certain uncertainty about it. Using more than one model will provide a better feel for that uncertainty.

11.4   Structured Packings

The recommendation to make estimates with more than one correlation given for random packings is equally applicable to structured packings.

It is interesting to observe that nobody seems to have published height equivalent to a theoretical plate (HETP) or HTU correlations for structured packings while for random packings they are used. The older thinking with use of HETP seems to be less used in the structured packing camp, but there is in principle no reason for the difference.

11.5   Packed Column Correlations

Wang, Yuan and Yu (2005) have reviewed considerable parts of the packed column papers and related mass transfer models and give a comprehensive list of equations. Tables 11.1 and 11.2 summarise available correlations presented by various people over more than 60 years. During this period the packings they are supposed to describe with respect mass transfer have changed. Logically the oldest correlations should not be able to deal with products being introduced later since their construction partially alter the flow behaviour in the packed section, for example by having a more grid-like structure rather than extensive surfaces. However, some of the old correlations do remarkably well. We shall discuss a select few. Note that these equations may have been rearranged relative to the original in order to have dimensionless groups. These dimensionless groups are defined differently by different people, mainly because of change in use of the characteristic length and for this reason they are defined here for each case. The exception is the Schmidt number (Sc) that is only dependent of physical properties of the fluid and the relevant diffusion coefficient.

$$Sc = \frac{\mu}{\rho D} \tag{11.1}$$

We choose to start with the correlations published by Onda, Takeuchi and Okumoto in 1968 based on work done at the University of Nagoya in Japan since it is often referred to. Their database included packings like Raschig rings, Berl saddles, Pall rings, spheres and rods. They gave a complete set of correlations for contact area and mass transfer coefficients.

$$\frac{k_G}{aD_G} = 5.23 \left(\frac{G_{wf}}{a\mu_G}\right)^{0.7} \left(\frac{\mu_G}{\rho_G D_G}\right)^{1/3} (ad_p)^{-2.0} \tag{11.2a}$$

$$\text{or} \quad Sh = 5.23 Re^{0.7} Sc^{1/3} (ad_p)^{-2.0} \tag{11.2b}$$

$$\left(k_L^0 \left(\frac{\rho_L}{\mu_L g}\right)^{1/3}\right) = 0.0051 \left(\frac{L_{wf}}{a_w \mu_L}\right)^{2/3} \left(\frac{\mu_L}{\rho_L D_L}\right)^{-1/2} (ad_p)^{0.4} \tag{11.3a}$$

$$\text{or} \quad \left(k_L^0 \left(\frac{\rho_L}{\mu_L g}\right)^{1/3}\right) = 0.0051 (Re)^{2/3} (Sc)^{-1/2} (ad_p)^{0.4} \tag{11.3b}$$

On inspection it is seen that for the gas side mass transfer coefficient the term on the left is the Sherwood number where the linear dimension is the inverse of 'a', while the dimensionless groups on the right are the Reynolds and Schmidt numbers followed by a dimensionless group of packing descriptors. The dimensionless group on the left hand side of the liquid side mass transfer coefficient is not the Sherwood number. Here the gravitational acceleration g has been brought in, and there is no diffusivity term. For the mass transfer interfacial area they give the following relation:

$$
\frac{a_e}{a} = 1 - \exp\left\{ -1.45 \left(\frac{\sigma_C}{\sigma_L} \right)^{0.75} \left(\frac{L_{wf}}{a\mu_L} \right)^{0.1} \left(\frac{L_{wf}a}{\rho_L^2 g} \right)^{-0.05} \left(\frac{L_{wf}^2}{\rho_L a\sigma_L} \right)^{0.2} \right\}
\tag{11.4a}
$$

$$
\text{or} \quad \frac{a_e}{a} = 1 - \exp\left\{ -1.45 \left(\frac{\sigma_C}{\sigma_L} \right)^{0.75} (Re_L)^{0.1} (Fr_L)^{-0.05} (We_L)^{0.2} \right\}
\tag{11.4b}
$$

In this relation the dimensionless groups within the exponential term is a ratio between a 'critical surface tension' and the liquid's surface tension followed by the Reynolds, Froude and Weber numbers. The critical surface tension is given as 61 for ceramic, 75 for steel and 33 for polyethylene all in dyn/cm. This size reflects the wetting properties of the material used to make the packing. These relations are much quoted, and seem to give good results also for more modern packings. (They have even been used in the context of Rotating packed beds (RPB)s!) It is quickly seen that the wetted area of the packing, or the effective interfacial area, cannot be bigger than the specific surface area of the packing. This represents a view of the liquid as trickling down the packing as an incomplete film.

Another correlation from this period is the one for interfacial area by Puranik and Vogelpohl (1974). It was used to model RPBs in a previous chapter along with the models of Onda, Takeuchi and Okumoto which is surprising but also shows the generality of these types of correlations that are based on the use of dimensionless groups and fitting exponents to them. In their work the area for mass transfer is divided into a number of categories. This aspect may be pursued in their original work. Next is their expression for predicted mass transfer area. It is observed that this relation can lead to interfacial areas higher than the specific surface area of the packing. This is different from the model in Equation 11.4.

$$
\frac{a_e}{a} = 1.045(Re)^{0.041}(We)^{0.133} \left(\frac{\sigma_C}{\sigma_L} \right)^{0.182}
\tag{11.5}
$$

$$
\text{where } Re = \frac{L_{vf}\rho_L}{a\mu_L} \text{ and } We = \frac{L_{vf}^2\rho_L}{a\sigma_L}
\tag{11.5a,b}
$$

The next group of correlations is due to work associated with professor Fair at University of Texas Austin where large pilot plants were set up at the Separations Research Program. His early work was actually done at Monsanto. For random packings the early Monsanto work resulted in correlations for HTUs and is passed by here. From the Austin work there is a correlation for interfacial area (Bravo, Rocha and Fair, 1982) that is:

$$
\frac{a_e}{a} = 0.310 \left(Ca_L Re_G \right)^{0.392} \frac{\sigma_L^{0.5}}{Z^{0.4}}
\tag{11.6a}
$$

$$
\text{where } Re_G = \frac{6\rho_G G_{Vf}}{a\mu_G} \text{ and the capillary number } Ca_L = \frac{L_{Vf}\mu_L}{\sigma_L}
\tag{11.6b}
$$

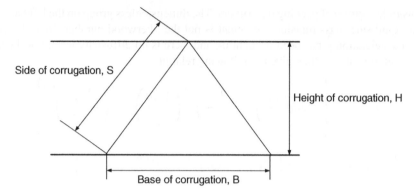

Figure 11.4   *Key sizes of the corrugation in structured packings.*

The last ratio is not dimensionless, and σ has to be in dyn/cm and Z is the height of packing in metres. (The coefficient 0.498 from the original work has been changed to accommodate Z in metres). There is reason to be careful with the use of Equation 11.6. The use of units for the dimension dependent ratio is different when referenced by Wang, Yuan and Yu (2005), and as will be seen in Section 11.7.3 the value obtained deviates from the others. A later work also presents correlations for mass transfer coefficients (Wagner, Stichlmair and Fair, 1997), but that is not included here because of its complexity. It may be pursued in the original publication (Figure 11.4).

The Austin work is better known for its correlations relating to structured packings. The early work was based on gauze packings and was presented as:

$$Sh_G = 0.0338\left(Re_G\right)^{0.8} Sc_G^{0.33} \tag{11.7}$$

$$\text{where } Re_G = \frac{\rho_G d_{eq}\left(u_{Le} + u_{Ge}\right)}{\mu_G} \quad Sh_G = \frac{k_G d_{eq}}{D_G} \tag{11.7a,b}$$

$$\text{and } u_{Ge} = \frac{G_{Vf}}{\varepsilon \sin(\alpha)} \quad \text{and } u_{Le} = \left(\frac{9\Gamma^2 g}{8\rho_L \mu_L}\right)^{1/3} \tag{11.7c,d}$$

$$\frac{k_L^0 s}{D_L} = 2\left(\frac{s^3}{D_L^3}\frac{9\Gamma^2 g}{8\rho_L \mu_L}\right)^{1/6} \tag{11.8}$$

The 'Austin' correlations have been refined by Rocha, Bravo and Fair (1996) to cover sheet metal packing, but the change is limited. Here α is the inclination angle of the corrugated packing and the characteristic dimension is s which represents the side dimension of the packing (Figure 11.4).

$$Sh_G = 0.054\left(Re_G\right)^{0.8} Sc_G^{0.33} \tag{11.9}$$

$$\text{where } Re_G = \frac{\rho_G s\left(u_{Le} + u_{Ge}\right)}{\mu_G} \quad Sh_G = \frac{k_G s}{D_G} \tag{11.9a,b}$$

$$\text{and } u_{Ge} = \frac{G_{Vf}}{\varepsilon\left(1 - h_L\right)\sin\left(\alpha\right)} \text{ and } u_{Le} = \frac{L_{Vf}}{\varepsilon\left(h_L\right)\sin\left(\alpha\right)} \qquad (11.9c,d)$$

$$\frac{k_L^0 s}{D_L} = 2\left(\frac{0.9 u_{Le} s}{\pi D_L}\right)^{1/2} \qquad (11.10)$$

For interfacial area in structured packings Fair and Bravo (1987, 1990) gave a seemingly simple relation where they related the effective packing area to the fractional flow related to the flooding condition. It is interesting to observe that they predict 'full wetting' of the packing from 85% load compared to flooding and upwards:

$$\frac{a_e}{a} = 0.5 + 0.58 F_{fraction} \text{ if } F_{fraction} < 0.85 \qquad (11.11a)$$

$$\frac{a_e}{a} = 1 \text{ if } F_{fraction} > 0.85 \qquad (11.11b)$$

Also the relation for effective contact area was refined in the work by Rocha, Bravo and Fair (1996). In this they introduce a packing enhancement factor and they give values for a few packings. This work takes the work of Shi and Mersmann (1985) as a starting point. Here, θ is the contact angle between liquid and surface, α is the angle between the horizontal and the inclination of the packing:

$$\frac{a_e}{a} = F_{SE}\frac{29.12\left(We_L Fr_L\right)^{0.15} S^{0.359}}{Re_L^{0.2}\varepsilon^{0.6}\left(1 - 0.93\cos\left(\theta\right)\right)\left(\sin\left(\alpha\right)\right)^{0.3}} \qquad (11.12)$$

Mersmann's group at TU München has also published widely in the field of packing technology over a long period. There is, for example a large influence of their work in Equation 11.12. In their laboratory, a large scale column is available (Mackowiak, 2010) but they have also looked into details of liquid flow over packing materials like studying rivulets when liquid is flowing down an inclined plane as an example. For random packings they have the following equations (Mersmann and Deixler, 1986; Shi and Mersmann, 1985).

$$a_e = 3.49 L_{vf}^{0.4}\left(\frac{\mu_L}{\rho_L}\right)^{0.2}\left(\frac{\rho_L}{\sigma_{L8}}\right)^{0.15}\left(\frac{a^2}{4\pi\varepsilon}\right)^{0.6}\left(1 - 0.93\cos\left(\theta\right)\right)^{-1} \qquad (11.13)$$

They also give this in a dimensionless form:

$$\frac{a_e}{a} = \frac{0.76\left(We_L Fr_L\right)^{0.15}}{\left(Re_L\right)^{0.2}\left(1 - 0.93\cos\left(\theta\right)\right)\varepsilon^{0.6}} \qquad (11.14)$$

where:

$$Re_L = \frac{L_{vf}\rho_L}{a\mu_L} \qquad We_L = \frac{L_{vf}^2\rho_L}{a\sigma_L} \qquad Fr_L = \frac{L_{vf}^2 a}{g} \qquad (11.14a,b,c)$$

In this correlation it is important to stick to the correct dimensions since it is not on a dimensionless form: 'g' is the gravitational acceleration, 9.81 m/s^2. In the work referenced they view the critical surface tension σ_C and the term $(1 - \cos(\theta))$ as the same.

They give a number of values for σ_C and we note that they give 0.043 for steel compared with 0.071 (they say) from Onda, Takeuchi and Okumoto (normally associated with 0.075), for polypropylene they give 0.033 (all in N/m).

$$Sh_G = K_{packing}\left(Re_G\right)^{2/3} Sc_G \tag{11.15}$$

$$\text{where } Sh_G = \frac{k_G d_{pe}}{D_G} \text{ and } Re_G = \frac{G_{Vf} \rho_G d_{pe}}{\mu_G} \tag{11.15a,b}$$

The factor $K_{packing}$ is 0.413 for rings, 0.730 for saddles and 0.673 for spheres.

Their correlation for the liquid side mass transfer coefficient introduces a new (in this context) dimensionless group in the form of the Galileo number:

$$Sh_L = 1.09\left(Re_L\right)^{1/3}\left(Sc_L\right)^{1/2}\left(Ga_L\right)^{1/6}\left(\frac{a}{a_e}\right)^{1/3} \tag{11.16}$$

Here

$$Sh_L = \frac{k_L^0 d_p}{D_L}; \quad Re_L = \frac{L_{vf}\rho_L}{a\mu_L}; \quad Ga_L = \frac{d_p^3 g \rho_L^2}{\mu_L^2} \tag{11.16a,b,c}$$

There is another group in Germany at the Ruhr University at Bochum under Billet who also authored a book on the subject (Billet, 1995). There are correlations due to this group for random packings, but there are coefficients needed that are packing specific. Such data are given in their papers, and the reader is referred to those to pursue this further (Billet and Schultes, 1992, 1993, 1999).

In recent years there have been a number of publications by Mackowiak also on mass transfer aspects of packed columns (in addition to his comprehensive book on fluid dynamics of the same). In this work the change in behaviour in a packed column when operated above and below the load line is pointed out, as more closely discussed in his book. In general, the loading point in this context is where the trend in pressure drop versus fluid flow starts to show a marked change in its slope. For the liquid side coefficient (Mackowiak, 2008, 2011):

$$k_L^0 = \frac{5.524}{\sqrt{\pi}} \frac{a^{1/12} D_L g^{1/6} L_{Vf}^{1/6}}{\left(1 - \phi_p\right)^{1/3} \varepsilon^{1/4}} \tag{11.17}$$

This is not given in dimensionless form but it will yield the mass transfer coefficient in 'm/s' as long as the sizes with dimensions are given in m, s and kg. The form factor ϕ is a packing related size. There is a tabulation in his paper and extracted values for a few well known packings are Raschig ring 0, Pall metal 0.28, Pall plastic 0.309, CMR metal 0.475. For the interfacial area there are two relations given. The first is for use with $Re_L > 2$ below the loading line:

$$\frac{a_e}{a} = 3.42 Fr_L^{1/3}\left(\frac{We}{Fr_L}\right)^{1/2} \tag{11.18}$$

and for $Re_L < 2$:

$$\frac{a_e}{a} = \frac{6.69}{a^{1/3}} \frac{\Delta\rho^{1/2} g^{1/6} L_{Vf}}{\sigma_L^{1/2}} \tag{11.19}$$

The notion has been that the effective mass transfer area, or interfacial area, in a packed column is provided by an incomplete liquid film flowing down the packing. Mackowiak points out that with the use of modern packings that are very open, not to say, grid or cage-like, the surface areas are increasingly determined by the drip points provided. Given the right conditions, the interfacial area could in principle exceed the nominal packing area.

11.6   Tray Columns

Tray columns are used extensively in gas treating. This is easy to forget since all the focus in flue gas treating is given to packed columns and this field is dominated by the volume of papers these days. The reason for this situation is simply the need for the lowest possible pressure drop in the case of flue gas where no pressure is really available to spend.

A tray represents a separation stage in the traditional thinking. Before the advent of computers it was to large extent a question of applying the McCabe–Thiele 'ladder-step' analysis to the problem in its simplest form, or by identifying key components as an example. A good starting point to brush up on this is *Perry's Chemical Engineers' Handbook* (Perry and Green, 1984) or some textbook like Coulson and Richardson's *Chemical Engineering*, Volume 2 (2002).

The problem is that full thermodynamic equilibrium is not achieved on any one tray. This is compensated for by defining a tray efficiency, often referred to as Murphree efficiency. There is also a point efficiency. If this tray efficiency was 60%, which would represent that the change in vapour phase composition was 60% complete, the number of theoretical trays determined would simply be divided by 0.6 to obtain the number of mechanical trays needed to be installed.

This has worked fine in distillation where tray efficiencies have been from 50% upwards. Although doubling the number of trays seems drastic, there was plenty of experience to substantiate these numbers. Not so with alkanolamine based absorbers, here we are talking about tray efficiencies from 5 to maybe 25%. Clearly it makes quite a difference if we divide by 0.05 (i.e. multiply by 20) or 0.25 (i.e. multiply by 4). The column height in the two cases would differ by a factor of 5. Thousands of plate based gas absorbers have been built over the years and there is experience in the industry that can be used for designing a new column. However, this experience is not freely available. It resides with technology providers and very large companies.

To help the chemical engineers deal with tray design and their separation efficiencies in a more educated manner, AIChE (American Institute of Chemical Engineers, 1959) launched their bubble cap tray manual in 1959. This manual has been further developed over the years, and refined versions may be found in 'Perry' and specialist books like Kister's (1992). Looking into the details of this manual's design approach, it is seen that it makes use of both gas side and liquid side film mass transfer coefficients, and these are combined in the way already discussed previously to provide an overall coefficient although the manual refers to it as a 'transfer unit'.

The traditional literature on tray design does not deal with mass transfer combined with chemical reaction, but this is straight forward to do. Tomcej, Otto and Nolte (1983) describes a column simulator for absorption of CO_2 and/or H_2S incorporating the enhancement of mass transfer by chemical reaction. It was called AMSIM and this is now adopted for tray style column estimates in at least a couple of the bigger flow sheeting packages.

11.7   Examples

11.7.1   Treatment of Natural Gas for CO_2 Content

Let us consider the relatively small natural gas treatment plant already considered for column diameter in Section 8.7.3, one treating 30 MMSCFD. This takes place at 40°C and 35 bar.

Composition is described in the table underneath:

Composition, mole fractions	Mole fraction	MW
CH_4	0.865	16
C_2H_6	0.050	30
CO_2	0.080	44
H_2S	0.005	32
Average MW calculated		19.02

The specification will be to reduce the contents of CO_2 to 2% and the H_2S to 4 ppm. For now we want to consider mass transfer specifics. What size of mass transfer coefficients could be expected?

- The linear gas velocity of this column is $= (0.3087 \, \text{m}^3/\text{s})/(1.423 \, \text{m}^2) = 0.22 \, \text{m/s}$.
- The linear liquid velocity is $= 0.04150 / 1.423 = 0.029 \, \text{m/s}$.

We shall first approach this by using the correlations due to Onda, Takeuchi and Okumoto. The first step is to calculate the effective interfacial area in the packing, which is 2″ Pall metal as before. To this effect there is a 'critical surface tension' related to the packing material (stainless steel: SS), which is $\sigma_c = 75 \, \text{dyn/cm}$. (Note that L_{wf} is the same as linear velocity multiplied with density). The inverse of surface area a represents a linear dimension normally seen in a Reynolds number. Inserting the appropriate values into Equation 11.4 we have:

$$\frac{a_e}{105} = 1 - \exp\left\{ -1.45 \left(\frac{75}{62}\right)^{0.75} \left(\frac{30.80}{(105)(0.0014)}\right)^{0.1} \right.$$

$$\left. \times \left(\frac{(30.80)(105)}{(1056)^2 (9.81)}\right)^{-0.05} \left(\frac{30.80^2}{(1056)(105)(0.062)}\right)^{0.2} \right\}$$

$$a_e = 99 \, \text{m}^2/\text{m}^3$$

Note that the Reynolds number used to estimate the mass transfer coefficient is based on the effective contact area, not '*a*' as previously. The diffusion coefficient of CO_2 in the liquid is obtained from Ying and Eimer (2012) and is $D_{CO2} = 2.82 \times 10^{-9} \, \text{m}^2/\text{s}$.

The packing size $d_p = 0.05$ m (2"). Using Equation 11.2, we obtain the liquid side mass transfer coefficient:

$$\left(k_L^0 \left(\frac{1056}{(0.0014)(9.81)}\right)^{1/3}\right) = 0.0051 \left(\frac{30.80}{(99)(0.0014)}\right)^{2/3}$$

$$\times \left(\frac{0.0014}{(1056)(2.82 \times 10^{-9})}\right)^{-1/2} ((105)(0.05))^{0.4}$$

$$k_L^0 = 0.00039 \, \text{m/s}$$

For the gas side mass transfer coefficient, we need the gas phase viscosity. This is obtained from Poling, Prausnitz and O'Connell (2001) assuming the gas is pure methane as an approximation. The viscosity is 117 μP (0.12×10^{-6} kg/m s). The diffusivity of CO_2 in the gas was estimated in the example in Chapter 6 and is 5.5×10^{-7} m^2/s. Entering the numbers into Equation 11.2 a value for the mass transfer coefficient for the gas side is found:

$$\frac{k_G}{(105)(5.5 \times 10^{-7})} = 5.23 \left(\frac{7.30}{(105)(0.12 \times 10^{-6})}\right)^{0.7}$$

$$\times \left(\frac{0.12 \times 10^{-6}}{(33.64)(5.5 \times 10^{-7})}\right)^{1/3} ((105)(0.05))^{-2.0}$$

$$k_G = 0.0041 \text{m/s}$$

We observe that the mass transfer coefficient for the liquid side is about 1 order of magnitude lower than that for the gas side. Note that the liquid side coefficient estimated is just that for physical absorption. It does not account for the chemical reaction that will take place. Coefficients for H_2S will be of the same order.

11.7.2 Atmospheric Flue Gas CO_2 Capture

Let us assume that the gas is CO_2 in N_2 and that physical properties for such a system may be used for estimating the mass transfer coefficients. The absorbent will again be 30% (wt) monoethanolamine (MEA) in water. We shall stick to using the equations from Onda, Takeuchi and Okumoto. Packings are the same as for the preceding example.

	Liquid	Gas
Density (kg/m^3)	1120	1.151
Viscosity (kg/m·s)	2	0.000017
Surface tension (N/m)	0.064	–
Diffusion coefficient of CO_2 (m^2/s)	2.82×10^{-9}	1.70×10^{-5}
Flow (kg/m^2·s)	1.23	1.08

From our data and the previous example and Section 8.7.4, it follows that the effective contact area is:

$$\frac{a_e}{105} = 1 - \exp\left\{-1.45\left(\frac{75}{64}\right)^{0.75}\left(\frac{2.81}{(105)(0.002)}\right)^{0.1}\left(\frac{(2.81)(105)}{(1120)^2(9.81)}\right)^{-0.05}\right.$$

$$\left.\times\left(\frac{2.81^2}{(1120)(105)(0.064)}\right)^{0.2}\right\}$$

$$a_e = 63\,\mathrm{m}^2/\mathrm{m}^3$$

The gas side mass transfer coefficient is:

$$\frac{k_G}{(105)(1.7\times10^{-7})} = 5.23\left(\frac{1.08}{(105)(0.17\times10^{-6})}\right)^{0.7}$$

$$\times\left(\frac{0.17\times10^{-6}}{(1.151)(1.7\times10^{-7})}\right)^{1/3}((105)(0.05))^{-2.0}$$

$$k_G = 0.053\,\mathrm{m/s}$$

And the liquid side mass transfer coefficient is:

$$\left(k_L^0\left(\frac{1120}{(0.002)(9.81)}\right)^{1/3}\right) = 0.0051\left(\frac{2.81}{(63)(0.002)}\right)^{2/3}$$

$$\times\left(\frac{0.002}{(1120)(2.82\times10^{-9})}\right)^{-1/2}((105)(0.05))^{0.4}$$

$$k_L^0 = 8.1\times10^{-5}\,\mathrm{m/s}$$

It is time to reflect on these results. We see that in the high pressure case the gas side mass transfer coefficient was 0.0041 compared with 0.053 m/s for the low pressure case. The latter value is 1 order of magnitude higher. The difference can in the main be attributed to the effect of pressure on the diffusion coefficient of CO_2 in the gas. For the liquid side mass transfer coefficient the result is the opposite. Here the value, 0.00039, is roughly five times higher than in the low pressure case. This is mainly due to the higher flow intensity of liquid in the high pressure case and is caused by the more dense gas needing less room for flow, and this leads to a higher liquid flux. The difference in contact area between gas and liquid, 99 for the high pressure case compared 63 m²/m³ for the low pressure case, also reflects this situation.

11.7.3   Treatment of Natural Gas for H_2O Content

The gas having been treated to reduce CO_2 content in Section 11.7.1 will need to be dehydrated, for example, to achieve a dew point of $-10°C$. It is now water that is the transferred species, and the absorbent will most likely be triethylene glycol (TEG). We can recycle the estimates in Section 11.7.1 as a start, but we need to use the diffusivities for water in the gas and liquid phases as applicable. The liquid flow will, however, be very low, and that will take us out of range of the correlations, but let us first see how bad or good estimates that can be made. The numerics are summarised next without going through listing all formulae with numbers as before.

Data changed relative to the CO_2 absorption case:

$D\left(H_2O\text{-gas}\right)$	6.60×10^{-7}	m^2/s
$D\left(H_2O\text{-TEG}\right)$	1.50×10^{-10}	m^2/s
Density, liquid	1100	kg/m^3
Surface tension	43	dyn/cm
Liquid rate, L_w	0.34	kg/s

Results using the model due to Onda, Takeuchi and Okumoto (1968):

	CO_2 case	H_2O case
Contact area, a_e (m^2/m^3)	99	29
$k_L^{\,0}$ (m/s)	0.00039	0.0000007
k_G (m/s)	0.0041	0.0043

Considering the numbers in the results table it is clear that the wetting of the packing is estimated to be poor which is not surprising given the low liquid flow. The gas phase mass transfer coefficient is slightly improved due to water being a faster diffusing molecule. The liquid side mass transfer coefficient is estimated to be really low. Although the diffusion coefficient of water in TEG is an order of magnitude lower than that of CO_2 in the amine solution, this cannot explain such a low number, three orders of magnitude lower. We observe that the effective contact area figures in the $k_L^{\,0}$ correlation, but again it is doubtful if that can explain all. The value estimated for the liquid side mass transfer coefficient must be treated with large pinch of salt. For any further work it is probably best to disregard this value on the basis that the flow of liquid in a glycol column is way outside the range of the Onda correlation. Instead it is probably better to use the value for CO_2 in lack of a more reasonable figure.

11.7.4   Comparison of Correlations

In Section 11.5 a number of correlations are given for mass transfer coefficients and effective interfacial areas. Which ones to choose? How do they compare? The only way to form an opinion to these questions is to try them out and get acquainted with their properties. There will be no exact answer however. To make this comparison it is convenient to make use of the column example already analysed in the example in Section 11.7.1.

Let us first assemble all data related to flows and physical properties of the example. They are summarised in the following table.

	Gas	Liquid
Density $\left(kg/m^3\right)$	33.64	1056
Viscosity $\left(kg/m^2 \cdot s\right)$	1.17×10^{-5}	0.002
Surface tension (N/m)	–	0.062
Diffusion coefficient $\left(m^2/s\right)$	5.40×10^{-7}	2.80×10^{-9}
Mass flux, G_{wf}, L_{wf} $\left(kg/m^2 \cdot s\right)$	7.29	30.80
Mass flow, G_w, L_w (kg/s)	10.38	43.83
Volume flux, G_{vf}, L_{vf} $\left(m^3/m^2 \cdot s\right)$	0.217	0.029
Volume flow, G_v, L_v $\left(m^3/s\right)$	0.309	0.0415
Schmidt number	0.644	676

Next we assemble data for the packings we shall use for this exercise. For the random packing the data are summarised in the following table.

Column diameter (m)	1.346
Column cross-section $\left(m^2\right)$	1.423
Packing type	Pall metal
Packing size, d_p (m)	0.05
Packing factor (F)	20
Packing surface area $\left(m^2/m^3\right)$	105
Critical surface tension, σ_c (dyn/m)	75
Packed height (m) (assumed)	8
Packing free fraction (ε)	0.97

Finally, for use with the structured packing oriented correlations, here is also a set of data for such a packing. We choose to use the same diameter without any further ado. This diameter will work since we are not operating in an extreme corner of packing capability, but beware that the pressure drops will be different using this approach. Packing data are from Kister (1992).

Column diameter (m)	1.346
Column cross-section $\left(m^2\right)$	1.423
Packing type	Mellapak 250.Y
Packing hydraulic diameter, d_p (m)	0.0072
Packing factor	–
Packing surface area $\left(m^2/m^3\right)$	250
Corrugation spacing (m)	0.012
Base of triangle (m)	0.024
Height of triangle (m)	0.017
Packing void fraction (ε)	0.9
channel inclination angle $(°)$	45

When using these parameters as appropriate in the various equations in Section 11.5, values are obtained for interfacial area and the mass transfer coefficients on the gas and liquid sides. This work is summarised in the following table:

Results summary	Interfacial area (m^2/m^3)	Gas side, $(k_G \, m/s)$	Liquid side, $(k_L \, m/s)$
Random packing, \2″ Pall metal	105	–	–
Onda *et al.* Equations 11.2–11.4	99	0.00407	0.00039
Bravo and Fair, Equation 11.6	442	–	–
Wagner, Stichlmair, Fair, equation	–	0.00440	0.00065
Mersmann *et al.*, Equations 11.13–11.15	88	0.00382	0.00030
Mackowiak, Equations 11.16 and 11.17	292	–	0.00001
Puranik and Vogelpohl, Equation 11.5	107	–	–
Structured packing, Mellapak 250.Y	250	–	–
Bravo, Rocha, Fair, Equations 11.7–11.8	–	0.00371	0.00050
Rocha, Bravo, Fair, Equations 11.9–11.12	137	0.00653	0.00029

Some of these numbers stand out by deviating from the rest and some look unreasonable. They are still included here on purpose since this is a situation we often meet when trying to utilise new (to us) correlations. Now for the discussion.

A few comments to these results are in order. The predictions following Mersmann, Deixler, Rocha, Bravo and Fair are very sensitive to liquid hold-up and the contact angle between packing material and liquid. There are hold-up correlations based on fluid dynamics behind their correlations and that is not fully exploited in the present work. This is clearly a shortcoming and particularly the liquid side coefficients and surface area will suffer from this. Inspecting the original publication of Shi and Mersmann it is clear that their value of contact angle, or the size $1 - \cos(\theta)$ which they use synonymously with the critical surface tension σ_C, that there is discrepancy with respect to what values are used between research groups. One should be extremely careful with 'cross-use' of such data. It should also be kept in mind that this example is based on data for pressurised gas (35 bar), and it seems that the more laboratory column based correlations do perhaps not cover such high pressures. It could perhaps explain the extreme estimate for random packing interfacial area based on the Bravo and Fair model.

When going back into the various publications, they claim that their various models represent their data basis within ±20% for most data. The deviations between the various correlations are clearly bigger than that. However, there are a few points that must be raised. In this case all these models have been used without considering the basis for them. The following points should be kept in mind.

- The interfacial area based on Bravo and Fair: There seems to be some confusion with respect to units to use for surface tension and packed height. (Different publications vary between ft and m for height and N/m and dyn/cm for surface tension. Units as per their own publication have been used.)
- Wagner, Stichlmair and Fair's work was based on liquids with low surface tensions, and a turbulence enhancement factor was set to 1 for the present estimates.

- Mackowiak's model for interfacial area stands out. He has published widely in this area and there is a need to go deeper into the background material.
- The structured packing model from Bravo, Rocha and Fair was developed based on gauze packings while that of Rocha, Bravo and Fair used sheet metal. It makes a difference. However, the gas side coefficient for the latter model seems high.

This exercise hopefully demonstrates the value of considering alternative models, and these are but a selection from those available. It also shows the value of the ability to validate a model against known columns where the flows and conditions are under control. Furthermore, the need for using safety factors in design should be clear. Hopefully, the risk involved with taking new and unfamiliar correlations into use has been highlighted. When new correlations are put into use, ensure the background is understood, check for misprints (that may occur from time to time) and test them against familiar correlations. It cannot be stressed too much that we must always fully understand what we are doing.

References

American Institute of Chemical Engineers (1959) *Bubble-Tray Design Manual*, American Institute of Chemical Engineers.

Billet, R. (1995) *Packed Towers (in Processing and Environmental Technology)*, VCH-Chemie Verlag, Weinheim.

Billet, R. and Schultes, M. (1992) Advantage in correlating packed column performance. *Inst. Chem. Eng. Symp. Ser.*, **128**, B129–B136.

Billet, R. and Schultes, M. (1993) Predicting mass transfer in packed columns. *Chem. Eng. Technol.*, **16**, 1–9.

Billet, R. and Schultes, M. (1999) Prediction of mass transfer columns with dumped and arranged packings. *Trans. IChemE*, **77** (part A), 498–504.

Bolles, W.L. and Fair, J.R. (1979) Performance and design of packed distillation columns. *Inst. Chem. Eng. Symp. Ser.*, **56** (2), 35.

Bolles, W.L. and Fair, J.R. (1982) Improved mass transfer model enhances packed column design. *Chem. Eng.*, **89**, 109–116.

Bravo, J.L. and Fair, J.R. (1982) Generalized correlation for mass transfer in packed distillation columns. *Ind. Eng. Chem. Process Des. Dev.*, **21**, 162–170.

Bravo, J.L., Rocha, J.A. and Fair, J.R. (1985) Mass transfer in gauze packings. *Hydrocarb. Proc.*, **64** (1), 91–95.

Brunazzi, E., Nardini, G., Paglianti, A. and Potarca, K. (1995) Interfacial area of mellapak packing: absorption of 1,1,1-trichloroethane by Genesorb 300. *Chem. Eng. Technol.*, **18**, 248–255.

Brunazzi, E. and Paglianti, A. (1997) Liquid-film mass-transfer coefficient in a column equipped with structured packings. *Ind. Eng. Chem. Res.*, **36**, 3792–3799.

Cornell, D., Knapp, W.G., Close, H.J. and Fair, J.R. (1960a) Mass transfer efficiency – packed columns. Part 2. *Chem. Eng. Prog.*, **56**, 48–53.

Cornell, D., Knapp, W.G. and Fair, J.R. (1960b) Mass transfer efficiency – packed columns. Part 1. *Chem. Eng. Prog.*, **56**, 68–74.

Costa Novella, E., Escudro, G.O., Zamorano, M.A.U. and Gomez, C.L. (1992) Mass transfer in liquids. Determination of effective, specific interfacial areas for packing. *Int. Chem. Eng.*, **32**, 292–301.

Fair, J.R., J.L. Bravo, (1987) Prediction of mass transfer efficiencies and pressure drop for structured tower packings in vapour/liquid service. *Int. Chem. Eng.*, Symposium Series no. **104**, A183–A201.

Fair, J.R. and Bravo, J.L. (1990) Distillation columns containing structured packings. *Chem. Eng. Prog.*, **86**, 19–29.

Gualito, J.J., Cerino, F.J., Cardenas, J.C. and Rocha, J.A. (1997) Design method for distillation columns filled with metallic, ceramic, or plastic structured packings. *Ind. Eng. Chem. Res.*, **36**, 1747–1757.

Hanley, B., Dunbobbin, B. and Bennett, D.A. (1994) Unified model for countercurrent vapor/liquid packed columns. 2. Equations for the mass-transfer coefficients, mass-transfer area, the HETP, and the dynamic liquid holdup. *Ind. Eng. Chem. Res.*, **33**, 1222–1230.

Henriques de Brito, M., von Stockar, U., Bangerer, A.M., Bomio, P. and Laso M. (1994) Effective mass transfer interfacial area in a pilot column equipped with structured packings and with ceramic packings. *Ind. Eng. Chem. Res.*, **33**, 647–656.

Henriques de Brito, M., von Stockar, U. and Bomio, P. (1992) Predicting the liquid-phase mass transfer coefficient $-k_L$- for the Sulzer structured packing Mellapak. *Inst. Chem. Eng. Symp. Ser.*, **128**, B137–B144.

Kister, H.Z. (1992) *Distillation – Design*, McGraw-Hill.

Kolev, N. (1976) Operational parameters of randomly packed columns. *Chem. Ing. Tech.*, **48**, 1105–1112.

van Krevelen, D.W. and Hoftijzer, P.J. (1948) Kinetics of simultaneous absorption and chemical reaction. *Chem. Eng. Prog.*, **44**, 529–536.

Linek, V., Petricek, P., Benes, P. and Braun, R. (1984) Effective interfacial area and liquid side mass transfer coefficients in absorption columns packed with hydrophilised and untreated plastic packings. *Chem. Eng. Res. Des.*, **62**, 13–21.

Mackowiak, J. (2008) Modellierung des flüssigkeitsseitigen Stoffüberganges in Kolonnen mit klassischen und gitterförmigen Füllkörpern. *Chem. Ing. Tech.*, **80** (1–2), 57–77.

Mackowiak, J. (2010) *Fluid Dynamics of Packed Columns*, Springer.

Mackowiak, J. (2011) Model for the prediction of liquid phase mass transfer of random packed columns for gas-liquid systems. *Chem. Eng. Res. Des.*, **89**, 1308–1320.

Mangers, R.J. and Ponter, A.B. (1980) Effect of viscosity on liquid film resistance to mass transfer in a packed column. *Ind. Eng. Chem. Process Des. Dev.*, **19**, 530–537.

Mersmann, A. and Deixler, A. (1986) Packed columns. *Ger. Chem. Eng.*, **9**, 265–276.

Morris, G.A. and Jackson, J. (1953) *Absorption Towers*, Butterworth Scientific Publications, London.

Nakajima, E.S., Maffia, M.C. and Meirelles, A.J.A. (2000) Influence of liquid viscosity and gas superficial velocity on effective mass transfer area in packed columns. *J. Chem. Eng. Jpn.*, **33**, 561–566.

Nawrocki, P.A., Xu, Z.P. and Chuang, K.T. (1991) Mass transfer in structured packing. *Can. J. Chem. Eng.*, **69**, 1336–1343.

Olujic, Z. (2002) Delft mode – a comprehensive design tool for corrugated sheet structured packings. AIChE Spring National Meeting, New Orleans, LA, March 10–14.

Olujic, Z., Kamerbeek, A.B. and de Graauw, J. (1999) A corrugation geometry based model for efficiency of structured distillation packing. *Chem. Eng. Proc.*, **38**, 683–695.

Onda, K., Takeuchi, H. and Okumoto, Y. (1968) Mass transfer coefficients between gas and liquid phases in a packed column. *J. Chem. Eng. Jpn.*, **1**, 56–62.

Perry, R.H. and Green, D. (1984) *Perry's Chemical Engineers' Handbook*, McGraw-Hill.

Poling, B.E., Prausnitz, J.M. and O'Connell, J.P. (2001) *The Properties of Gases and Liquids*, 5th edn, McGraw-Hill.

Puranik, S.S. and Vogelpohl, A. (1974) Effective interfacial area in irrigated packed columns. *Chem. Eng. Sci.*, **29**, 501–507.

Rizutti, L. and Brucato, B. (1989) Liquid viscosity and flow rate effects on interfacial area in packed columns. *Chem. Eng. J.*, **41**, 49–52.

Rocha, J.A., Bravo, J.L. and Fair, J.R. (1996) Distillation columns containing structured packings: a comprehensive model for their performance. 2. Mass-transfer models. *Ind. Eng. Chem. Res.*, **35**, 1660–1667.

Sherwood, T.K. and Holloway, F.A.L. (1940) Performance of packed towers – liquid film data for several packings. *Trans. Am. Inst. Chem. Eng.*, **36**, 39–69.

Shetty, S. and Cerro, R.L. (1997) Fundamental liquid flow correlations for the computation of design parameters for ordered packings. *Ind. Eng. Chem. Res.*, **36**, 771–783.

Shi, M.G. and Mersmann, A. (1985) Effective interfacial area in packed columns. *Ger. Chem. Eng.*, **8**, 87–96.

Shulman, H.L. and de Grouff, J.J. (1952) Mass transfer coefficients and interfacial areas for 1-inch Raschig rings. *Ind. Eng. Chem.*, **44**, 1915–1922.

Shulman, L., C.F. Ullrich, A.Z. Proulx, J.O. Zimmerman, (1955) II. Wetted and effective – interfacial areas, gas – and liquid-phase mass transfer rates. *A.I.Ch.E. J.*, 1, 252–264.

Siminiceanu, I., A. Friedl, M. Dragan, (2002), as referenced by Wang *et al.* (2005). Presentation at Scientific Conference Meeting "35 years of Petroleum-Gas University Activity", Ploiesti, Romania, November 27–29.

Spiegel, L. and Meier, W. (1987) Correlations of the performance characteristics of the various Mellapak types (capacity, pressure drop and efficiency). *Inst. Chem. Eng. Symp. Ser.*, **62**, A202.

Strigle, R.F. Jr., (1987) *Random Packings and Packed Towers. Design and Applications*, Gulf Publishing, Houston, TX.

Tomcej, R.A., F.D. Otto, F.W. Nolte, (1983) Computer simulation of amine treating units. Laurance Reid Gas Conditioning Conference, Norman, OK.

Wagner, I., Stichlmair, J. and Fair, J.R. (1997) Mass transfer in beds of modern, high-efficiency random packings. *Ind. Eng. Chem. Res.*, **36**, 227–237.

Wang, G.Q., Yuan, X.G. and Yu, K.T. (2005) Review of mass-transfer correlations for packed columns. *Ind. Eng. Chem. Res.*, **44**, 8715–8729.

Xu, Z.P., Afacan, A. and Chuang, K.T. (2000) Predicting mass transfer in packed columns containing structured packings. *Chem. Eng. Res. Des.*, **78**, 91–98.

Ying, J. and Eimer, D.A. (2012) Measurements and correlations of diffusivities of nitrous oxide and carbon dioxide in monoethanolamine + water by laminar liquid jet. *Ind. Eng. Chem. Res.*, **51**, 16517–16524.

Zech, J.B. and Mersmann, A.B. (1979) Liquid flow and liquid-phase mass transfer in irrigated packed columns. *Inst. Chem. Eng. Symp. Ser.*, **56**, 39.

Further Reading

van Krevelen, D.W. and Hoftijzer, P.J. (1947) Studies of gas absorption. 1. Liquid film resistance to gas absorption in scrubbers. *Recl. Trav. Chim. Pays-Bas*, **66**, 49–66.

Olujic, Z. (1997) Development of a complete simulation model for predicting the hydraulic and separation performance of distillation columns equipped with structured packings. *Chem. Biochem. Eng. Q.*, **11**, 31–46.

Richardson, J.F., Harker, J.H. and Backhurst, J.R. (2002) *Coulson & Richardson's Chemical Engineering*, 5th edn, vol. **2**, Butterworth-Heinemann.

Rizutti, L., Augugliaro, V. and Lo Cascio, G. (1981) Influence of liquid viscosity on the effective interfacial area in packed columns. *Chem. Eng. Sci.*, **36**, 973–978.

12

Chemistry and Mass Transfer

12.1 Background

The use of absorbents reacting chemically with the absorbate shot ahead in the 1930s when Bottoms (1930, 1931) introduced alkanolamines for gas treating. Since then many thousands of such absorption plants have been built for various applications. It is an important process. The use of chemicals meant that the absorbents could carry much more absorbate per unit volume circulated. The downside was a higher energy consumption associated with the desorption part of the process. Energy was cheap, however.

There were materials issues involved with such plants and to a lesser degree there still are. Early designs were limited to 15% (wt) of monoethanolamine (MEA) due to corrosion issues. In the 1960s inhibitors were introduced and up to 30% (wt) could be used. The alkanolamines were also subject to degradation. That will be further discussed in a later chapter.

12.2 Equilibrium or Kinetics

What is the main benefit of using a chemical absorbent? Has it to do with a higher solubility per unit volume or has it to do with higher mass transfer rates? It is important to understand the fundamentals related to this. The answer is that both aspects are important. Let the absorption of CO_2 into 30% (wt) aqueous MEA absorbent be compared to absorption into water.

Figure 12.1 shows the solubility of CO_2 into MEA solution. Assume that the partial pressure of CO_2 is 100 kPa and the system temperature is 25°C. Under these circumstances the liquid's capacity for CO_2 is a so-called 'loading' of α mole CO_2/mole MEA and $\alpha = 0.6$ (the symbol y is also sometimes used in the literature). An MEA solution of 30% is roughly the same as 5000 mol/m³. Hence the solution's capacity for CO_2 is:

$$= \left(5000\,\text{mol MEA/m}^3\right) \left(0.6\,\text{mol CO}_2/\text{mol MEA}\right) = 3000\,\text{mol CO}_2/\text{m}^3$$

Gas Treating: Absorption Theory and Practice, First Edition. Dag Eimer.
© 2014 John Wiley & Sons, Ltd. Published 2014 by John Wiley & Sons, Ltd.

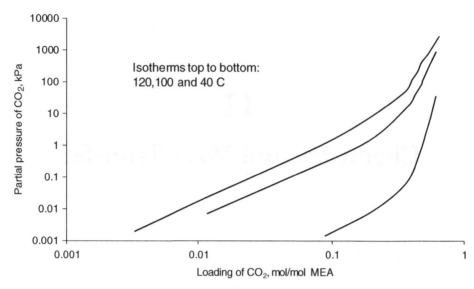

Figure 12.1   *Absorption isotherms for the system CO_2-MEA-water. Based on data from Jou, Otto and Mather (1995). Solution strength is 30% (wt).*

This value is reduced to 2000 mol CO_2/m^3 net effective loading if it is assumed that the lean absorbent has a lean loading of 0.2 mol CO_2/mol MEA.

From Chapter 7 we know that the solubility of CO_2 in water at 20°C is $m_i = 1.1$. At the present temperature we shall use 1.2 since the solubility is less at a higher temperature. Hence, when assuming full desorption of CO_2, the carrying capacity of water for CO_2 is

$$= C_i^G/m_i = \left(p_{CO_2}\right)\left(RTm_i\right)^{-1}$$
$$= (100\,\text{kPa})\left[\left(0.0083\,\text{kPa}\cdot\text{m}^3\cdot\text{mol}^{-1}\cdot\text{K}^{-1}\right)(298\,\text{K})(1.2)\right]^{-1}$$
$$= 34\,\text{mol}\cdot\text{m}^{-3}$$

The difference in CO_2 carrying capacity is about 60 to 1. If water were used, 60 times more liquid would need to be pumped around the cycle. Given a circulation rate for the MEA solution at 500 m^3/h and an absorption pressure of 50 bar, the energy needed to pump the solution would be around 1 MW. In the case of water this would be 60 MW, which is really prohibitive. The reduced absorbent circulation is obviously of great value. The higher solubility is indeed very attractive. (The energy needed for regeneration shall be left alone for now.)

As if this higher affinity for CO_2 was not enough, there is also an advantage with respect to the mass transfer rate. It shall be postulated that the mass transfer rate in the case as stated is in the order of 100 times faster for the MEA solution relative to water. How this comes about is the subject of this chapter.

12.3 Diffusion with Chemical Reaction

There are numerous articles on this subject to be found in the literature. Some discuss the principles based on mathematics and many are concerned with specific systems. Overviews have been made by Danckwerts (1970) and Astarita, Savage and Bisio (1983). Their books discuss extensively the subject of diffusion combined with chemical reaction. Later works have not particularly added to the fundamental knowledge in this field beyond the occasional detail that would be for specific problems beyond the scope of this text.

It is always a discussion point to decide which mass transfer mechanism model is to be used for absorption analysis. Here the penetration model due to Higbie will be used. The reason is purely practical. Danckwerts (1970) makes his fundamental analysis of the problem based on absorption into quiescent liquids, and he provides analytical solutions to the equations. Since this is as good as any approach to achieve a fundamental understanding, this is the path that will be followed. There may be nuances in answers if the analysis was made based on the Whitman film theory. Some of these nuances will be commented upon at the end as Astarita, Savage and Bisio (1983) make use of this fundament as well as using penetration theory and even surface renewal theories.

A further reason for focusing on the penetration theory is that experimental techniques for measuring reaction rates often involve the use of liquid surface stable over a defined, determinable time.

It will be remembered that Higbie's penetration model assumes that the liquid surface is renewed at fixed time intervals, and that within those time intervals the absorbed component 'i' will diffuse into the liquid as if this was quiescent. To analyse diffusion with chemical reaction we reuse an earlier figure (here Figure 12.2) to show how a component

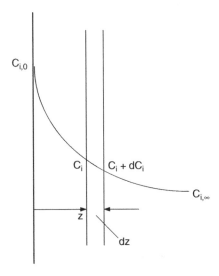

Figure 12.2 *Control element for diffusion into a quiescent liquid.*

mass balance is made with allowance for chemical reaction. Most engineers struggle to understand this concept. Hopefully the progression here is slow enough to allow the matter to mature.

The component mass balance is based on the same principle as always:

$$\text{Accumulated} = \text{In} - \text{Out} + \text{Produced}$$

In this case there will be accumulation, and this term is not zero. Hence:

$$A dz \frac{\partial C_i}{\partial t} = \left(-A D_i \frac{\partial C_i}{\partial z} \right) - \left(-A D_i \frac{\partial \left(C_i + \frac{\partial C_i}{\partial z} dz \right)}{\partial z} \right) + r_i A . dz \tag{12.1}$$

which becomes:

$$\frac{\partial C_i}{\partial t} = D_i \frac{\partial^2 C_i}{\partial z^2} + r_i \tag{12.1a}$$

This equation is essentially Fick's second law with a reaction term added. The boundary conditions will be discussed on a case to case basis as solutions are provided. It is worth noting that if the component 'i' is consumed in the reaction, the term r_i will be negative.

Assumptions implicit in the analysis done in this chapter include constant temperature and constant physical properties.

12.4   Reaction Regimes Related to Mass Transfer

To solve the Equation 12.1(a) analytically specific cases must be considered. These cases have to do with the speed of the chemical reaction relative to the diffusion process. It is convenient to divide this mathematically oriented problem into regimes that may be referred to as:

- Physical absorption
- First order irreversible reaction (first order w.r.t. component 'i' reacting)
- Instantaneous irreversible reaction (where reaction order is immaterial)
- Second order reaction.

Reversible reactions and simultaneous absorption will be dealt with in Section 12.6 and separate from this list in an attempt to simplify the task of achieving an initial overview.

12.4.1   Absorption with Slow Reaction

Slow reaction in this context means that the extent of reaction taking place in the mass transfer zone is negligible. It is so slow that there is no enhancement of mass transfer rate. This means that the absorption can be analysed as physical absorption. This has already been discussed in Chapter 6 on diffusion and in Chapter 10 on mass transfer models. It is just brushed upon here as a reminder of the basis it is coming from. The rate of mass transfer may be described by either:

$$N_i = \frac{D_i}{l}\left(C_i^{Li} - C_i^{Lb}\right) \tag{10.1}$$

or

$$N_i = 2\sqrt{\frac{D_i}{\pi\theta}}\left(C_{i0} - C_{i\infty}\right) \tag{10.9b}$$

The expression 10.1 comes from the film model, and Equation 10.9(b) comes from the penetration model. The surface renewal model shall be left alone for now.

This description is sufficient as far as the rate of mass transfer is concerned. However, the regime of slow reaction may be further sub-divided (Astarita, Savage and Bisio, 1983). In chemical engineering at large there are gas–liquid reactions that take place in the liquid bulk. An example is the reaction between carbon monoxide and methanol to form methyl formate in the presence of, for example the sodium or potassium salt of methanol.

Since the subject of 'slow reaction' is of little interest to gas treating, it will not be further discussed in any detail. Clearly the liquid residence time will be important since in this case the reaction rate determines the conversion.

12.4.2 Fast First Order Irreversible Reaction

In the case of absorption with first order chemical reaction the general Equation 12.1(a), which is based on the penetration theory, becomes:

$$\frac{\partial C_i}{\partial t} = D_i\frac{\partial^2 C_i}{\partial z^2} - k_1 C_i \tag{12.2}$$

This is a very important case in gas treating. Although the reaction of CO_2 with alkanolamines is second order, this reaction may often be approximately described as a first order reaction. The regime is then referred to as pseudo first order. The implications shall become clear as the discussion progresses. Danckwerts (1970) gives an analytical solution for Equation 12.2 when the boundary conditions are:

$$C_i = C_i^{Li} \quad z = 0 \quad t > 0 \tag{12.3}$$

$$C_i = C_i^{Lb} = 0 \quad z > 0 \quad t = 0 \tag{12.4}$$

$$C_i \rightarrow C_i^{Lb} = 0 \quad z \rightarrow \infty \quad t > 0 \tag{12.5}$$

At time zero there will be no 'i' present anywhere in the liquid, and this will also be the case for the bulk of the liquid at any later time. The concentration of 'i' at the interface at all times after time zero will be saturated physically at the partial pressure of component 'i' at the gas side. Danckwerts' solution for the concentration profile for component 'i' is, on this basis, given as:

$$\frac{C_i}{C_i^{Li}} = \frac{1}{2}e^{-z\sqrt{\frac{k_1}{D_i}}}\,erfc\left\{\frac{z}{2\sqrt{D_i t}} - \sqrt{k_1 t}\right\} + \frac{1}{2}e^{-z\sqrt{\frac{k_1}{D_i}}}\,erfc\left\{\frac{z}{2\sqrt{D_i t}} + \sqrt{k_1 t}\right\} \tag{12.6}$$

The rate of absorption at any time t is as before:

$$N_i = -D_i \left(\frac{dC_i}{dz} \right)_{z=0} \tag{10.5}$$

The expression for C_i from Equation 12.6 may, as when developing the penetration model, be substituted into Equation 10.5. Danckwerts (1970) gives the result of this as:

$$N_i = C_i^{Li} \sqrt{D_i k_1} \left\{ erf \left[\sqrt{k_1 t} \right] + \frac{e^{-k_1 t}}{\sqrt{\pi k_1 t}} \right\} \tag{12.7}$$

As before, this has to be integrated over the time period the stagnant element spends at the surface. That is:

$$Q_i = \int_0^\theta |N_i|_{z=0} dt \tag{10.8}$$

The result according to Danckwerts is:

$$Q_i = C_i^{Li} \sqrt{\frac{D_i}{k_1}} \left\{ \left[k_1 t + \frac{1}{2} \right] erf \left[\sqrt{k_1 t} \right] + \sqrt{\frac{k_1 t}{\pi}} e^{-k_1 t} \right\} \tag{12.8}$$

When Q_i is divided by the time t, the average absorption rate in the time period t becomes:

$$N_i|_{average} = \frac{Q_i}{t} = C_i^{Li} \sqrt{\frac{D_i}{k_1}} \left\{ \left[k_1 + \frac{1}{2t} \right] erf \left[\sqrt{k_1 t} \right] + \sqrt{\frac{k_1}{\pi t}} e^{-k_1 t} \right\} \tag{12.9}$$

Analysing the effect of chemical reactions on the rate of mass transfer is very much about using good approximations. A very interesting possibility is that either the chemical kinetics are fast or the contact time is very short, or both. In this case it is pointed out by Danckwerts that when:

$$k_1 t \gg 1 \tag{12.10}$$

then Equation 12.6 can be approximated by:

$$\frac{C_i}{C_i^{Li}} = e^{-z\sqrt{\frac{k_1}{D_i}}} \tag{12.11}$$

And:

$$N_i = C_i^{Li} \sqrt{D_i k_1} \tag{12.12}$$

$$Q_i = C_i^{Li} \sqrt{D_i k_1} \left(t + \frac{1}{2k_1} \right) \tag{12.13}$$

$$N_i|_{average} = C_i^{Li} \sqrt{D_i k_1} \left(1 + \frac{1}{2k_1 t} \right) \tag{12.14}$$

Alternative approximations may be made in the case where:

$$k_1 t \ll 1.$$

In that case:

$$N_i = C_i^{Li} \sqrt{\frac{D_i}{\pi t}} \left(1 + k_1 t\right) \tag{12.15}$$

$$Q_i = 2C_i^{Li} \sqrt{\frac{D_i t}{\pi}} \left(1 + \frac{k_1 t}{3}\right) \tag{12.16}$$

$$N_i\big|_{average} = C_i^{Li} \sqrt{\frac{D_i}{\pi t}} \left(1 + \frac{k_1 t}{3}\right) \tag{12.17}$$

When there is a fast irreversible reaction taking place when the component 'i' is absorbed, this component disappears in the mass transfer zone. See Figure 12.3. This makes its concentration profile steeper than if there were no reaction. In view of Equation 10.5 this clearly makes the mass transfer rate faster.

In gas treating there is another approximation that is of great importance and is often used. It is known from Chapter 4 that CO_2 reacts with alkanolamines according to a second order reaction scheme, one that is first order with respect to each of the reactants. Very often the alkanolamines are present in such a surplus that the reaction becomes pseudo first order with CO_2 as the limiting reactant. This condition is strived for in experimental set-ups because of the analytical solutions available to interpret the measured data.

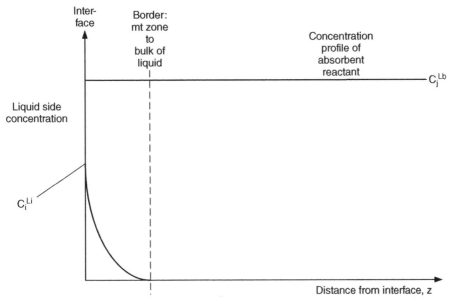

Figure 12.3   *Concentration profile of component 'i' when it undergoes a fast first order irreversible reaction when it is absorbed.*

In case of a pseudo first order reaction, the previous equations may be used when we set:

$$k_1 = k_2 C_j = k_{ps1} = k_2 \, [Absorbent] \tag{12.18}$$

Here k_{ps1} is the pseudo first order reaction constant and component j is the absorbent assumed to be present in great surplus. The condition for a second order reaction to be treated as a pseudo first order reaction is further discussed in Section 12.5.

12.4.3   Instantaneous Irreversible Reaction

When a component being absorbed is taking part in an instantaneous reaction, it is implied that there is no co-existence between the absorbed species and the reactant in the absorbent. Hence the reaction must take place in a plane parallel to the interface. There is no unreacted reactant between this plane and the interface. This is illustrated in Figure 12.4.

Since the reaction is instantaneous, the reaction mechanism and mathematics to describe it is of no importance as long as the consumption of the reactants is correctly described. The position of the reaction plane is influenced by the diffusion rates of the reactants. The concentration of the reactant in the absorbent influences its diffusion through the provision of a concentration gradient. The solubility of the absorbed species, that determines the value of C_i^{Li}, is a physical parameter that is given by nature.

Given that the instantaneous chemical reaction can be described by:

$$A + \nu B \rightarrow P \tag{12.19}$$

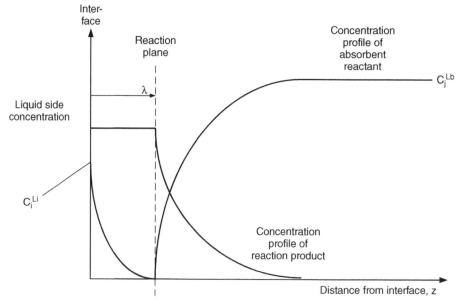

Figure 12.4   *Concentration profiles in the liquid phase for a typical case of absorption with irreversible instantaneous chemical reaction. (Based on application of the penetration theory.)*

Mass balances for both species and a coupling via the reaction stoichiometry is needed to determine the concentration profiles and absorption rates. An example stoichiometry is:

$$H_2S + (1.0) \; MEA \; \rightarrow \; (MEAH^+)(HS^-)$$

For the absorbing component 'i', when considering the region where $z < \lambda$, the mass balance equation when using the penetration theory as the fundament is:

$$\frac{\partial C_i}{\partial t} = D_i \frac{\partial^2 C_i}{\partial z^2} \tag{12.20}$$

The boundary conditions are:

$$C_i = C_i^{Li} \; z = 0 \; t > 0 \tag{12.21}$$

$$C_i = C_i^{Lb} = 0 \; z \geq \lambda \tag{12.22}$$

The mass balance for the absorbent reactant component 'j', the corresponding equations for the region $z > \lambda$, where 'j' can exist, are:

$$\frac{\partial C_j}{\partial t} = D_j \frac{\partial^2 C_j}{\partial z^2} \tag{12.23}$$

The boundary conditions are:

$$C_j = C_j^{Lb} \; t = 0 \tag{12.24}$$

$$C_j = 0 \; z \leq \lambda \tag{12.25}$$

Since the penetration theory is the basis, the reaction plane position will vary with time, that is $\lambda = \lambda(t)$. The reaction plane is moving because the penetration theory represents an unsteady state situation. The two differential Equations 12.20 and 12.23 are coupled by the stoichiometric requirement at the reaction plane:

$$D_i \frac{d^2 C_i}{dz^2} = v D_j \frac{d^2 C_j}{dz^2} \; \text{(i.e. at z = } \lambda(t)) \tag{12.26}$$

The reaction plane moves with time, and a relation is needed to describe this motion. There is no net chemical reaction taking place except at the reaction plane. It should be noted that the reaction plane is much nearer the gas–liquid interface than the so-called mass transfer film thickness when there is physical absorption only, and the concentration gradient at the interface is thus steeper for the case with instantaneous chemical reaction.

An expression is needed to describe the rate of movement of the reaction plane. By inspection of Figure 12.5 it is seen that at a given $z = \lambda$, then:

$$-\frac{\partial C_i}{\partial z} = \frac{\dfrac{\partial C_i}{\partial t} dt}{\dfrac{d\lambda}{dt} dt} \tag{12.27}$$

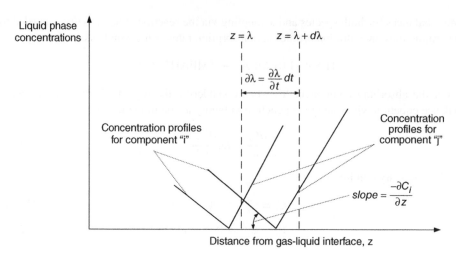

Figure 12.5 *Illustration of concentration profiles at the reaction plane and how they move with time in consideration of λ being a function of time.*

This equation, after elimination of dt over and under the RHS fraction, may be rearranged to:

$$\frac{d\lambda}{dt} = \left[-\frac{\dfrac{\partial C_i}{\partial t}}{\dfrac{\partial C_i}{\partial z}} \right]_{z=\lambda(t)} \tag{12.27a}$$

The solution to Equations 12.20, 12.23 and 12.27a is given by Danckwerts (1970) and Astarita, Savage and Bisio (1983). The concentration profiles are, first for the absorbed component 'i':

$$\frac{C_i}{C_i^{Li}} = \frac{erfc\left[\dfrac{z}{2\sqrt{D_i t}}\right] - erfc\left[\dfrac{\beta}{\sqrt{D_i}}\right]}{erf\left[\dfrac{\beta}{\sqrt{D_i}}\right]} \quad \text{when } 0 < z < \lambda \tag{12.28}$$

$$\frac{C_i}{C_i^{Li}} = 0 \text{ when } z > \lambda \tag{12.29}$$

$$\lambda = 2\beta\sqrt{D_i} \tag{12.30}$$

For the absorbent chemical 'j':

$$\frac{C_j}{C_j^{Li}} = 0 \text{ when } 0 < z < \lambda \tag{12.31}$$

$$\frac{C_j}{C_j^{Lb}} = \frac{erf\left[\dfrac{z}{2\sqrt{D_j t}}\right] - erf\left[\dfrac{\beta}{\sqrt{D_j}}\right]}{erfc\left[\dfrac{\beta}{\sqrt{D_j}}\right]} \quad \text{when } z > \lambda \tag{12.32}$$

The rate of absorption of component 'i' is as before determined by:

$$N_i = -D_i \left(\frac{dC_i}{dz} \right)_{z=0} \tag{10.5}$$

The function for C_i is obtained from Equation 12.28 where β is a constant yet to be determined. The mathematical procedure to obtain an expression for the rate of absorption at any time t is the same as before. Danckwerts (1970) gives the result as:

$$N_i = \sqrt{\frac{D_i}{\pi t}} \frac{1}{erf\left[\frac{\beta}{\sqrt{D_i}}\right]} C_i^{Li} \tag{12.33}$$

By inspecting the rate Equation 12.33 it is seen, by comparing it to the rate equation for physical absorption (Equation 10.9b), that the rate of absorption is altered by a factor defined as:

$$E_i = \frac{1}{erf\left[\frac{\beta}{\sqrt{D_i}}\right]} \tag{12.34}$$

The factor E_i is referred to as the enhancement factor for absorption with instantaneous irreversible reaction. The absorption rate equation may now be written as:

$$N_i = \sqrt{\frac{D_i}{\pi t}} E_i C_i^{Li} = k_L^0 E_i C_i^{Li} \tag{12.35}$$

The size β introduced in the previous equations is defined by, and must be determined by trial and error, from the expression:

$$\frac{C_j^b}{\upsilon C_i^{Li}} \sqrt{\frac{D_j}{D_i}} \cdot \left\{ \exp\left[\frac{\beta^2}{\sqrt{D_i}} - \frac{\beta^2}{\sqrt{D_j}} \right] \right\} \cdot erf\left[\frac{\beta}{\sqrt{D_i}} \right] + erf\left[\frac{\beta}{\sqrt{D_j}} \right] = 1 \tag{12.36}$$

In an absorption situation in gas treating it is likely that:

$$\frac{C_j^b}{\upsilon C_i^{Li}} \gg 1 \tag{12.37}$$

The condition defined by Equation 12.37 implies that:

$$\sqrt{\beta} \ll D \tag{12.38}$$

(Here, D covers both diffusion coefficients)

When this is true, the transcendental functions involved can be series expanded. Doing this, and using the definition of E_i (Equation 12.34), Equation 12.36 becomes:

$$E_i = \sqrt{\frac{D_i}{D_j}} + \frac{C_j^b}{\upsilon C_i^{Li}} \sqrt{\frac{D_j}{D_i}} \tag{12.39}$$

This equation allows the estimation of the enhancement factor for absorption with an instantaneous reaction from fundamental properties of the system. As seen, it turns out that the constant β does not need to be determined to estimate E_i, but it would be needed to calculate the concentration profiles.

A similar result may be obtained using the film theory where the reaction plane would be stationary. Astarita, Savage and Bisio (1983) give the result as:

$$E_i = 1 + \frac{C_j^b \, D_j}{v C_i^{Li} \, D_i} \tag{12.40}$$

It is worth noting that both mass transfer models' starting points lead to related results. The essential difference is the square root dependency on the ratio of diffusion coefficients predicted by the penetration theory compared with the linear dependency of the same ratio predicted by the film theory. This is the same difference as for physical absorption. The first term of Equations 12.39 and 12.40 will usually be much smaller than the second term.

12.4.4   Instantaneous Reversible Reaction

The previous chapter discussed the situation when it was reasonable to assume an irreversible chemical reaction in combination with mass transfer. In gas treating the main reactions are reversible. The reason why the analysis of irreversible reactions has been elaborated is that this is often a good approximation and it also serves to highlight potentials. Furthermore the foregoing material is a key factor in setting up experiments to measure key kinetic variables and interpret the results as will be discussed in Chapter 15.

The key reactions in catching CO_2 and H_2S are obviously reversible, or the consumption of reactants would be prohibitive. When a reaction is reversible and instantaneous, it will proceed very fast in both directions. During desorption when heavily loaded solutions are desorbed, it is intuitively obvious that reversibility must be considered. However, also in absorption reversibility must be accounted for when the back pressure from the solution is no longer negligible. The term 'reversible' must here be seen in this context.

Astarita and Savage (1980a,b) extended the previous work to deal with instantaneous reversible reactions. The material in this section is based on this work. These authors' starting point is that under these circumstances equilibrium prevails everywhere in the solution because the reaction(s) are infinitely fast. The absorption reaction of H_2S with a generic alkanolamine R_3N (one or two of the Rs could be an H) is ionic and thus infinitely fast:

$$H_2S + R_3N = (R_3NH^+)(HS^-) \tag{12.41}$$

Or in terms of a general volatile component A that reacts with a base B to form a product P:

$$A + vB = P \tag{12.42}$$

If there is more than one non-volatile compound B, that also include the products, then it is possible to describe this by:

$$A + \sum_j v_j B_j = 0 \tag{12.43}$$

where the stoichiometric coefficient v_j is negative for the products.

In this section it shall be understood that all concentrations except where otherwise stated are given as generic for the liquid phase. The equilibrium may be described by the thermodynamic equilibrium constant K for Reaction 12.43:

$$K = \frac{\prod_j C_j^{-v_j}}{C_i} \tag{12.44}$$

To develop the mathematics it is convenient to introduce a few definitions and new variables. The first is the extent of reaction, ξ, relative to the values in the bulk of the liquid. Here we have that:

$$C_i = C_i^b - \xi \tag{12.45}$$

$$C_j = C_j^b - v_j \xi \tag{12.46}$$

The restriction on ξ is that it may not be a value that makes any concentration negative. A further concept is a new coefficient μ_j. It is defined by the following relations:

$$\sum_j \mu_j v_j = 0 \tag{12.47}$$

This relation comes about because there is a need to make a balance on the atoms involved in the molecules participating in reactions. Next we define the molarity of the solution as:

$$m = \sum_j \mu_j C_j \tag{12.48}$$

For this development it is chosen to base it on the transport equations for the film theory, which is different from using the penetration theory as a base as in the preceding sections. In the case of the film theory the time dependent accumulation term is equal to zero in Equation 12.1(a), the diffusion equations for i and j become:

$$0 = D_i \frac{\partial^2 C_i}{\partial z^2} + r_i \tag{12.49}$$

$$0 = D_j \frac{\partial^2 C_j}{\partial z^2} + r_j \tag{12.50}$$

Since:

$$r_j = v_j r_i \tag{12.51}$$

the Equations 12.49 and 12.50 can be combined to eliminate the reaction rates:

$$D_j \frac{\partial^2 C_j}{\partial z^2} = v_j D_i \frac{\partial^2 C_i}{\partial z^2} \tag{12.52}$$

It is furthermore defined that:

$$r_{Dj} = \frac{D_j}{D_i} \tag{12.53}$$

And using this ratio in Equation 12.52 leads to:

$$r_{Dj} \frac{\partial^2 C_j}{\partial z^2} = v_j \frac{\partial^2 C_i}{\partial z^2} \tag{12.54}$$

Combining Equations 12.44 and 12.54 closes this problem since Equation 12.54 represents one equation for the concentrations of each of the js and the equilibrium relation 12.44 add another to deal with C_i. To solve these equations the boundary conditions must be defined. Since this is based on the film theory based equations, it is clear that:

$$z = \delta; C_i = C_i^b, \text{ and } C_j = C_j^b,$$
(12.55)

At the interface there may still be a chemical reaction (r_o) going on and the flux of a non-volatile component j diffusing towards the interface is given by:

$$z = 0; N_j = D_j \left[\frac{dC_j}{dz} \right]_{z=0} = v_j r_o$$
(12.56)

For the volatile component 'i' it is, as usual, governed by it being in physical equilibrium with its concentration on the gas side of the interface:

$$z = 0; C_i = C_i^{Li}$$
(12.57)

The mass flux of component 'i' is that caused by the reaction at the interface plus its diffusion into the liquid:

At:

$$z = 0; N_i = r_o - D_i \left[\frac{dC_i}{dz} \right]_{z=0}$$
(12.58)

As usual it is desired to express the flux of i in the form:

$$N_i = E_\infty k_L^o \left(C_i^{Li} - C_i^{Lb} \right)$$
(12.59)

Since this is film theory based:

$$k_L^o = \frac{D_i}{\delta}$$
(12.60)

So far, so good. There is still a need to arrive at expressions that can be utilised to analyse the mass transfer behaviour of this reversible system. N_i may be eliminated between Equations 12.58 and 12.59:

$$N_i = E_\infty k_L^o \left(C_i^{Li} - C_i^{Lb} \right) = r_o - D_i \left[\frac{dC_i}{dz} \right]_{z=0}$$
(12.61)

$$r_o = D_i \left[\frac{dC_i}{dz} \right]_{z=0} + E_\infty k_L^o \left(C_i^{Li} - C_i^{Lb} \right)$$
(12.61a)

From Equation 12.56:

$$r_o = \frac{D_j}{v_j} \left[\frac{dC_j}{dz} \right]_{z=0}$$
(12.56a)

The reaction rate at the interface, r_o, may now be eliminated between Equations 12.56(a) and 12.61(a):

$$r_o = \frac{D_j}{v_j} \left[\frac{dC_j}{dz} \right]_{z=0} = D_i \left[\frac{dC_i}{dz} \right]_{z=0} + E_\infty k_L^o \left(C_i^{Li} - C_i^{Lb} \right)$$
(12.62)

Enter Equation 12.60 for k_L^0 into Equation 12.62:

$$\frac{D_j}{v_j}\left[\frac{dC_j}{dz}\right]_{z=0} = D_i\left[\frac{dC_i}{dz}\right]_{z=0} + E_\infty \frac{D_i}{\delta}\left(C_i^{Li} - C_i^{Lb}\right) \tag{12.62a}$$

Divide through by D_i, multiply by v_j and use of the definition from Equation 12.53. This yields:

$$r_{Dj}\left[\frac{dC_j}{dz}\right]_{z=0} = v_j\left\{\left[\frac{dC_i}{dz}\right]_{z=0} + \frac{E_\infty}{\delta}\left(C_i^{Li} - C_i^{Lb}\right)\right\} \tag{12.62b}$$

Next the Equation 12.54 is integrated from $z = 0$ to a generic position z:

$$\int_0^z r_{Dj}\frac{\partial^2 C_j}{\partial z^2}dz = \int_0^z v_j\frac{\partial^2 C_i}{\partial z^2}dz \tag{12.63}$$

$$r_{Dj}\frac{dC_j}{dz} - r_{Dj}\left[\frac{dC_j}{dz}\right]_{z=0} = v_j\frac{dC_i}{dz} - v_j\left[\frac{dC_i}{dz}\right]_{z=0} \tag{12.63a}$$

The second term on the left-hand side is defined by Equation 12.62(b) and this is substituted into Equation 12.63(a):

$$r_{Dj}\frac{dC_j}{dz} - v_j\left\{\left[\frac{dC_i}{dz}\right]_{z=0} + \frac{E_\infty}{\delta}\left(C_i^{Li} - C_i^{Lb}\right)\right\} = v_j\frac{dC_i}{dz} - v_j\left[\frac{dC_i}{dz}\right]_{z=0} \tag{12.64}$$

$$r_{Dj}\frac{dC_j}{dz} = v_j\left\{\left[\frac{dC_i}{dz}\right]_{z=0} + \frac{E_\infty}{\delta}\left(C_i^{Li} - C_i^{Lb}\right)\right\} + v_j\frac{dC_i}{dz} - v_j\left[\frac{dC_i}{dz}\right]_{z=0} \tag{12.64a}$$

Two terms cancel out on the right-hand side, and:

$$r_{Dj}\frac{dC_j}{dz} = v_j\left\{\frac{E_\infty}{\delta}\left(C_i^{Li} - C_i^{Lb}\right) + \frac{dC_i}{dz}\right\} \tag{12.64b}$$

This equation is next integrated from $z = 0$ to δ:

$$\int_0^\delta r_{Dj}\frac{dC_j}{dz}dz = \int_0^\delta v_j\left\{\frac{E_\infty}{\delta}\left(C_i^{Li} - C_i^{Lb}\right) + \frac{dC_i}{dz}\right\}dz \tag{12.65}$$

So:

$$r_{Dj}\left(C_j^b - C_j^i\right) = v_j\left\{\frac{E_\infty}{\delta}\left(C_i^{Li} - C_i^{Lb}\right)(\delta - 0) + \left(C_i^{Lb} - C_i^{Li}\right)\right\} \tag{12.65a}$$

$$C_j^b - C_j^i = \frac{v_j}{r_{Dj}}\left\{E_\infty\left(C_i^{Li} - C_i^{Lb}\right) + \left(C_i^{Lb} - C_i^{Li}\right)\right\} \tag{12.65b}$$

$$C_j^i = C_j^b - \frac{v_j}{r_{Dj}}\left\{E_\infty C_i^{Li} - E_\infty C_i^{Lb} + C_i^{Lb} - C_i^{Li}\right\} \tag{12.65c}$$

$$C_j^i = C_j^b - \frac{v_j}{r_{Dj}}\left\{(E_\infty - 1)\left(C_i^{Li} - C_i^{Lb}\right)\right\} \tag{12.65d}$$

Equation 12.65(d) is the expression that was sought for. This allows the calculation of the interface concentration of non-volatile component j while accounting for differing diffusion coefficients for the various components j. For convenience it is defined that:

$$\zeta = (E_\infty - 1) \left(C_i^{Li} - C_i^{Lb} \right) \tag{12.66}$$

and

$$E_\infty = 1 + \frac{\zeta}{C_i^{Li} - C_i^{Lb}} \tag{12.66a}$$

and introducing Equation 12.66 into Equation 12.65(d):

$$C_j^i = C_j^b - \frac{v_j \zeta}{r_{Dj}} \tag{12.67}$$

Comparing Equation 12.67 with the expression in Equation 12.46 it is clear that the value obtained in Equation 12.67 could not be arrived at by the more trivially derived expression in Equation 12.46. This is the case in spite of the fact that both chemical and physical equilibria prevail at the interface. The chemical equilibrium at the interface is described by:

$$K = \frac{\prod_j \left[C_j^b - \frac{v_j \zeta}{r_{Dj}} \right]^{-v_j}}{C_i^{Li}} \tag{12.68}$$

There is also chemical equilibrium in the bulk of the liquid and this is described by:

$$K = \frac{\prod_j \left[C_j^b \right]^{-v_j}}{C_i^{Lb}} \tag{12.69}$$

As will become clear next, it is useful to divide Equation 12.69 by Equation 12.68 and this yields a ratio that defines a new variable ψ:

$$\psi = \frac{C_i^{Li}}{C_i^{Lb}} = \prod_j \left[1 - \frac{v_j \zeta}{C_j^b r_{Dj}} \right]^{-v_j} \tag{12.70}$$

Equation 12.70 may be solved for the variable ζ, and E_∞ can thereafter be found from Equation 12.66a. This was the target aimed for (although Equation 12.70 represents a polynomial, only one of the roots will make all concentrations positive and this is the true root to be used.)

To obtain a feel for this parameter E_∞ it is useful to consider how its value may change when we alter the concentrations of the non-volatile components and the mass transfer driving force. As stated by Astarita and Savage (op. cit):

$$\frac{d\psi}{d\zeta} > 0 \tag{12.71}$$

And: when $\psi = 1$ then $\zeta = 0$. See Equations 12.66 and 12.70. In this case there is no driving force and thus no mass transfer taking place. Clearly from Equation 12.71 it follows that:

$$\psi > 1 \rightarrow \zeta > 0 \text{ for absorption and}$$

$$\psi < 1 \rightarrow \zeta < 0 \text{ for desorption.}$$

It follows from Equation 12.66a that $E_\infty > 1$ for both absorption and desorption, that is, mass transfer is enhanced by the chemical reaction. This is as we would expect.

In the case of *absorption* the lower bound of ζ has been established as zero. What about the upper bound? Since $K > 0$ (a thermodynamic constant), it follows from Equation 12.68 that:

$$C_j^b > \frac{v_j \zeta}{r_{Dj}} \text{ and hence } \zeta < \frac{r_{Dj} C_j^b}{v_j} \text{ and } 0 < \zeta < \frac{r_{Dj} C_j^b}{v_j} \tag{12.72}$$

It is the component j with $v_j > 0$ (i.e. a reactant for absorption, not a product) that has the lowest value of ζ that is important to consider. This component is referred to as the diffusionally limiting reactant for absorption (DLRA). It is not necessary that the stoichiometrically limiting component is the same. From Equation 12.72 it follows that when:

$$\psi \rightarrow \infty \text{ then } \zeta \rightarrow \frac{r_{Dj} C_j^b}{v_j} \text{ as the upper bound} \tag{12.73}$$

Further to this, the enhancement factor tends to:

$$E_\infty \rightarrow 1 + \frac{r_{Dj} C_j^b}{v_j \left(C_i^{Li} - C_i^{Lb} \right)} \tag{12.74}$$

and the flux of the volatile component tends to:

$$N_i \rightarrow k_L^0 \left\{ 1 + \frac{r_{Dj} C_j^b}{v_j \left(C_i^{Li} - C_i^{Lb} \right)} \right\} \left(C_i^{Li} - C_i^{Lb} \right) \tag{12.75}$$

Or:

$$N_i \rightarrow k_L^0 \left\{ \left(C_i^{Li} - C_i^{Lb} \right) + \frac{r_{Dj} C_j^b}{v_j} \right\} \tag{12.75a}$$

and since $C_i^{Li} \gg C_i^{Lb}$ because of the condition Equation 12.73:

$$N_i \rightarrow k_L^0 \left\{ C_i^{Li} + \frac{r_{Dj} C_j^b}{v_j} \right\} \tag{12.75b}$$

In the case of *desorption* the upper bound of ζ has been established as 0. To make it easier to distinguish the mathematical expressions associated with desorption, the index j for this component will be replaced by 'k'. It is the component k with $v_k < 0$ (i.e. a reactant for *desorption*) that has the lowest value of $|\zeta|$ that is important to consider. This component is referred to as the diffusionally-limiting reactant for desorption (DLRD). It is not necessary that the stoichiometrically limiting component is the same. The bounds for ζ during desorption will be:

$$\frac{r_{Dk}C_k^b}{v_k} \leq \zeta \leq 0 \tag{12.76}$$

The defined ratio ψ can only be as low as 0, and this would happen when the value of the concentration of the volatile component at the interface is negligible. For ζ this corresponds to the lower bound in Equation 12.76. When this information is used in Equations 12.59 and 12.66(a), it is obtained that when $\psi \to 1$, then:

$$E_\infty \to 1 + \frac{1}{C_i^{Li} - C_i^{Lb}} \frac{r_{Dk}C_k^b}{v_k} \tag{12.77}$$

The enhancement factor E_∞ will be positive since $v_k < 0$ and $C_i^{Li} - C_i^{Lb} < 0$ and the flux when written as before will tend to:

$$N_i \to k_L^o \left\{ 1 + \frac{1}{C_i^{Li} - C_i^{Lb}} \frac{r_{Dk}C_k^b}{v_k} \right\} (C_i^{Li} - C_i^{Lb}) \tag{12.78}$$

$$N_i = k_L^o \left\{ (C_i^{Li} - C_i^{Lb}) + \frac{r_{Dk}C_k^b}{v_k} \right\} \tag{12.78a}$$

Here, N_i will be negative. This signifies that the flux of component 'i' is going from the liquid to the gas. It should be noticed that this result comes about automatically without the need to rewrite the flux equation.

A third limiting situation will arise when the driving force tends to zero. At zero driving force for component 'i' there will of course be no mass transfer. It is seen from Equations 12.66(a) that an infinitely large enhancement factor is predicted. Hence a very small driving force suggests a very large enhancement factor. Remembering the conditions used for deriving the previous equations, there is clearly a need to look at this situation specifically.

The line of argument is as follows (Astarita and Savage, 1980b). The first derivative $\frac{d\psi}{d\zeta}$ is finite also at zero driving force, at which point this is equal to 1 and $\zeta = 0$. It follows that:

$$\psi = 1 + \left[\frac{d\psi}{d\zeta} \right]_{\zeta=0} \zeta \tag{12.79}$$

The Equation 12.70 is expanded, and the following is obtained:

$$\zeta = \frac{\psi - 1}{\sum_j \frac{v_j^2}{C_j^b r_{Dj}}} \tag{12.80}$$

This latter result is taken from the referenced paper without further ado. When this result is used in Equations 12.66(a) and 12.59, the following expressions result:

$$E_\infty = 1 + \frac{1}{\sum_j \frac{C_i^b v_j^2}{C_j^b r_{Dj}}} \tag{12.81}$$

If the definition:

$$\epsilon = \frac{c_i^b}{c_j^b} \tag{12.82}$$

is used in Equation 12.81, then it becomes:

$$E_\infty = 1 + \frac{1}{\sum_j \frac{\epsilon v_j^2}{r_{Dj}}} \tag{12.81a}$$

And using this relation for E_∞ in the flux equation, it is obtained that:

$$N_i = k_L^0 \left[1 + \frac{1}{\sum_j \frac{\epsilon v_j^2}{r_{Dj}}} \right] \left(C_i^{Li} - C_i^{Lb} \right) \tag{12.83}$$

The ratio ϵ is interesting. Since the concentration of the volatile component is a small number while the absorbent's reactant j is likely to be large, ϵ will be a small number. Hence the enhancement factor will be large when calculated from Equation 12.81(a) but it will be a finite number also for zero driving force. Figures 12.6 and 12.7 illustrate the

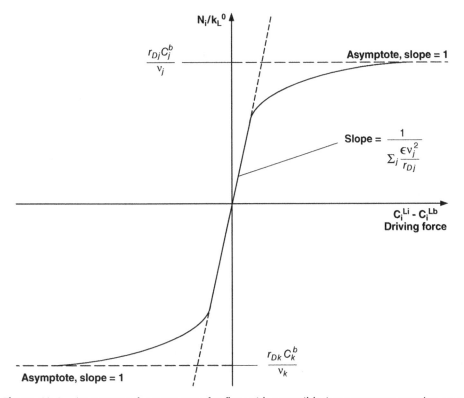

Figure 12.6 *Asymptotes for mass transfer flux with reversible instantaneous reaction as a function of mass transfer driving force.*

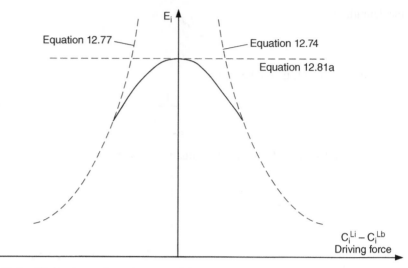

Figure 12.7 *Comparison of predictions of the instantaneous enhancement factor from the various relations derived.*

newfound insight into the bounds of instantaneous reversible reaction and mass transfer in both absorption and desorption.

The mass transfer flux for the volatile component may in parts of a typical absorption–desorption plant also be limited by the gas side mass transfer resistance.

12.4.5 Second Order Irreversible Reaction

If the reactant present in the absorbent is not available in sufficient surplus to make the reaction pseudo first order, a more complex mathematical analysis is needed. Simultaneous mass balances for the absorbate and the reactant absorbent must be solved. These equations are as usual developed by taking the mass balance over a control element stretching from z to $z + dz$. If the analysis is again based on the penetration theory, the equations are, first for the absorbed component (absorbate):

$$\frac{\partial C_i}{\partial t} = D_i \frac{\partial^2 C_i}{\partial z^2} - k_2 C_i C_j \tag{12.84}$$

The boundary conditions are:

$$C_i = C_i^{Li} \quad z = 0 \, t \geq 0 \tag{12.85}$$

$$C_i \rightarrow C_i^{Lb} = 0 \quad z \rightarrow \infty \tag{12.86}$$

The mass balance for the absorbent reactant component 'j', is the same as the corresponding equations for the region $z > \lambda$ for instantaneous reaction but now the reaction term needs to be introduced:

$$\frac{\partial C_j}{\partial t} = D_j \frac{\partial^2 C_j}{\partial z^2} - k_2 C_i C_j \tag{12.87}$$

The boundary conditions are:

$$C_j = C_j^{Lb} \quad t = 0 \tag{12.88}$$

$$C_j \rightarrow C_j^{Lb} \quad z \rightarrow \infty \quad t > 0 \tag{12.89}$$

$$\frac{dC_j}{dz} = 0 \quad z = 0 \quad t > 0 \tag{12.90}$$

The solution to these equations may be found in Danckwerts (1970) but he refers to work by Perry and Pigford (1953), Brian, Hurley and Hasseltine (1961) and Pearson (1963). These authors have solved the previous equations and simplified the solution by introducing the following dimensionless groups:

$$E = \frac{Q}{2C_i^{Li}} \sqrt{\frac{\pi}{D_i t}} \tag{12.91}$$

$$M = \frac{\pi}{4} k_2 C_j^b t \tag{12.92}$$

The result can be summarised by the following approximate formula:

$$E = \frac{\sqrt{M \dfrac{E_i - E}{E_i - 1}}}{\tanh \sqrt{M \dfrac{E_i - E}{E_i - 1}}} \tag{12.93}$$

The E_i is as defined in Section 12.4.3 dealing with instantaneous irreversible reaction.

12.5 Enhancement Factors

The enhancement factor and its role in mass transfer rate analysis have been explained in the previous sections. It may be related to reaction kinetics and mass transfer coefficients. This will be discussed here. The relation between the enhancement factor and the dimensionless Hatta number is shown in Figure 12.8 with the enhancement factor for mass transfer with instantaneous reaction as a parameter. The graphical representation given in this figure is a classic in the literature on absorption with chemical reaction. A particularly nice version is given by Levenspiel (1999) where the limits between the reaction regimes are given. This approach is adopted here.

The Hatta number is defined as:

$$Ha = \frac{\sqrt{D_i k_2 C_j^b}}{k_L^0} \tag{12.94}$$

If Equation 10.10 is used to substitute for k_L^0, it is easily demonstrated that $(Ha)^2 = M$ or $Ha = \sqrt{M}$. The size M is defined by Equation 12.92. Some authors, notably Danckwerts

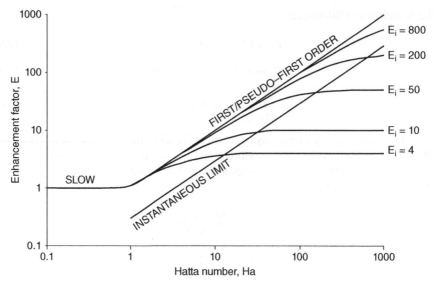

Figure 12.8   *Enhancement factor for absorption with chemical reaction versus the Hatta number with a few* E_i *values as asymptotes.*

(1970), use $\sqrt{M}$ as abscissa instead of *Ha* for the graph in Figure 12.8. There is no need for confusion on this point. Insert the expression for k_L^0 from Equation 10.10 into Equation 12.94:

$$Ha = \frac{\sqrt{D_i k_2 C_j^b}}{2\sqrt{\dfrac{D_i}{\pi t}}} = \sqrt{\frac{k_2 C_j^b}{4/(\pi t)}} = \sqrt{\frac{\pi}{4} k_2 C_j^b t} = \sqrt{M} \qquad (12.95)$$

As discussed under 'slow reactions', the enhancement factor is then equal to one, that is there is no enhancement of the mass transfer rate. Since the reaction does not take place to a significant degree in the mass transfer zone, it does not have an effect on the mass transfer. This is the situation for Hatta numbers less than one. Strictly speaking the criterion is Ha $\ll$ 1. There is a transition region between the slow and the first order reaction regimes. The enhancement factor could be around 1.3 when Ha $= 1$.

It must be understood that there are a number of approximations implied in the foregoing mathematical treatment, and the following criteria are also approximate. However, in general there is (Levenspiel, 1999).

- Pseudo first order reaction regime when $E_i > 5$ (Ha)   (12.96)
- Instantaneous reaction regime when $E_i < $ (Ha) $/5$ (12.97)
- When $5E_i > $ Ha $> E_i/5$ the system will be in the transition between the first order and the instantaneous reaction regimes, and the second order reaction kinetics need to be accounted for when estimating E.   (12.98)

There are other ways of judging the regime situation. For example Danckwerts (1970) states that the pseudo first order reaction regime will limited by:

$$1 \ll Ha \ll E_i. \tag{12.99}$$

The 'Levenspiel criteria', although just approximations, are more specific and by inspection of the curves in Figure 12.8 they are seen to be accurate enough for engineering purposes.

It is worthwhile considering some limiting behaviours associated with the graphical description given in Figure 12.8. The effect of the Hatta number for slow reaction regime has already been mentioned. However, both a very short exposure time and a slow reaction could cause a slow reaction regime behaviour (see Equation 12.92). This is more easily seen from the dimensionless group M than from the Hatta number. In this regime the rate of reaction has a very limited effect on the rate of mass transfer.

In the pseudo first order reaction regime, the liquid phase reactant, for example MEA, diffuses to the gas–liquid interface fast enough to avoid a mathematically significant depletion of this reactant in the mass transfer zone. This is indeed why the reaction is referred to as pseudo first order. The concentration of this reactant is approximately the same in the mass transfer zone as in the bulk of the liquid. Here, it is the concentration of the free reactant that must be used. It must be remembered that in a typical system the absorbent's reactant is consumed as the absorbent travels along the column's length. This must be accounted for when estimating enhancement factors.

As the reaction becomes faster and faster it will approach the instantaneous reaction situation where the rate of reaction no longer determines the rate of absorption. When this stage is reached, the limiting process is the rate of diffusion of reactants towards the plane of reaction. However, it is the presence of the instantaneous reaction that positions the reaction plane near to the surface such that the absorbed component, for example H_2S, will have a shorter distance to diffuse. This makes its concentration profile steeper compared to physical absorption and thus the absorption rate is very much increased by the presence of the reaction.

A faster reaction as depicted by the size of its reaction kinetic constant will influence the absorption rate more than a slower reaction no matter the contact time t. This is evident from the group M and Figure 12.8. It is also worth noting that a faster reaction can influence the absorption rate at smaller contact times, t, than a slower reaction. When considering process intensification units, this may be important to consider.

12.5.1 Transition from Slow to Fast Reaction

There is one situation that is bound to crop up when working with slower reacting species like the tertiary amines. This is when substantial parts of it is loaded with either CO_2 or H_2S in which case the Hatta number may actually be lower than 1. Considering Figure 12.8 it is clear that there is no sharp point or discontinuity at $Ha = 1$. In Figure 12.9 this area is blown up to highlight this. For practical numerical work it is of course possible to simply say that:

$$\text{If } Ha < 1, \text{ then } E = 1 \tag{12.100}$$

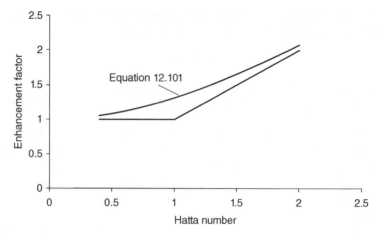

Figure 12.9 *The transition from slow to fast reaction. The upper curve represents equation 12.101.*

It is, however, more satisfactory to have a mathematical representation depicting a smooth transition. Referring to Astarita, Savage and Bisio (1983), this may be expressed as:

$$E = \frac{Ha}{\tanh(Ha)} \tag{12.101}$$

It is, however, important to remember that the relations relating to calculating the enhancement factor as summarised by Figure 12.8, are approximations. They are good but it is always necessary to consider the assumptions made in relation to the results obtained.

12.6　Arbitrary, Reversible Reactions and/or Parallel Reactions

The foregoing discussion may seem to address only a simple situation with single and irreversible reactions. This may or may not be realistic. The theory discussed has certainly had a profound effect on how data for these reactions are gathered. However, it is realised that there is a back pressure of CO_2 and/or H_2S from the absorbent solutions as this becomes more and more loaded with these gases.

It is important to remember that the basic transport equations developed to analyse absorption with chemical reaction have a reaction term that is, in principle, general in the sense that any reaction could be accounted for, including a reversible reaction. When there is a reversible reaction to account for, mass balances must also be made for the additional compounds involved, like carbamate and the protonated amine molecule as examples. The number of mass balances to solve has thus increased to 4 from 1 or 2.

When both CO_2 and H_2S are present over an alkaline absorbent, there will also be parallel reactions since they both will react with the alkali, as they are acids. Here the gases being absorbed will compete for the same absorbent molecules. This means that their transport

Table 12.1 *Past theoretical studies of absorption with chemical reaction.*

Author (year)	Order	Reversible/ irreversible	Parallel	Consecutive
Danckwerts (1950)	–	–	–	–
Perry and Pigford (1953)	2nd	Irrev	–	–
Brian *et al.* (1961)	2nd	Irrev	–	–
Pearson (1963)	2nd	Irrev	–	–
Hikita and Asai (1964)	mth,nth	Irrev	–	–
Onda *et al.* (1970a,b,c)	2nd	Rev	–	–
Onda *et al.* (1970a,b,c)	–	–	–	consecutive
Sada and Kumazawa (1973)	Varying	–	–	–
Hikita *et al.* (1973)	2nd	Irrev	–	–
Hikita and Asai (1976)	2nd	Irrev/rev	–	2 step
Sada *et al.* (1976)	Computational Study	–	–	–
Hikita *et al.* (1977)	mth,nth	Irrev	Simulation, absorption	–
Sada *et al.* (1977a)	Chemical Reaction n both phases			
Sada *et al.* (1977b)	Approx.	Analytical	Solutions	–

rate equations are coupled and must be solved simultaneously. As pointed out in chapter 4, CO_2 can also enter into parallel reactions on its own as it may react both directly with an amine and via hydroxide ions. This is accounted for in the reaction term as an adjustment, and is easier to deal with.

A number of authors have analysed a number of situations over the years. Table 12.1 gives a list of such publications. These papers may be worth looking up for further discussions. A detailed discussion of all the various possibilities is beyond the scope of the present text. As computers have become more powerful and available, it is likely that the more complex situations alluded to will be analysed numerically for direct application in process design.

12.7 Software

As evident from the preceding discussions a lot of effort has been expended in finding more or less approximate solutions for mass transfer with chemical reaction that could be used with a calculator, pen and paper. This work has brought a lot of insight into the field and it is well worth a study. Often it may be possible to use the results of past research such as this to obtain quick overviews of the problem at hand. Another aspect is that it is always good practice to obtain a thorough understanding of the challenge before expending a lot of work on computations to take care of the details.

This said, and given the present day status of computers and software, it is becoming common to make numerical analysis of concentration profiles and relate them to the rate of mass transfer. The software used tends to be a mixture of custom programming and use of mathematical libraries for solving equations. An example would be a Fortran compiler with the IMSL (International Mathematical and Statistical Library).

There is also a trend that the major flowsheet simulators sold commercially offer rate based column models that may be applied to absorption with chemical reaction. The more dedicated the software, the less in-house work is needed. The downside of everything being sold as a specific solution is that there is less flexibility. One of the bigger suppliers sells a framework for such columns that is very flexible but this requires more user input including a bit of programming.

This is a comparatively specialised field with a limited number of people doing in-depth process analysis of this type. The range of software products offered is also limited. Organisations having the right know-how in general can make more money from selling engineering services than from selling a few program licenses.

12.8   Numerical Examples

12.8.1   Natural Gas Problem with MEA

In the example in Section 11.7.1 the mass transfer coefficients for a natural gas treating absorber were calculated. A question remaining for this problem is, how will the chemical reaction between CO_2 and the free MEA in the absorbent affect the mass transfer? How will this differ between the rich and lean end of the column?

The column operated at 35 bar, 40°C with an aqueous 5 M MEA solution. The lean amine had a loading (α) of 0.20 mol/mol while the rich loading was 0.45 mol/mol. Gas feed had 8% CO_2 while the treated gas had 2%. We also remember that $CO_2 + 2$ MEA $->$ adduct, that is the stoichiometric coefficient (v) is equal to two.

To evaluate the effect of chemical reaction we need:

- Kinetic constant for the reaction
- Solubility of CO_2 (will be via N_2O analogy)
- Diffusivity of CO_2 (will be via N_2O analogy)
- Diffusivity of MEA.

These may be obtained from Ying and Eimer (2012a,b, 2013) and Snijder *et al.* (1993). The obtained values, without further ado are:

- Kinetic constant, $k_2 = 13\,359\ \mathrm{m^3/kmol \cdot s}$

- $H_{CO_2,C} = 4303$ kPa·m^3/kmol $\gg m_{CO2} = 1.65$ (dimensionless solubility)
- $D_{CO_2} = 1.85 \times 10^{-9}$ m^2/s
- $D_{MEA} = 1.06 \times 10^{-9}$ m^2/s.

Free MEA available for reaction with CO_2 is $[MEA]_{free} = [MEA]_{no\,min\,al}\,(1 - v\alpha)$.

The partial pressures of CO_2 in the rich and lean end are $(35\,bar)(0.08) = 2.8\,bar$ and $(35\,bar)(0.02) = 0.7\,bar$, respectively.

For evaluating the mass transfer resistances and the overall mass transfer coefficient it is known from Chapter 7 (Equation 7.4, but now with the enhancement factor added) that:

$$\frac{1}{K_G} = \frac{1}{k_G} + \frac{m_i}{k_L^0 E}$$

where the three terms are overall resistance, gas side resistance and liquid side resistance.

From the example in Section 11.7.1 we have that:

- $k_G = 0.00039$ m/s
- $k_L^0 = 0.0041$ m/s

The Hatta number is adopted from Equation 12.94:

$$Ha = \frac{\sqrt{D_{co_2}k_2\,[MEA]_{free}}}{k_L^0}$$

We also repeat from Equation 12.40 that

$$E_i = 1 + \frac{C_j^b\,D_j}{vC_i^{Li}\,D_i}$$

And in case of we are in the fast second order reaction regime, we may need to calculate the enhancement factor E via the expression (Equation 12.93):

$$E = \frac{\sqrt{M\dfrac{E_i - E}{E_i - 1}}}{\tanh\sqrt{M\dfrac{E_i - E}{E_i - 1}}}$$

The technique is to guess a value for E, use the estimate we have for E_i, then calculate E from the expression. If needed, we guess a new value for E, and so on. We also remember that:

$$M = Ha^2$$

Inserting the numbers into the expressions given earlier, it is obtained that:

	Rich end	Lean end
Concentration of free MEA (kmol/m³)	0.5	3.0
Concentration of CO_2 at interface (kmol/m³)	0.065	0.016
E_i, the instantaneous reaction enhancement factor	3.2	54
Ha number	9.0	22
$5 \times Ha$	45	270
Is $5 \times Ha < E_i$?	No	No
$Ha/5$	1.8	5.4
Is $E_i < Ha/5$?	No	No
Regime	Fast second order	Fast second order
Guess E	2.967	18.2
Calculate E from the expression previously	2.966	18.1
Gas side resistance, $1/k_G$ (s/m)	244	244
Liquid side resistance, $1/k_L^0 \cdot E$ (s/m)	1429	234
Total resistance to mass transfer, $1/K_G$ (s/m)	1673	478
Estimate for K_G, (m/s)	0.0006	0.0021
% of resistance in the gas film	15	51
% of resistance in the liquid film	85	49

The numbers in the table speak for themselves, but it is interesting to note that in neither case can we assign all the resistance to either the gas or liquid side, although at the rich end a liquid side assumption would not be way out.

Note also that we check for which regime with respect to reaction related to mass transfer that we are in by comparing Ha and E_i as has been discussed in relation to Figure 12.8. In this context we also observe that Ha is big enough to look away from the possibility of being in the transition region between slow and fast reaction discussed in Section 12.5.1.

12.8.2   Flue Gas Problem

This problem is very much like the previous in Section 12.8.1 but is included to illustrate the different results ensuing form different operation pressure and a tougher specification for CO_2 in treated gas. Like in the previous example, in the example in Section 11.7.2 the mass transfer coefficients for a natural gas treating absorber were calculated. A question remaining for this problem is, how will the chemical reaction between CO_2 and the free MEA in the absorbent affect the mass transfer? How will this differ between the rich and lean end of the column?

The column operated at 1.05 bar, 40°C with an aqueous 5 M MEA solution. The lean amine had a loading (α) of 0.20 mol/mol while the rich loading was 0.45 mol/mol. Gas feed had 4% CO_2 while the treated gas had 0.4%. We also remember that $CO_2 + 2$ MEA $->$ adduct, that is the stoichiometric coefficient (v) is equal to 2.

The partial pressures of CO_2 in the rich and lean end are $(1.05\,\text{bar})(0.04) = 0.042\,\text{bar}$ (4.2 kPa) and $(1.05\,\text{bar})(0.004) = 0.0042\,\text{bar}$ (0.42 kPa), respectively.

The mass transfer coefficients are different for this problem. The gas and liquid fluxes are different and the diffusivity of CO_2 in the gas phase is higher. See the relevant examples in Chapters 6 and 11.

The physical data and reaction kinetics are the same as in the previous example as are the method of numerical analysis. The numbers are summarised next.

Inserting the numbers into the previous expressions given, it is obtained that:

	Rich end	Lean end
Concentration of free MEA ($kmol/m^3$)	0.5	3.0
Concentration of CO_2 at interface, $kmol/m^3$	0.00098	0.00010
E_i, the instantaneous reaction enhancement factor	148	8827
Ha number	43.3	106.2
$5 \times Ha$	215.5	531
Is $5 \times Ha < E_i$?	No	Yes
$Ha/5$	8.7	21.2
Is $E_i < Ha/5$?	No	No
Regime	Fast second order	Pseudo first order
Guess E	37.6	105.5
Calculate E from the previous expression	37.6	105.5
Gas side resistance, $1/k_G$ (s/m)	19	19
Liquid side resistance, $1/k_L^0 E$ (s/m)	543	193
Total resistance to mass transfer, $1/K_G$ (s/m)	562	212
Estimate for K_G (m/s)	0.0018	0.0047
% of resistance in the liquid film	97	91
% of resistance in the gas film	3	9

We notice that in this case a pseudo first order reaction regime may be assumed for the lean end. This mean that $E = Ha$. The E in the table has, however, been estimated as if there was a second order regime. There is nothing wrong with that, only more cumbersome. It was done on purpose to illustrate that if we approximate to pseudo first order regime, there will often be an observable difference, in this case 105.5 instead of 106.2. This is less than 1% different and illustrates that choosing to go with the pseudo first order regime would have been perfectly acceptable.

We also notice that at both ends of the column the resistance is mainly on the liquid side.

12.8.3 Natural Gas Problem Revisited with MDEA

In this example we shall revisit the example in Section 12.8.1 and exchange the amine from MEA to MDEA. The concentration will still be 5 $kmol/m^3$.

The column is operated at 35 bar, 40°C as before. The lean amine had a loading (α) of 0.20 mol/mol while the rich loading was 0.60 mol/mol. We also remember that $CO_2 + 1$ MDEA $\rightarrow$ adduct, that is the stoichiometric coefficient (ν) is equal to 1. Although the new liquid would lead to a somewhat changed liquid side mass transfer coefficient, the old value is kept since the change is not expected to be large enough to interfere significantly with the values obtained when analysing allocation of mass transfer resistances.

We will, however, obtain new values for diffusivity and gas solubilities as well the kinetic constant to be used for estimating Ha number and the enhancement factors.

- CO_2 diffusivity and solubility is obtained from Haimour and Sandall (1984)
- MDEA diffusivity is obtained from Snijder *et al.* (1993)
- The kinetic constant, k_2, is obtained from Haimour, Bidarian and Sandall (1987).

Some interpolation and extrapolation has been necessary in this process.
The obtained values, without further ado, are:

- Kinetic constant, $k_2 = 9.4\,m^3/kmol\cdot s$
- $H_{CO2,C} = 5020\,kPa\cdot m^3/kmol$ --$\gg$ $m_{CO2} = 1.93$ (dimensionless solubility)
- $D_{CO2} = 0.726 \times 10^{-9}\,m^2/s$
- $D_{MDEA} = 0.31 \times 10^{-9}\,m^2/s$.

In this case we will need to consider the transition between slow and fast reaction which means that:

$$E = \frac{Ha}{\tanh{(Ha)}}$$

Inserting the numbers into the expressions given previously, it is obtained that:

	Rich end	Lean end
Concentration of free MDEA ($kmol/m^3$)	2	4
Concentration of CO_2 at interface ($kmol/m^3$)	0.056	0.014
E_i, the instantaneous reaction enhancement factor	16.3	123.5
Ha number	0.3	0.4
Is $Ha < 1$?	Yes	Yes
Regime	Slow-fast transition	Slow-fast transition
Calculate E from the expression earlier, Equation 12.101	1.0	1.1
Gas side resistance, $1/k_G$ (s/m)	244	244
Liquid side resistance, $1/k_L^0\,E$ (s/m)	4803	4670
Total resistance to mass transfer, $1/K_G$ (s/m)	5047	4913
Estimate for K_G (m/s)	0.0002	0.0002
% of resistance in the gas film	5	5
% of resistance in the liquid film	95	95

There is no appreciable acceleration of mass transfer in this case. It is easy to see why so much research has gone into attempts at finding good 'activators'.

References

Astarita, G. and Savage, D.W. (1980a) Theory of chemical desorption. *Chem. Eng. Sci.*, **35**, 649–656.

Astarita, G. and Savage, D.W. (1980b) Gas absorption and desorption with reversible instantaneous chemical reaction. *Chem. Eng. Sci.*, **35**, 1755–1764.

Astarita, G., Savage, D.W. and Bisio, A. (1983) *Gas Treating with Chemical Solvents*, John Wiley& Sons, Inc.

Bottoms, R.R. (1930) Process for separating acid gases. US Patent 1,783,901.

Bottoms, R.R. (1931) Organic bases for gas purification. *Ind. Eng. Chem.*, **23**, 501–504.

Brian, P.L.T., Hurley, J.F. and Hasseltine, E.H. (1961) Penetration theory for gas absorption accompanied by a second order chemical reaction. *AIChE J.*, **7**, 226–231.

Danckwerts, P.V. (1950) Unsteady-state diffusion or heat-conduction with moving boundary. *Trans. Faraday Soc.*, **46**, 701–712.

Danckwerts, P.V. (1970) *Gas-Liquid Reactions*, McGraw-Hill.

Haimour, N., Bidarian, A. and Sandall, O.C. (1987) Kinetics of the reaction between carbon dioxide and methyldiethanolamine. *Chem. Eng. Sci.*, **42**, 1393–1398.

Haimour, N. and Sandall, O.C. (1984) Absorption of carbon dioxide into aqueous methyldiethanolamine. *Chem. Eng. Sci.*, **39**, 1791–1796.

Hikita, H. and Asai, S. (1964) Gas absorption with (m,n)-th order irreversible chemical reaction. *Int. Chem. Eng.*, **4**, 332–340.

Hikita, H. and Asai, S. (1976) Gas absorption with a two-step chemical reaction. *Chem. Eng. J.*, **11**, 123–129.

Hikita, H., Asai, S. and Ishikawa, H. (1973) Lévêque model for mass transfer with an irreversible second-order chemical reaction. *Bull. Univ. Osaka Prefecture, Ser. A: Eng. Nat. Sci.*, **22** (1), 57–67.

Hikita, H., Asai, S. and Ishikawa, H. (1977) Simultaneous absorption of two gases which react between themselves in a liquid. *Ind. Eng. Chem. Fundam.*, **16**, 215–219.

Jou, F.-Y., Otto, F.D. and Mather, A.E. (1995) The solubility of CO_2 in a 30 mass percent monoethanolamine solution. *Can. J. Chem. Eng.*, **73**, 140–147.

Levenspiel, O. (1999) *Chemical Reaction Engineering*, 3rd edn, John Wiley & Sons, Inc.

Onda, K., Sada, E., Kobayashi, T. and Fujine, M. (1970a) Gas absorption accompanied by complex chemical reactions – I reversible chemical reactions. *Chem. Eng. Sci.*, **25**, 753–760.

Onda, K., Sada, E., Kobayashi, T. and Fujine, M. (1970b) Gas absorption accompanied by complex chemical reactions – II consecutive chemical reactions. *Chem. Eng. Sci.*, **25**, 761–768.

Onda, K., Sada, E., Kobayashi, T. and Fujine, M. (1970c) Gas absorption accompanied by complex chemical reactions – III parallel chemical reactions. *Chem. Eng. Sci.*, **25**, 1023–1028.

Pearson, J.R.A. (1963) Diffusion of one substance into a semi-infinite medium containing another with second-order reaction. *Appl. Sci. Res.*, **A11**, 321–328.

Perry, R.H. and Pigford, R.L. (1953) Kinetics of gas-liquid reactions. Simultaneous absorption and chemical reaction. *Ind. Eng. Chem.*, **45**, 1247–1253.

Sada, E., Ameno, T. and Kondo, M. (1976) Computational study of the simultaneous chemical absorption of two gases based on the penetration model. *J. Chem. Eng. Jpn.*, **9**, 409–410.

Sada, E. and Kumazawa, H. (1973) Gas absorption accompanied by a complex chemical reaction: variation of the enhancement factors with the orders of the reaction. *Chem. Eng. Sci.*, **28**, 1903–1905.

Sada, E., Kumazawa, H. and Butt, M.A. (1977a) Mass transfer with chemical reaction in both phases. *Can. J. Chem. Eng.*, **55**, 475–476.

Sada, E., Kumazawa, H. and Butt, M.A. (1977b) Gas absorption with complex chemical reaction: approximate analytical solutions. *Can. J. Chem. Eng.*, **55**, 623–625.

Snijder, E.D., te Riele, M.J.M., Versteeg, G.F. and van Swaaij, W.P.M. (1993) Diffusion coefficients of several aqueous alkanolamine solutions. *J. Chem. Eng. Data*, **38**, 475–480.

Ying, J., Eimer, D.A. and Yi, W. (2012) Measurements and correlation of physical solubility of carbon dioxide in (monoethanolamine + water) by a modified technique. *Ind. Eng. Chem. Res.*, **51**, 6958–6966.

Ying, J. and Eimer, D.A. (2012) Measurements and correlations of diffusivities of nitrous oxide and carbon dioxide in monoethanolamine + water by laminar jet. *Ind. Eng. Chem. Res.*, **51**, 16517–16524.

Ying, J. and Eimer, D.A. (2013) Determination and measurements of mass transfer kinetics of CO_2 in concentrated aqueous monoethanolamine solutions in a stirred cell. *Ind. Eng. Chem. Res.*, **52**, 2548–2599.

Further Reading

Cussler, E.L. (2009) *Diffusion*, 3rd edn, Cambridge University Press, Cambridge.

Richardson, J.F., Harker, J.H. and Backhurst, J.R. (2002) *Coulson and Richardson's Chemical Engineering*, 5th edn, vol. **2**, Butterworth-Heinemann.

13

Selective Absorption of H$_2$S

13.1 Background

H$_2$S poses a different challenge to gas treating than CO$_2$. It is a very poisonous gas with an obnoxious smell and paralyses the human olfactory function after a while, which means there is no natural warning of danger. Combusting it to let it out as SO$_2$ is not allowed because it would be harmful to the environment. H$_2$S cannot be caught and released as has been the practice with CO$_2$.

Smaller quantities of H$_2$S may be dealt with by using scavengers. For larger quantities, up to levels of 5–20 t/day, there are direct oxidation processes that may be used (Tennyson and Schaaf, 1977) (see Sections 2.3.1 and 2.3.3). Both these approaches are selective with respect to H$_2$S relative to CO$_2$. However, for large gas fields H$_2$S is converted to sulfur in so-called Claus plants. It is advantageous for these plants to have as little CO$_2$ as possible to dilute the H$_2$S. It is important to keep the partial pressure of H$_2$S as high as possible, not the least since Claus plants are atmospheric. This is to ensure the best possible conversion. Next, a higher gas volume means that the Claus plant will be physically bigger. Finally, there are challenges with respect to achieving a sufficiently high and stable process temperature if the gas is too dilute.

Claus plants have also an expensive tail gas treatment unit needed to deal with unconverted H$_2$S. The size and cost of this unit is very sensitive to non-H$_2$S in the feed to the Claus plant.

This is the backdrop that provides an incentive to make the H$_2$S absorption process as selective with respect to H$_2$S as possible.

The interest in, and the development of, selective absorption processes seemed to gather momentum in the 1970s at a time when many new fields containing both CO$_2$ and H$_2$S were developed and the cost of energy started to be significant. However, the desire for selectivity was not new as evidenced by older papers (Frazier and Kohl, 1950). They tested 50 amines before they settled on doing further experiments with methyldiethanolamine (MDEA). At that time the theoretical fundament to work from was not very developed.

Gas Treating: Absorption Theory and Practice, First Edition. Dag Eimer.
© 2014 John Wiley & Sons, Ltd. Published 2014 by John Wiley & Sons, Ltd.

Computers were not yet available for such work. Progress must have been more laborious to achieve with a lot of experimentation needed. This field still draws a lot of interest.

13.2   Theoretical Discussion of Rate Based Selectivity

Absorption with chemical reaction is discussed at length in Chapter 12. The key element exploited to make the process selective is that H_2S, being a Brønsted acid, reacts instantaneously with the alkaline alkanolamine usually used as an absorption reactant, while CO_2 has a finite reaction rate and is thus slower to be absorbed.

The selectivity can be further improved by using tertiary alkanolamines as absorbents as these cannot react directly with CO_2 and must use the slower reaction route via hydrolysis of CO_2. Some so-called sterically hindered amines with unstable carbamates show the same behaviour.

When looking at this from a theoretical point of view, the starting point is the equations for rate based mass transfer, but this time there are two compounds being absorbed in parallel. Hence an extra equation is needed. In the text that follows, the penetration theory is used as a basis. The reaction between H_2S and the amine is still instantaneous and takes place at a plane as before.

For CO_2 that undergoes a fast reaction the rate equation becomes:

$$\frac{\partial C_{CO_2}}{\partial t} = D_{CO_2}\frac{\partial^2 C_{CO_2}}{\partial z^2} - k_1 C_{CO_2} C_j \tag{13.1}$$

A boundary condition is that:

$$C_{CO_2} \to C_{CO_2}^{Lb} \text{ when } z \to \infty \tag{13.2}$$

Notice that there is no assumption of pseudo first order reaction in Equation 13.1. It will become clear below that this complication is necessary. Between the gas–liquid interface and the reaction plane the free amine concentration C_j is zero, but this is not the case on the 'bulk side' of the reaction plane, that is $z > \lambda$.

For H_2S, where the reaction is instantaneous, it is a little more complicated. Since the reaction is instantaneous, H_2S cannot co-exist with the absorbent reactant. Hence, there have to be two coupled equations as explained in Chapter 12:

$$\frac{\partial C_{H_2S}}{\partial t} = D_{H_2S}\frac{\partial^2 C_{H_2S}}{\partial z^2} \tag{12.20/13.3}$$

The boundary conditions are:

$$C_{H_2S} = C_{H_2S}^{Li}; \ z = 0 \ t > 0 \tag{12.21/13.4}$$

$$C_{H_2S} = C_{H_2S}^{Lb} = 0; \ z \geq \lambda \tag{12.22/13.5}$$

The mass balance for the absorbent reactant component 'j', the corresponding equations for the region $z > \lambda$, where 'j' can exist, are:

$$\frac{\partial C_j}{\partial t} = D_j\frac{\partial^2 C_j}{\partial z^2} - k_1 C_{CO_2} C_j \tag{13.6}$$

The boundary conditions are:

$$C_j = C_j^{Lb}; \ t = 0 \qquad (12.24/13.7)$$

$$C_j = 0; \ z \leq \lambda \qquad (12.25/13.8)$$

Equation 13.6 differs from Equation 12.23 in that there is now a reaction term due to CO_2 being available as a reactant where there is no H_2S. The Equations 12.20/13.3 and 13.6 are coupled as already explained in Chapter 12 for Equations 12.20 and 12.23/13.6.

What is new here compared to the situations discussed in Chapter 12? The key point is that there are now two gases being absorbed and they compete for the attention of the absorption chemical reactant. Furthermore, there is CO_2 available to react with the amine on the bulk side of the reaction plane. Whether this reaction will influence the absorption of H_2S or not depends on its rate. If it slow enough to be considered pseudo first order, there should be negligible effect. On the other hand, a fast reaction would deplete the amine concentration, and this should lead to the reaction plane moving marginally away from the gas–liquid interface and slow down the rate of H_2S absorption.

It is here easier to proceed by reference to the film theory. In the region $0 < z < \lambda$ there can be no free absorbent chemical reactant as the H_2S would 'devour' it on the spot. Hence both CO_2 and H_2S must propagate by diffusion only. In Equation 13.6 the value of C_j would simply become zero for $z < \lambda$, and the situation for H_2S is as described by Equation 13.3 with its boundary conditions. This is illustrated in Figure 13.1 where the reaction

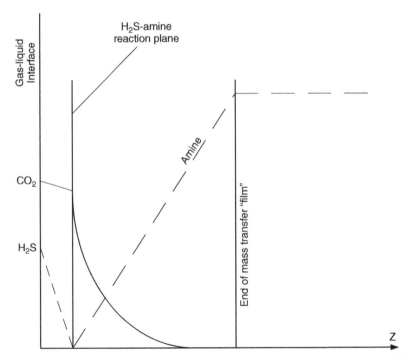

Figure 13.1 *Concentration profiles in the mass transfer zone for co-absorption of CO_2 and H_2S in the presence of, for example an alkanolamine based on profiles expected from film theory.*

plane for H_2S and amine is at a position $z = \lambda$. All chemical reaction between H_2S and the amine takes place at this plane. For $z < \lambda$, there is no amine for CO_2 to react with, and hence the concentration profile is linear as for physical absorption. However, once past the reaction plane, there is amine available for CO_2 to react with, and there is then an abrupt downturn in its concentration profile. The concentration profile for the amine is roughly the same as for instantaneous reaction since we assume that CO_2 would not particularly deplete the solution of amine towards the interface, that is we assume a pseudo first order reaction for the CO_2 profile to the right of $z = \lambda$. The full solution would need to consider equilibria as well since CO_2 would start to 'kick' H_2S out of the solution given enough time and sufficiently high absorption level. Figure 13.3 shows an equilibrium relationship for the two gases in aqueous diethanolamine (DEA) solution. Clearly in this figure, there is a shortage of DEA to chemically bind both CO_2 and H_2S in the higher loading region.

13.3   What Fundamental Information is Available in the Literature?

13.3.1   Equilibrium Data

Absorption equilibria have been measured for both CO_2 and H_2S and for their mixtures for a number of absorbents. The data provided here represent only a small selection to illustrate a few points. Figure 13.2 shows that CO_2 tends to have a higher equilibrium pressure than H_2S at the same loading when only one of them is present at a time, but the difference is

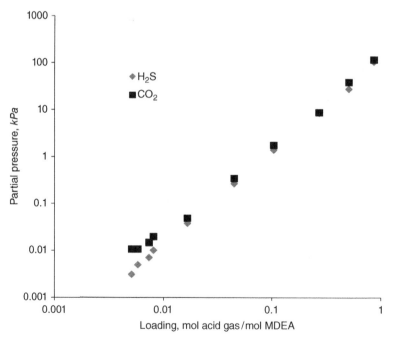

Figure 13.2   *Absorption equilibria between H_2S, CO_2 and 4.28 M aqueous MDEA at 40°C. Based on data from Jou, Mather and Otto (1982). (Note: These are data for CO_2 and H_2S one by one.)*

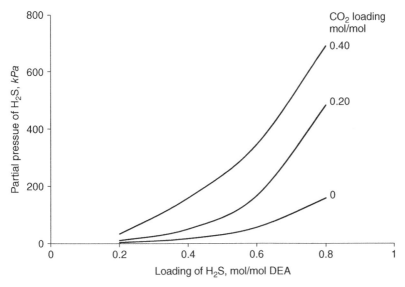

Figure 13.3 *Simultaneous absorption equilibria of CO_2 and H_2S over aqueous DEA. Based on data from Kent and Eisenberg (1975).*

not very great and becomes smaller at higher loadings. The difference is too small to be useful for selective absorption. In many cases there is more CO_2 than H_2S in the gas to be treated which means that equilibrium absorption would be dominated by CO_2.

The data shown in Figure 13.2 do not account for the relative acid strengths since the measurements were done with only one gas present at a time. There are data available measured for gas mixtures. Figure 13.3 represents such measurements and shows that these gases compete for the alkaline chemicals in the absorbent.

Acid strengths for CO_2 and H_2S are shown as functions of temperature in Figure 4.1. It is seen that the CO_2 has a lower pK_a value than H_2S and is thus a stronger acid.

13.3.2   Rate and Selectivity Research Data

There is a general consensus that modelling selective H_2S absorption should be done using rate based models. Bullin and Polasek (1982) used TSWEET with special options. Weiland and co-workers used a self-made model, and in the 1990s they developed a program called Pro-Treat that is rate based. They also present a number of case studies, but these were only to a small extent compared to real plant data (Chakravarty, Phukan and Weiland, 1985; Weiland, 1996; Weiland, Sivasubramanian and Dingman, 2003). ASPEN Ratefrac (a rate based column model) was used by Bolhàr-Nordenkampf *et al.* (2004). These authors were also able to compare their predictions with industrial columns. Better predictions were made with Ratefrac than the equilibrium based Radfrac column model (the standard ASPEN option for columns) but the industrial performance was underestimated. Ratefrac was also used by Rochelle and co-workers (Jassim and Rochelle, 2006) but their attention was only focused on CO_2 absorption. MacKenzie *et al.* (1987) points out that most amine based gas treatment absorbers have 20 trays or an equivalent packed height. In view of the uncertainty of data and the risk adversity of field owners, it will be difficult to convince them that columns should be shorter to achieve selectivity. On this basis it is advocated that columns should be fitted with extra liquid entry nozzles in a few positions down the column. If these are blinded and not piped in, they need not be prohibitively expensive. Such an approach is also discussed by Weiland and Dingman (1995).

Experimental studies have been reported by Haimour and Sandall (1983, 1987a,b) and Al-Ghawas and Sandall (1991) using laminar jet and stirred cell, Saha, Bandyopadhyay, Saju and Biswas (1993), Mandal, Biswas and Bandyopadhyay (2004) and Mandal and Bandyopadhyay (2005) using wetted wall columns, and by Xu, Zhang and Zheng (2002), Lu, Zheng and He (2006), and Godini and Mowla (2008) using packed columns. There is also a study by Bendall, Aiken and Mandas (1983) where spray nozzles were used. The results certainly support the notion that residence times should be kept low to achieve selective absorption of H_2S. This has taken the forms of varying liquid hold-up and varying the gas rates. Increased amine concentration leads to more CO_2 being absorbed. The CO_2 absorption was also faster during drop formation. It was observed that the selectivity in favour of H_2S was improved once the (50 μm) drops were formed and became stagnant inside. Xu, Zhang and Zheng also investigated the effect of a few non-aqueous solutions. In general better selectivity was achieved when no or little water was present. In view of the polar nature of the reactions involved in CO_2 absorption and the need for hydration, this is as might be expected. Amines studied included MDEA, aminomethyl propanol (AMP), DEA and monoethanolamine (MEA). Even MEA can provide some selectivity, but in general it is the tertiary amines that do not form carbamates that show the best potential for selective absorption. A so-called sterically hindered amine like AMP behaves like a tertiary amine in this respect. Cadours, Roquet and Perdu (2007) have measured absorption rates of both H_2S and CO_2 into preloaded solutions of aqueous DEA. There is obvious competition between these gases for being absorbed to the extent that one is desorbed when the other is absorbed. Huffmaster (1997) have analysed removal of H_2S down to 4 ppm, and they point out that the lean absorbent must have an equilibrium H_2S pressure no more than a quarter of the H_2S partial pressure in the treated gas if such a gas specification is to be achieved. They also point out that Sulfolan, when added, will help in the removal of the organic sulfur components. This is in accordance with the discussion in Chapter 4.

The work by Bendall, Aiken and Mandas points out that more CO_2 relative to H_2S was absorbed while the drops were formed in the atomisation step rather than later when the drops had become internally stagnant. This observation is natural in view of the concentration profiles of H_2S and CO_2 illustrated in Figure 13.1.

Tests run in a real gas treating plant are commented on next.

13.4   Process Options and Industrial Practice

Any of the commercial alkanolamines could provide some kinetic selectivity towards H_2S, but it was aqueous MDEA solutions that became the backbone in the development of kinetically selective absorption processes. A tertiary amine like TEA could be expected to match MDEA and has a longer history in gas treating, but it does not seem to figure in the development that took place from the 1970s onwards. By 1990 the best part of this development had been done. Focus turned to amine mixtures in order to tune the processes to absorb also CO_2 to the level necessary for further use of the gas.

Industrial practice in this area seems to have settled on MDEA with or without another alkanolamine to tailor the solution to absorb the desired mix of H_2S and CO_2. Claus technology has, after all, been developed and can now be built to operate stably at lower H_2S concentrations than earlier.

Ammons and Sitton (1981) from Phillips Petroleum report data from a real MDEA based gas treater placed in Waveland, Mississippi. This unit treated $93\,000\,Sm^3/h$ of gas having up to 4.8% CO_2 and 40–60 ppm of H_2S. The CO_2 slippage was normally 50–60% of that in the gas feed. The operating pressure was 65 bar. The absorber had 20 trays and lean amine could be fed to the top tray and every two trays down to the middle. Running tests within the constraints of the plant operations they observed that:

- CO_2 slippage increased when the MDEA concentration increased.
- CO_2 slippage decreased when the temperature increased.
- CO_2 slippage increased when the lean amine feed point was lowered, that is, when the column was shortened.
- CO_2 slippage decreased when the absorbent to CO_2 feed ratio increased.

These observations are logical in that the viscosity increases and the diffusivity coefficient decreases, and thus the liquid side mass transfer coefficient is expected to decrease when the MDEA concentration is increased. An increased concentration of MDEA would also be expected to push the reaction plane towards the interface and thus make the H_2S concentration slope steeper and therefore speed up its absorption. Furthermore, CO_2 will react faster and lead to a higher enhancement factor when the temperature goes up. A shorter column will logically imply less absorption, but for H_2S this had negligible effect. The column should be just high enough to catch the H_2S required to be removed and preferably let CO_2 slip through. Finally more absorbent would be expected to pick up more CO_2.

There is, however, a conflict between statements made concerning the effect of the concentration of amine on the mass transfer rate of CO_2. It is stated that increased concentration can cause both an increase and a decrease in mass transfer. The first could be expected because the pseudo first order reaction constant becomes higher when the concentration is increased. The viscosity of the solution will, however, increase leading to a lower value of the CO_2 diffusion coefficient. This point will not be argued further here but the former effect was clearly shown in experiments with CO_2 only (Jassim *et al.*, 2007).

There is also a report of a gas treating plant in Wyoming where generic MDEA was used successfully for selective H_2S removal from the gas (Harbison and Handwerk, 1987). Selective H_2S removal is the subject of presentations regularly at the annual Laurance Reid Gas Conditioning Conference. A proprietary solvent, UCARSOL 111, was run with 60–78% CO_2 slip in tests (Thomas, 1988). Slippage is, however, very dependent on the gas' content of CO_2 and H_2S. Little CO_2 and a lot of H_2S means that CO_2 slippage is next to impossible. In 1991 Taylor, Hugill, van Kessel and Verburg (1991) described selective removal of H_2S in a complex plant in Emmen in the Netherlands using Shell's Sulfinol M process which uses MDEA. This process was also used successfully in a solvent replacement case in Canada (Grant, Sourisseau and Weiss, 2007). Better selectivity was achieved in one case by implementing interstage cooling in the absorber which is equivalent to lowering the temperature (Robertson *et al.*, 2004). This curbs the CO_2 absorption rate the most due to reduced temperature and hence reaction rate. More examples may be found.

13.5   Key Design Points

A selective absorption process is more demanding to design technically than the more conventional cycles. The normal safety margins in dimensioning the equipment, particularly column height, must be avoided because they are detrimental to the object of the process design. This means that the absorption column must be no higher than absolutely necessary and the absorbent circulation rate must also be limited. Removal of these safety margins means that there is a greater need to fully understand the chemical engineering design of the process, and also to have good data available. There is obviously a need for good simulation tools to do process analysis if such designs are to be successful. The use of such techniques have been discussed by Weiland, Sivasubramanian and Dingman (2003) who used their ProTreat simulator to analyse possibilities for letting CO_2 slip. Vickery, Adams and Wright (1992) described the use of ChemShare's GasPlant software for the same purpose.

13.6   Process Intensification

Rotating packed beds were discussed in Chapter 9 but it is prudent to draw special attention to possibilities of process intensification in the context of rate based selective absorption of H_2S. Rotating packed beds represent the ultimate in short exposure times. They also provide very high liquid side mass transfer coefficients and can have very high specific contact areas. These features must be used wisely. Tests run in the late 1980s do support the notion that process intensification can lead to better selectivity. Recent work in China on this subject done in a pilot plant certainly gives clear evidence to this line of thought (Qian, Li and Guo 2012). Their results showed that co-absorbed CO_2 was reduced from 80 to 10% by going from a conventional column to a rotating packed bed. The absorbent was MDEA.

13.7   Numerical Example

In this example we look again at the natural gas treating example last visited in Sections 12.8.1 and 12.8.3. The amine to be studied here is MDEA. This time around it is considered with H_2S added to the feed. The analysis will only be done at the rich end point of the column. All data from Sections 12.8.1 and 12.8.3 are the same but the rich loading has been increased to 0.7 mol/mol. For this analysis it does not matter if it is loaded with H_2S or CO_2 as the estimates only need to know the concentration of free amine available for reaction in the mass transfer film. Pressure was 35 bar and the temperature was 40°C.

   To keep the estimates simple and suitable for a compact development the CO_2 absorption enhancement is estimated as if H_2S was not involved. That is to say, the amine zone nearest to the interface is neglected. The real mass transfer coefficient estimated for CO_2 in the end will thus be overestimated. Hence, the selectivity with respect to H_2S will be higher than suggested here.

The solubility of H_2S in the solution is estimated based on the ratio of solubilities between CO_2 and H_2S being the same in water and the solution. In this case H_2S is 3.2 times more soluble. A similar exercise has been made to estimate the diffusivity but here the ratio is around 1.08. Henry's law is used to estimate concentration of the gases in the solution. Data for diffusivity is from Haimour and Sandall (1984, 1988) while solubilities may be found in Table 5.2.

For CO_2 we will need to estimate the Hatta number. This will also be estimated by a lower bound value of the kinetics constant given by Haimour and Sandall (1987a).

The Hatta number is adopted from Equation 12.94: $Ha = \dfrac{\sqrt{D_{CO_2} k_2 [MEA]_{free}}}{k_L^0}$

We also repeat from Equation 12.40 that: $E_i = 1 + \dfrac{C_j^b}{\upsilon C_i^{Li}} \dfrac{D_j}{D_i}$

And in the transition between slow and fast reaction: $E = \dfrac{Ha}{\tanh(Ha)}$

Summary of estimates:

	CO_2	H_2S
MDEA concentration, nominal (kmol/m³)	5	–
Acid gas loading	0.70	–
Stoichiometric coefficient	1	1
Concentration of free MEA (kmol/m³)	1.5	1.5
Mole fraction of component in the gas	0.08	0.02
Partial pressure (bar)	2.8	0.7
Solubility (kPa m³/kmol)	5020	1569
Dimensionless solubility	1.93	0.60
Diffusion coefficients (m²/s)	0.726×10^{-9}	0.673×10^{-9}
Concentration of component in liquid at interface (kmol/m³)	0.056	0.045
E_i	12.5	16.5
Ha	0.3	3.5×10^{10}
E based on slow-fast transition and instantaneous	1.0	16.5
Resistance in liquid phase (s/m)	4838	94
Resistance in gas phase (s/m)	244	244
Total mass transfer resistance (s/m)	5082	338
% resistance in liquid	95	28
% resistance in gas	5	72
Overall mass transfer coefficient, K_G (m/s)	0.0002	0.0030

It may be observed that the ratio of mass transfer coefficient in favour of H_2S is:

$$0.0030/0.0002 = 15$$

On the other hand the ratio of driving forces assuming no appreciable back pressure from the liquid is:

$$0.02/0.08 = 0.25$$

The real selectivity may then be stated as: $15 \times 0.25 = 4$.

As stated initially, the selectivity is actually better because of the non-existence of free MDEA at and near the gas-liquid interface. This deliberation is, however, just a taster with respect to the intricacies involved in estimating this fully. The message is that although a full analysis borders on being realistic, it is possible to do a relatively quick estimate to get a feel for the problem. This is probably the level most engineers will get to.

References

Al-Ghawas, H.A. and Sandall, O.C. (1991) Simultaneous absorption of carbon dioxide, carbonyl sulphide and hydrogen sulphide in aqueous methyldiethanolamine. *Chem. Eng. Sci.*, **46**, 665–676.

Ammons, H.L., D.M. Sitton, (1981) Operating data from a commercial MDEA gas treater. Gas Conditioning Conference, Norman, OK.

Bendall, E., Aiken, R.C. and Mandas, F. (1983) Selective absorption of H_2S from larger quantities of CO_2 by absorption and reaction in fine sprays. *AIChE J.*, **29**, 66–72.

Bolhàr-Nordenkampf, M., Friedl, A., Koss, U. and Tork, T. (2004) Modelling selective H_2S absorption and desorption in an aqueous MDEA-solution using a rate-based non-equilibrium approach. *Chem. Eng. Proc.*, **43**, 701–715.

Bullin, J.A., J. Polasek, (1982) Selective absorption using amines. 61st GPA Annual Convention, Tulsa, OK.

Cadours, R., Roquet, D. and Perdu, G. (2007) Competitive absorption-desorption of acid gas into water-DEA solutions. *Ind. Eng. Chem. Res.*, **46**, 233–241.

Chakravarty, T., Phukan, U.K. and Weiland, R.H. (1985) Reaction of acid gases with mixtures of amines. *Chem. Eng. Progr.*, **81**, 32–36.

Frazier, H.D. and Kohl, A.L. (1950) Selective absorption of hydrogen sulphide from gas streams. *Ind. Eng. Chem.*, **42**, 2288–2292.

Godini, H.R. and Mowla, D. (2008) Selectivity study of H_2S and CO_2 absorption form gaseous mixtures by MEA in packed beds. *Chem. Eng. Res. Des.*, **86**, 401–409.

Grant, J., K. Sourisseau , M. Weiss, (2007) Conversion of Sulfinol-D to MDEA at the Shell Canada Burnt Timber facility. Laurance Reid Gas Conditioning Conference, Norman, OK.

Haimour, N. and Sandall, O.C. (1983) Selective removal of hydrogen sulphide from gases containing hydrogen sulphide and carbon dioxide using diethanolamine. *Sep. Sci. Technol.*, **18**, 1221–1249.

Haimour, N. and Sandall, O.C. (1984) Molecular diffusivity of hydrogen sulphide in water. *J. Chem. Eng. Data*, **29**, 20–22.

Haimour, N. and Sandall, O.C. (1987) Simultaneous absorption of H_2S and CO_2 into aqueous methyldiethanolamine. *Sep. Sci. Technol.*, **22**, 921–947.

Haimour, N. and Sandall, O.C. (1988) Absorption of H_2S into aqueous methyldiethanolamine. *Chem. Eng. Commun.*, **59**, 85–93.

Harbison, J.L., G.E. Handwerk, (1987) Selective removal of H_2S utilizing generic MDEA, Paper H. Laurance Reid Gas Conditioning Conference, Norman, OK.

Huffmaster, M.A. (1997) Stripping requirements for selective treating with Sulfinol and amine systems. Laurance Reid Gas Conditioning Conference, Norman, OK.

Jassim, M. and Rochelle, G. (2006) Innovative absorber/stripper configurations for CO_2 capture by aqueous monoethanolamine. *Ind. Eng. Chem. Res.*, **45**, 2465–2472.

Jassim, M., Rochelle, G., Eimer, D. and Ramshaw, C. (2007) Carbon dioxide absorption and desorption in aqueous monoethanolamine solutions in a rotating packed bed. *Ind. Eng. Chem. Res.*, **46**, 2823–2833.

Jou, F.-Y., Mather, A.E. and Otto, F.D. (1982) Solubility of H_2S and CO_2 in aqueous methyldiethanolamine. *Ind. Eng. Chem. Process. Des. Dev.*, **21**, 539–544.

Kent, R.L., B. Eisenberg, (1975) Equilibrium of H_2S and CO_2 with MEA and DEA solutions. Gas Conditioning Conference, Norman, OK.

Lu, J.-G., Zheng, Y.-F. and He, D.-L. (2006) Selective absorption of H_2S form gas mixtures into aqueous solutions of blended amines of Methyldiethanolamine and 2-tertiarybutylamino-2-ethoxyethanol in a packed column. *Sep. Purif. Technol.*, **52**, 209–217.

MacKenzie, D.H., Prambil, F.C., Daniels, C.A. and Bullin, J.A. (1987) Design and operation of a selective sweetening plant using MDEA. *Energy Prog.*, **7**, 31–36.

Mandal, B.P. and Bandyopadhyay, S.S. (2005) Simultaneous absorption of carbon dioxide and hydrogen sulphide into aqueous blends of 2-amino-2-methyl-1-propanol and diethanolamine. *Chem. Eng. Sci.*, **60**, 6438–6451.

Mandal, B.P., Biswas, A.K. and Bandyopadhyay, S.S. (2004) Selective absorption of H_2S from gas streams containing H_2S and CO_2 into aqueous solutions of N-methyldiethanolamine and 2-amino-2-methyl-1-propanol. *Sep. Purif. Technol.*, **35**, 191–202.

Qian, Z., Li, Z.-H. and Guo, K. (2012) Industrial applied and modelling research on selective H_2S removal using a rotating packed bed. *Ind. Eng. Chem. Res.*, **51**, 8108–8116.

Robertson, K., L. Stern, M. Tonjes, L. Dreitzler, D. Stevens, (2004) Increase H_2S/CO_2 selectivity with absorber interstage cooling. Laurance Reid Gas Conditioning Conference, Norman, OK.

Saha, A.K., Bandyopadhyay, S.S., Saju, P. and Biswas, A.K. (1993) Selective removal of H_2S form gases containing H_2S and CO_2 by absorption into aqueous solutions of 2-amino-2-methyl-1-propanol. *Ind. Eng. Chem. Res.*, **32**, 3051–3055.

Taylor, N.A., J.A. Hugill, M.M. van Kessel, R.P.J. Verburg, (1991) Sulfinol-M provides the solution to a tough treating challenge. Laurance Reid Gas Conditioning Conference, Norman, OK.

Tennyson, R.N. and Schaaf, R.P. (1977) Guidelines can help proper process for gas-treating plants. *Oil Gas J.*, **10**, 78–80 and 85–86.

Thomas, J.C. (1988) Improved selectivity achieved with UCARSOL INNOVATOR solvent 111. Laurance Reid Gas Conditioning Conference, Norman, OK.

Vickery, D.J., J.T. Adams, R.D. Wright, (1992) The effect of tower parameters on amine based gas sweetening systems. Laurance Reid Gas Conditioning Conference, Norman, OK.

Weiland, R. (1996) Mass-transfer rate-based simulation of amine treating – applying GRI/GPA research results to plant operations. Proc GPA Annual Convention, Denver, CO, March 11–13.

Weiland, R., M.S.J.C. Dingman, (1995) Effect of solvent blend formulation on selectivity in gas treating. Laurance Reid Gas Conditioning Conference, Norman, OK.

Weiland, R., M.S. Sivasubramanian , J.C. Dingman, (2003) Effective amine technology: controlling selectivity, increasing slip, and reducing sulphur. Proceeding of the 53rd Laurance Reid Gas Conditioning Conference, Norman, OK, Feburary 24.

Xu, H.-J., Zhang, C.-F. and Zheng, Z.-S. (2002) Selective H_2S removal by nonaqueous methyldiethanolamine solutions in an experimental apparatus. *Ind. Eng. Chem. Res.*, **41**, 2953–2956.

14

Gas Dehydration

14.1 Background

Natural gas is water contaminated when it comes out of the reservoir. As the gas pressure drops, the gas becomes cooler and water will precipitate so it must be knocked out. The dominant components of the natural gas, particularly methane, have the ability to form hydrates if the temperature and pressure allows hydrates to form. In the case of temperature it happens when the gas is chilled enough, and for pressure when it increases above a certain level. This is extensively dealt with in the literature as gas hydrates represent a speciality field of their own with extensive activity. For now we simply conclude that water removal is a must.

The field of water removal has been extensively reviewed (e.g. Kohl and Nielsen, 1997; Pearce and Sivalls, 1993; Wagner and Judd, 2006). There is also an interesting comparison made by Netusil and Ditl (2011). Generally the water must be removed to a level where the gas' water dew point is below the temperature seen as the lowest the gas will be exposed to. Within the process this is decided by the process engineer but there will also be an export gas specification. Gas is normally exported into a pipeline and the pipeline operator will have his own specification to ensure that water will not condense in the pipeline. In some places the ambient temperature might drop as low as −40°C or perhaps even further. Figure 14.1 is given to provide a feel for the problem. Part of this equation is that the gas temperature will drop when its pressure is reduced due to the Joule–Thompson effect. There is a pressure drop in a pipeline due to flow induced friction, and there could be pressure variations due to change in gas off-take from the pipeline.

Free water and a potential for hydrate formation in a pipeline spell trouble. At worst the whole pipeline might be blocked by a hydrate plug. This will normally require significant effort to remove with much downtime and lost production as a result. Typically the plug will eventually disintegrate when the pipeline pressure is reduced significantly. In this process there is a hazard associated with the loosened plug or bits thereof flying along the pipeline with the gas (Canadian Association of Petroleum Producers, 2007).

Gas Treating: Absorption Theory and Practice, First Edition. Dag Eimer.
© 2014 John Wiley & Sons, Ltd. Published 2014 by John Wiley & Sons, Ltd.

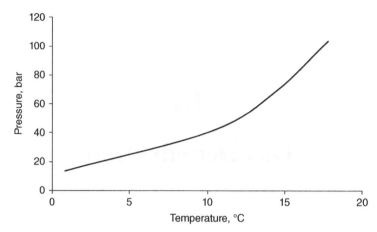

Figure 14.1 *Hydrate formation curve for 90% (mol) methane and 10% ethane based on estimates by the Katz technique (Campbell, 1976). Hydrates will form to the left of the curve.*

14.2   Dehydration Options

Calcium chloride is often mentioned as a desiccant. It is said to be applicable if water removal demand is not too big and the plant is on the low capacity side.

Adsorbents like silica gel, activated alumina and molecular sieve zeolites are all good desiccants. Very low dew points may be achieved. However, the adsorbed water is hard to desorb. Energy is needed to recycle the desiccant and the temperature needed to regenerate, for example the zeolites is in the order of 200°C and higher. There will also be a need to use stripping gas with the inconvenience of handling this. In general these adsorption processes are seen as expensive, and their use is limited to applications where very low dew points are needed. A typical flowsheet is shown in Figure 14.2. Here, some stripping gas is needed and this is rarely available on site. In the flowsheet a portion of the dry product gas is extracted for regeneration purposes as a stripping gas is needed. This stripping gas is at first heated and the hot gas is then used to heat up the offline adsorption column while also carrying released gas out of said column. The gas is subsequently cooled and the condensed gas separated and removed from the process. In a subsequent process step the stripping gas is not heated and in this period used to cool the adsorption column back to its process temperature. Also in this period water will be carried out and condensed. During both these periods the gas is recompressed and recycled back to the process upstream of the adsorption unit. A typical cycle configuration could be 8 h on adsorption, 5 h off line being heated and 3 h offline being cooled. Similar flowsheets would be needed if other desiccants were used but the temperature and cycle times would be different. The cycle could be made shorter to reduce adsorber volume but most likely more parallel units would be needed. For big units more than two units might be preferred anyway.

Liquid desiccants have come to mean glycols. This field is today dominated by triethylene glycol (TEG) although monoethylene glycol (MEG) diethylene glycol (DEG) now and again are advocated as more cost effective. TEG has become the preferred tool because it is more resistant to thermal degradation and partially as an indirect result of this it allows

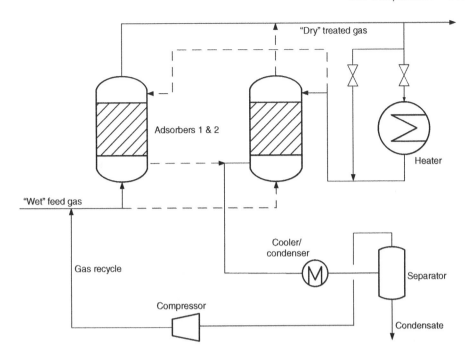

Figure 14.2 *Typical flowsheet of an adsorption dehydration unit.*

higher dew point depressions for the gas to be treated because it may be regenerated at higher temperatures. MEG and DEG are stilled used for injections into pipelines where the cost of buying the glycol make-up is much more important. TEG is more expensive than DEG and MEG with the latter normally being the cheaper. Glycol prices do shift due to market forces.

Dew point depression is a much used term in dehydration. It is simply expressed as the number of degrees centigrade (or Fahrenheit) by which the dew point is lowered by the treatment. This approach is due to the apparent simplicity of dehydration plants, and developed to give quick designs. It has served the industry well for a long time.

The rest of this chapter will be confined to discussing the liquid desiccants.

14.3 Glycol Based Processes

Water removal by the glycol based absorption–desorption process is the work horse in the gas treating industry. A typical flowsheet is shown in Figure 14.3. The gas is always scrubbed before it enters the glycol column (absorber). The term 'scrub' is often used in the gas industry. It simply means that any droplets and particles carried by the gas shall be removed, usually by settling in an empty column shell. Someone once said that if you can't afford the scrubber, you can't afford the water removal plant. That describes the importance of this unit.

The absorption column is where the water is removed from the gas. It may be based on trays or packings. The choice is by and large circumstantial and is based on local economics

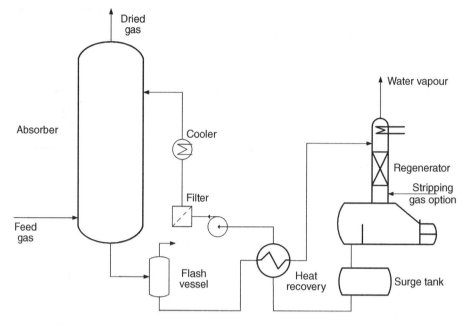

Figure 14.3   *The main features of a glycol dehydration plant flowsheet.*

and tradition. Tradition should not be knocked as it represents the operator's experience from past units. Smooth operation is the main target for the operator as there is more money to be made from this than a marginal saving on investment cost. Tray columns used to be the norm but are now challenged by packed columns. Kean Turner and Price (1991a,b) reported on extensive and successful tests made by ARCO where they investigated the use of structured packings in the water absorber. Such packed columns can be designed with a lower diameter and seems to be preferred for large offshore plants where the associated weight reduction is important.

The flow of glycol is very low in the context of column liquid load. For tray columns this is not difficult to handle, especially with bubble cap trays. A packed column, however, will need a fairly advanced liquid distributor. It would typically involve small diameter drip tubes that could easily be fouled by particulate matter or waxy material in the liquid. These tubes could be as close as 10 cm apart even for very large columns. Even very big plants could circulate as little as 1–3 m³/h of TEG.

There are rules of thumb for specifying the number of trays in the absorber. A simple unit, that would require about one theoretical stage, would be built with four trays given 25% tray efficiency. Kean, Turner and Price (1991a,b) state that six to eight trays are common for such simple separations. Øi (2003) recommends 50% Murphree tray efficiency used when accurate equilibrium data are available. The availability of such data is an issue. For lower dew points 10 trays may be chosen (Campbell, 1976). Rigorous calculations are done when a closer look is wanted. There is similar guidance to be found for packings. Kean, Turner and Price (1991a,b) report on height equivalent to a theoretical plates (HETPs) for structured packings as 3–6 ft (1–2 m). They also claim a 50% weight reduction compared

Table 14.1 Relative volatilities, $(y_{H_2O}/x_{H_2O})/(y_{glycol}/x_{glycol})$, of water and glycols at a pressure of 600 mm Hg (Dow, 1992).

Glycol	Water's relative volatility	Molecular weight of the glycol
MEG	15	62.1
DEG	51	106.1
TEG	81	150.2

to tray columns based on both reduced height and diameter. The diameter in particular may be reduced by 30%. This is significant in an offshore context where it is often preferred to use structured packings as the column diameter can be shrunk. The effect of this is important when the pressure is high and the weight of the unit takes on importance. For big fields it may be a question of feasibility of a one train approach due to a combination of capacity and pressure and their influence on shell thickness. Shell thickness is proportional to both pressure and diameter. Quickie estimates based on *Chemical Engineers' Handbook* (Perry, Chilton and Kirkpatrick, 1963) indicate that a 3.5 m diameter column at pressures of 50–100 bar would have wall thicknesses from 75 to 160 mm. If the column shell was 10 m long the shell (excluding dished ends) would weigh 60–120 tons. Salamat (2012) suggests that the packed section should normally be around 5 m.

Some natural gas is co-absorbed, and a flash unit is added to allow this to be recovered for fuel gas purposes. The residence time for liquid in this flash vessel should be no less than 10–12 minutes if light gas is processed, and 20 minutes for heavy gases (Salamat, 2012). The solution is then pre-heated before going to the still (alias regenerator) (see Figure 14.3). The still is mostly a small column that ensures a minimum loss of glycol with the water vapour leaving. The relative volatilities of water and glycol are very high, see Table 14.1. Removal of water from the glycol is mainly accomplished in the so-called regenerator; that is, essentially a boiler or evaporator. The solution runs into the regenerator from the still. From the regenerator the vapour goes back up the still while the regenerated liquid flows to a surge tank below. The preheating of water rich glycol is typically done by this liquid being routed through the surge tank in a subsequent spiral pipe for heat recovery. Due to the low quantity circulating there is a fairly small demand for heat exchange surface. Hence, this seemingly inefficient heat exchanger design is adequate. Such a unit may be operated with or without stripping gas. The stripping gas would usually be natural gas that could otherwise have been sold and after such use as likely as not will need to be flared unless there is a use for it as fuel gas. Operators would prefer to operate without stripping gas but the situation may be forced by the water dew point specification.

The regenerator alias still may be run with a little stripping gas to make a water leaner glycol to enable lower dew points to be achieved for the gas. Rich glycol is typically 95% (by weight) glycol, that is 5% water. (This corresponds to a water mole fraction of 0.30 in TEG). Lean glycol must have a water content low enough to allow the gas to be dried to specification. This is discussed by Piemonte, Maschietti and Gironi (2012) who state that lean glycol concentration should be 'well above 99.0%' for modern applications. Actually 99.95% is often quoted for achieving very low dew points. An appropriate height for the still is 2.5 m (8 ft) unless the reboiler duty is small (Salamat, 2012).

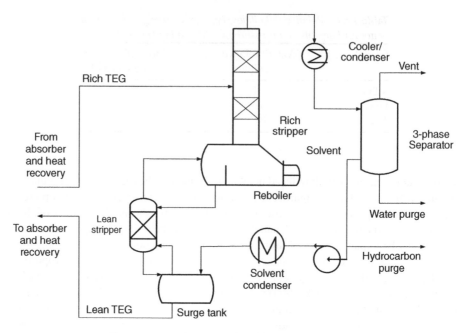

Figure 14.4   *The principle of the DRIZO process.*

The regenerated glycol is pumped from the surge tank back to the absorber. This pump is usually a special design that recovers energy from the water rich glycol being expanded and uses this energy to drive the pumping of the water lean glycol. Since the rich glycol stream is roughly 5% bigger by weight than the lean glycol, there is enough recovered energy to drive the pump. It is often referred to as a balanced pump. Its energy efficiency is accordingly better than 95%. Care must be taken in the design and operation of this unit. It avoids an electricity installation and consumption and it is commonly used.

The glycol will pick up contamination from the gas, and filters are normally installed in the process. Contamination and glycol management are discussed next.

There are varieties on the flowsheet described. The most well-known deviation is the so-called DRIZO process (Smith, 1997) that may lower the dew point to −95°C or even −100°C. It also deals with some heavier hydrocarbons that would otherwise be a contamination problem to the standard process. In this process a hydrocarbon solvent is used in a closed loop in the regeneration section of the process as shown Figure 14.4. This solvent acts as a stripping gas in the rich and lean regenerator stills and the process can in this way achieve TEG purities as high as 99.999 wt%. In turn, gas dew points as low as −100°C can be achieved. Another feature of the DRIZO process is its ability to handle heavy hydrocarbons, particularly aromatics, without these being allowed to accumulate in the TEG. As seen in the figure, it is possible to extract a purge stream from the solvent loop in the regeneration section.

There is also a variation known as the 'Cold Finger' process (Piemonte, Maschietti and Gironi 2012; Rahimpour, Jokar, Feyzi and Asghari, 2013; Reid, 1971) that can also lower the dew point further than the standard process. The special feature is the Coldfinger

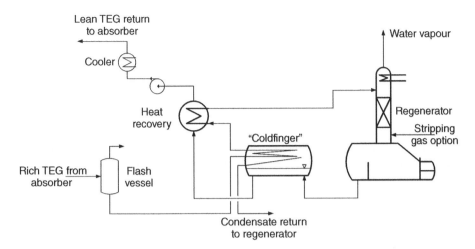

Figure 14.5 *The Coldfinger process concept. Absorber omitted.*

apparatus. Its position in the process is shown in Figure 14.5. The hot lean TEG from the reboiler (or evaporator as the case might be) is passed into a vessel as shown. This hot liquid will partially evaporate, and the vapour will be much richer in water than in TEG. There is a cooling tube, or tubes, in the top of the vessel that will partially condense this vapour, and the condensate thus produced will essentially be water with some TEG in it. This condensate is suitable for being returned to the regenerator where it will lead to more water leaving over the top. In this way the lean TEG will be made leaner before being returned to the absorber. The leaner TEG will enable the achievement of a lower water dew point in the treated gas. How much water will be removed in the Coldfinger? There is no heat added in this vessel and there is mainly water vapour to make up the pressure. The water vapour will approach the equilibrium with the solution flowing through. Obviously the quantity of water evaporated will be determined by the evaporation-caused cooling but there are other limits that may be influenced by the shape and design of the Coldfinger vessel like depth of liquid, liquid surface area and hold-up. There is no absolute answer to the question of how much extra water is removed.

All fields producing gas will need a dehydration unit and there are many small variations on this process going around. The differences are not significant unless there is a particular challenge that needs to be solved for the application of interest.

14.4 Contaminants and Countermeasures

Inevitably contaminants will appear in the glycol solution. There is salt carryover from upstream units as prior scrubbing is not 100% effective. When the gas is treated for acid gas, this is done upstream of the water removal unit. This provides protection for the glycol unit but absorbent carryover becomes a new consideration.

Solution filters are installed, but serious salt contamination must be specially dealt with. An option is ion exchange. Salt build-up will lead to precipitation in the regenerator where water content is at its lowest. The glycol itself is nowhere as good a solvent for salts as water.

It is common that a glycol solution becomes dark as in black. If this is seen along with a fine precipitate, it may be due to corrosion problems with FeS or Fe_3O_4 as products. Black solutions are also caused by heavy hydrocarbons absorbed from the feed gas accumulating. Thermal degradation or oxidation may also lead to a black solution. The thermal degradation products have a burnt sugary smell associated with them.

Surfactants represent another problem, and they may lead to foaming. These may come from upstream use of chemicals, example given in the CO_2/H_2S removal plant.

Hydrocarbons will also be absorbed in the solution and will accumulate if left to their own devises. An active carbon filter on the glycol circulation circuit to deal with this is necessary. That would also deal with any compressor oils appearing. The importance of such filtration of the glycol absorbent cannot be overstressed (Salamat, 2012).

A thorough discussion of alternative solutions to these problems will not be addressed here. The aim of this introduction is merely to create awareness of these possible process problems.

14.5   Operational Problems

Glycol units do not need to be troublesome to operate. There is a need to maintain the solution, but otherwise there are few problems. Examples of trouble are described in the literature, see example given the book by Lieberman (1987) and the conference paper by Rueter, Beitler, Sivalls and Evans (1997).

There will be glycol losses from the operation. Benchmarking such losses is important. Salamat (2012) suggests that total losses should not be much larger than 3.3 kg TEG per million Sm^3 gas treated. (0.025 US gal per MMSCF). The losses overhead is around $1/2$ of these.

14.6   TEG Equilibrium Data

Equilibrium data in the form of dew point charts are provided by glycol vendors like Dow (1992). These charts are based on available data, but they are not documented in the academic sense. Published equilibrium data for water in TEG were reviewed by Parrish, Won and Baltatu (1986) along with publication of new data measured. They concluded that there was a significant spread between sources and chose to exclude some of the data for their new dew point charts. The 'Parrish' dew point charts have since been adopted by the Gas Processors Association (GPA) and is published in the Gas Processors Supplier Association (GPSA) *Engineering Data Book*, 12th edition. (GPSA, 2004). Bestani and Shing (1989) extended the measurements of Parrish, Won and Baltatu, to a wider range of conditions. These data were further reviewed by Clinton, Hubbard and Shah, in 2008 in an in-depth work where they reviewed all data published on the water-TEG equilibrium as well as discussing the calculation of dew points in the gas phase where pressure must be accounted for. A different kind of approach was taken by Twu, Tassone, Sim and Watanasiri (2005) who used to an equation of state to model the TEG-water system with a view to use it

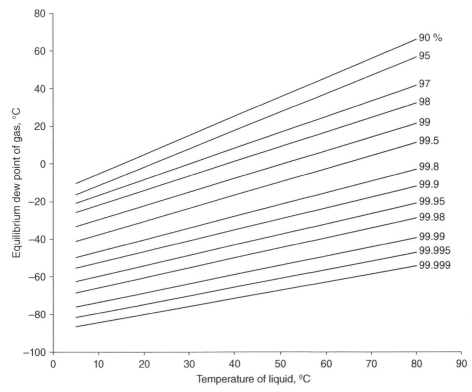

Figure 14.6 *Dew point chart based on the regression model of Bahadori and Vuthaluru (2009).*

for this type of process design. It appears to be a good way forward, but a good bit of programming is necessary to exploit these correlations that are shown to fit literature data very well. Bahadori and Vuthaluru (2009) have made a representation of the same data with good results using a system of multiple third order polynomials. Unlike Twu, Tassone, Sim and Watanasiri, they did not attempt to deal with gas phase non-idealities, but they were able to reproduce the standard dew point chart as widely published. Their version matches the data from Twu, Tassone, Sim and Watanasiri. The Bahadori and Vuthaluru, correlations have been used to create the dew point chart in Figure 14.6.

A variety of techniques have been used over the years to determine absorption equilibria for water over glycols. There is no such thing as a right or wrong technique, given that the user knows what he is doing, but there are a number of experimental difficulties that need to be overcome. Equilibrium must, of course, be achieved and that may be difficult enough. However, when the water dew point is low say −10°C, there is little water present in the vapour phase. Sampling then becomes a challenge as the surrounding air is likely to have a much higher water content. It will not take much contamination of a sample to seriously distort the result. Measuring the water content also becomes a challenge.

In view of the difficulties obviously involved in establishing equilibria, it would be good to have a quantified treatment of the uncertainties associated with these data. Such information is not available. However, the uncertainties are taken into account through conservatism in process design where recommendations and experience figures are widely applied. Long history proves that this approach works but it is open to challenge. Having said that, the work done by Clinton, Hubbard and Shah (2008) goes a long way towards resolving issues on the equilibrium data front. However, data for water in MEG or DEG have received little attention with in-house data published by the glycol vendors as the standard source of information. Since these glycols are mainly used for less precise work as hydrate inhibitors, this is not quite as critical.

14.7   Hydrate Inhibition in Pipelines

Glycols and sometimes methanol are injected into pipelines to inhibit hydrate formation. See Figure 14.1. They will simply dissolve water condensate such that there is no free water to form hydrates. Hydrate formation in pipelines has, for example been discussed by Dendy Sloan (1993).

The use of inhibitors has allowed untreated gas to be piped over long distances. In off-shore gas production this is important as it represents one enabling factor to develop fields with sub-sea installations and direct piping to shore for treatment.

There are even big schemes with very long pipelines where the liquid is recovered at the end and piped back in a separate pipeline.

In the context of pipeline schemes the price of the desiccant is more important than its ability to achieve high dew point depressions. For this reason MEG and DEG are usually preferred over TEG in such applications.

14.8   Determination of Water

Quality control of the glycol solution has been discussed by Pearce and Sivalls (1993) and will not be repeated here.

Determination of water in the gas and the liquid is an interesting subject. For analysing the liquid for water there is no real alternative to the Karl–Fischer (KF) titration in its original form or the KF coulometer derivation. They are both commercially available.

There are a number of instruments available to measure the water content of the gas. Sampling should preferably be done by direct piping to the instrument. If low contents are to be measured, a significant time must be allowed to enable the system to reach its steady state. It must be remembered that even metal surfaces in pipes will adsorb some water from the gas above the surface. It takes a long time to remove this if the surface has been previously exposed to high water concentrations. Measuring principles used include dew formation at cooled mirrors, adsorption that influences the adsorbents electrical conductivity or capacitance. Co-adsorption or condensation of hydrocarbons may lead to false dew points. The pitfalls are many.

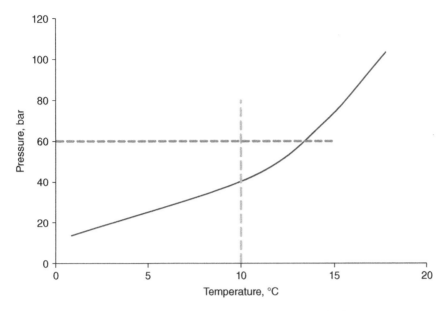

Figure 14.7 *Figure 14.1 with conditions marked.*

14.9 Example Problems

14.9.1 Example 1: Check for Hydrate Potential

If the gas pressure is 60 bar and the gas is subjected to temperatures below 10°C, is there a danger of hydrate formation if water condenses?

Using Figure 14.1 we find that the point we have been asked to investigate lies to the left of the hydrate curve, and the conclusion is therefore that hydrate formation will be possible (Figure 14.7).

14.9.2 Example 2: TEG and Water Balance

The gas stream coming from the acid gas treatment outlined in Section 11.7.1 (again a development from Section 8.7.3) needs to be dehydrated. It is saturated with water at 40°C. The gas pressure is 35 bar. What is its water content before and after dehydration if we assume ideal gas behaviour? How pure will the lean TEG need to be to bring its dew point down to −10°C? How big a flow of TEG is needed? Assume the cooling facilities for lean TEG allows cooling to 40°C.

The saturation pressure of water at 40°C is 0.07375 bar (see any steam table). The mole fraction of water in the feed is accordingly 0.07375 bar / 35 bar = 0.002107. In the treated gas the water dew point is specified to be −10°C, and its vapour pressure will then be 0.00260 bar. This means the water mole fraction in treated gas is 0.00260 bar / 35 bar = 0.000075 (or 75 ppm if you like).

Using Figure 14.6, we find the water equilibrium loadings, or rather the weight percent TEG in the absorbent.

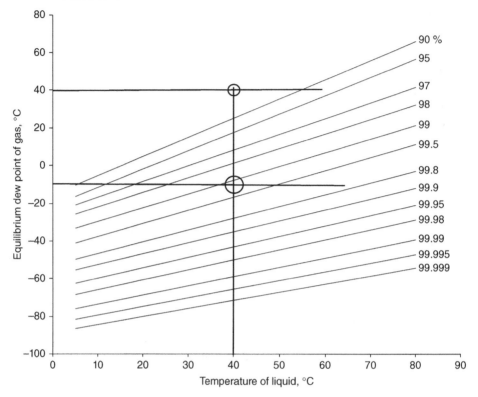

We see from the lower point indicated that the lean TEG must be between 99 and 99.5% (wt).

From the water rich end, upper point, we see that the rich TEG could be less than 90% (wt). However, we would probably choose to let the rich TEG be 95% (wt), that is 5% (wt) water.

The amount of gas to be treated is 1494 kmol/h (we do not adjust for acid gas removal). From the other example we also have that $G_w = 10.38$ kg/s.

The amount of water to be removed is:

$$(1494 \text{ kmol/h}) (0.002107 - 0.000075) = 3.0358 \text{ kmol/h} = 54.64 \text{ kg/h}$$

Let the TEG concentration swing from 99.5 to 95% (wt). An approximate value for the flow of TEG may be found from:

$$L_w (0.995 - 0.95) = 54.64 \text{ kg/h} \text{ -- >> } L_w = 1214 \text{ kg/h} = 0.337 \text{ kg/s}$$

The density of this liquid is a little above that of water (it is about 1100 kg/m³). Formally the volumetric flow is accordingly (1214 kg/h) / (1100 kg/m³) = 1.1 m³/h. We choose to set this TEG flow to 1.3 m³/h.

This is a small volumetric flow.

14.9.3   Example 3: Tower Diameter

What will be the diameter of the tower outlined in Example 2?

The diameter of this tower will be the same as that estimated for Section 8.7.3. The much lower flow of glycol compared with the amine solution to remove CO_2 in the referred example will of course reduce the 'flow parameter' but the ensuing 'capacity parameter' is not sensitive to this in the relevant part of the diagram. The new flow parameter will be:

$$\frac{0.337}{10.38} \sqrt{\frac{33.64}{1100}} = 0.0057$$

An inspection of the generalised pressure drop chart in Figure 8.2 tells us that we are off the chart and that is due to the extremely low liquid flow rate. Using random packings for such a tower is not widely applied. Most such towers are probably using trays, but structured packings are also used, especially where large gas quantities are treated and it is attractive to reduce the column diameter. Any packed column for this application faces a challenge with respect to liquid distribution. A 'high tech' distributor with closely placed drip points will need to be used.

14.9.4   Example 4: Mass Transfer Resistances

What can we say about mass transfer resistances for the tower discussed in Sections 14.9.2 and 14.9.3?

Although we would not use a random packed column for this application, we can still use the mass transfer coefficients derived in Section 11.7.3 to evaluate the relative sizes of the mass transfer resistances in a glycol tower.

From the dew point chart in Figure 14.6 and information of water vapour pressures over sub-cooled water (Perry and Chilton, 1983), it is possible to derive the dimensionless Henry's coefficient needed for this purpose. An estimate is 0.0016.

The mass transfer coefficients derived in Section 11.7.3 were:

- $k_G = 0.0043$ m/s
- $k_L^0 = 0.00039$ m/s (since we concluded to use the estimate for CO_2 reasoning that the water value was based on using the correlation way out of range).

$$\text{The gas side resistance is } \frac{1}{k_G} = \frac{1}{0.0043} = 233$$

$$\text{The liquid side resistance is } \frac{m}{k_L^0} = \frac{0.0016}{0.00039} = 4.1$$

The total resistance is 237, and the overall gas side mass transfer coefficient:

$$K_G = 0.0042 \text{ m/s}$$

In this case the gas side part of the resistance is $(100)(233)/(237) = 98\%$ which is reasonable in view of the high solubility of water in TEG.

References

Bahadori, A. and Vuthaluru, H.B. (2009) Rapid estimation of equilibrium dew point of natural gas in TEG dehydration systems. *J. Nat. Gas Sci. Eng.*, **1**, 68–71.

Bestani, B. and Shing, K.S. (1989) Infinite dilution activity coefficients of water in TEG, PEG, glycerol and their mixtures in the temperature range 50 to 140°C. *Fluid Phase Equilib.*, **50**, 209–221.

Campbell, J.M. (1976) *Gas Conditioning and Processing*, Campbell Petroleum Series, vol. **2**, John M. Campbell and Company, Norman, OK.

Canadian Association of Petroleum Producers (2007) Guide. Prevention and Safe Handling of Hydrates, www.capp.ca (accessed 26 March 2014).

Clinton, P., R.A. Hubbard , H. Shah, (2008) A review of TEG-water equilibrium data and its effect on the design of glycol dehydration units. Proceedings of Laurance Reid Gas Conditioning Conference, Norman, OK.

Dendy Sloan, E. (1993) A kinetics means of hydrate pluggage prevention. Proceedings of Laurance Reid Gas Conditioning Conference, Norman, OK.

Dow (1992) *A Guide to Glycols*, Dow Chemical Co., Midland, MI.

Dow (2007) *Triehtylene Glycol*, Dow Chemical Co, Midland, MI.

GPSA (Gas Processors Supplier Association) (2012) *GPSA Engineering Data Book*, 13th edn, GPSA.

Kean, J.A., Turner, H.M., Price, B.C. (1991a) Structured packing in triethylene glycol dehydration service. Proceedings of Laurance Reid Gas Conditioning Conference, Norman, OK (Note: Names of packing revealed in this reference, but not in 1991b).

Kean, J.A., Turner, H.M. and Price, B.C. (1991b) How packing works in dehydrators. *Hydrocarbon Process.*, **70**, 47–52.

Kohl, A. and Nielsen, R. (1997) *Gas Purification*, 5th edn, Gulf Publishing, Company.

Lieberman, N.P. (1987) *Troubleshooting Natural Gas Processing*, PennWell Books, Tulsa, OK.

Netusil, M. and Ditl, P. (2011) Comparison of three methods for natural gas dehydration. *J. Nat. Gas Chem.*, **20**, 471–476.

Øi, L.E. (2003) Estimation of tray efficiency in dehydration absorbers. *Chem. Eng. Process. Process Intensif.*, **42**, 867–878.

Parrish, W.R., K.W. Won , M.E. Baltatu , (1986) Phase behaviour of the TEG-water system and dehydration / regeneration design for extremely low dew point requirements. Proceedings of the Annual Convention of the GPA, San Antonio, TX, pp. 202–210.

Pearce, R.L., C.R. Sivalls, (1993) Fundamentals Manual, 2nd edn, "Fundamentals of gas dehydration design and operation of glycol solutions". Proceedings of Laurance Reid Gas Conditioning Conference, Norman, OK.

Perry, R.H. and Chilton, C.H. (1983) *Chemical Engineers' Handbook*, 6th edn, McGraw-Hill.

Perry, R.H., Chilton, C.H. and Kirkpatrick, S.D. (1963) *Chemical Engineers' Handbook*, 4th edn, McGraw-Hill.

Piemonte, V., Maschietti, M. and Gironi, F. (2012) A Triethylene glycol-water system: a study of the TEG regeneration processes in natural gas dehydration plants. *Energy Sources, Part A*, **34**, 456–464.

Rahimpour, M.R., Jokar, S.M., Feyzi, P. and Asghari, R. (2013) Investigating the performance of dehydration unit with Coldfinger technology in gas processing plant. *J. Nat. Gas Sci. Eng.*, **12**, 1–12.

Reid, L.S. (1971) Apparatus for dehydrating organic liquids. US Patent 3,589,984.

Rueter, C., C. Beitler, C.R. Sivalls, J.M. Evans, (1997) Design and operation of glycol dehydrators and condensers. Laurance Reid Gas Conditioning Conference, Norman, OK, March 2–5.

Salamat, R. (2012) Gas dehydration offshore or onshore, how, how much and design tips. SPE 154134, mtg QATAR, May 14–16.

Smith, R.S. (1997) Enhancement of DRIZO dehydration. Laurance Reid Gas Conditioning Conference, Norman, OK, USA, March 2–5.

Twu, C.H., Tassone, V., Sim, W.D. and Watanasiri, S. (2005) Advanced equation of state method for modelling TEG-water for glycol gas dehydration. *Fluid Phase Equilib.*, **228–229**, 213–221.

Wagner, R. and Judd, B. (2006) Fundamentals – gas sweetening. Laurance Reid Gas Conditioning Conference, Norman, OK, February 26–March 01.

15

Experimental Techniques

15.1 Introduction

The way absorption columns have been developed and designed has changed profoundly since the early days of gas treating. Since the pioneering days of chemical solvent theory in the 1950s (or even perhaps late 1940s) there has been a trend towards a more theoretically based estimation of absorption with chemical reaction. Although empiricism and use of past experience is by no means dead and still valuable, the trend is for a theoretically rate based approach in design. This necessitates the provision of fundamental data. This chapter will summarise the equipment that may be used to measure reaction rates and solubilities needed for their interpretation and use. Physical data is not discussed here, nor is the plethora of other apparatus for measurement of mass transfer rates.

15.2 Experimental Design

The problem associated with measurements of reaction rates in a complex system may seem to be on the complicated side but good planning of experiments goes a long way toward making it easy. There are many publications in the past describing how this may be done. Summaries are given by Danckwerts (1970) and Astarita, Savage and Bisio (1983). Commonly used equipment is discussed next one-by-one highlighting key points but not giving full reviews.

A key aspect of such experiments is to ensure that the reaction conditions allow the system to be mathematically treated as pseudo first order reaction rather than second order (or seemingly of fractional order). In the latter case the mathematics becomes much more complicated and numerical solutions will most likely be needed. The more complicated the interpretation of data, the higher the uncertainty in the derived value. When fitting parameters in regression analysis, it is important that there is sensitivity between parameters and the object function. It is also an advantage to simplify to avoid unnecessary distortion of parameters due to interference when regressing data. Pseudo first order means that the

Gas Treating: Absorption Theory and Practice, First Edition. Dag Eimer.
© 2014 John Wiley & Sons, Ltd. Published 2014 by John Wiley & Sons, Ltd.

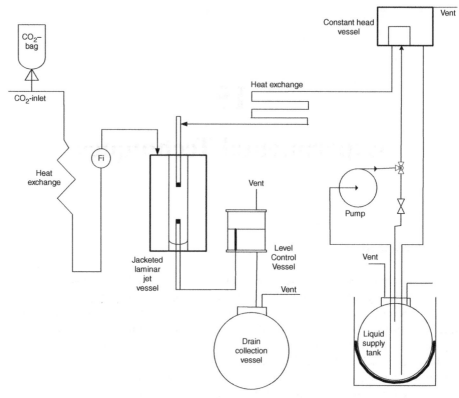

Figure 15.1 *Typical flowsheet for a laminar jet rig. This flowsheet is also used for single sphere and wetted wall rigs.*

absorbent concentration can be eliminated as a variable and this helps with respect to determining the kinetic constant.

It goes without saying that the temperature must be controlled. To resolve this, the equipment may be jacketed with a thermal fluid (most likely water) circulating through, or the equipment may be submerged in a temperature controlled bath. Feed streams should also be tempered. This is basic experimental technique and there will be no further discussion of temperature control here although it can be challenging enough.

The overall flowsheets of the experimental apparatus are very similar for laminar jet, wetted wall and single sphere. The stirred cell is different and is more like the equilibrium cells used. Typical flowsheets are shown in Figures 15.1 and 15.2.

As always there are operational problems associated with experimental research. The present equipment is no exception. The devil lies in the attention to detail. Unfortunately these details are often not discussed when people publish and a potential new user is well advised to make a thorough literature search before starting. Even so, difficult and essential details are hard to come by. When it comes to write-up, there is a tendency to forget the troubles that seem trivial in retrospect. And the reader's ability to reproduce the work is given less priority than it used to.

On the subject of details, most of the equipment types discussed here are dependent on being leak tight as they depend on measurements of either pressure changes or volumetric

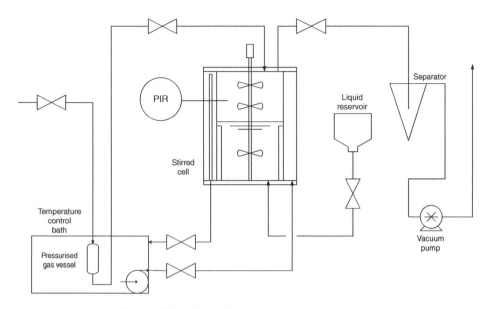

Figure 15.2   *Typical flowsheet for a stirred cell rig.*

measurement of a small gas make-up stream. This kind of measurement is based on a mass balance approach and any leakage would spoil this. Trouble free operation demands that due attention is paid to all joints and seals when engineering and mounting the equipment. This challenge is often underestimated.

A further point when wanting to close a mass balance is interference from other components. Water is a particularly relevant example as it is significantly volatile to influence the composition of the gas phase.

When the equipment is finally commissioned, the first trial runs should always be attempts at trying to reproduce trusted data. This ought to be repeated for every new person intending to use the instrument. Once this is accomplished, the experimenter is ready to explore new territory.

Interfacial resistance was once a debated issue. However, studies at the universities in Delaware, USA and Cambridge, UK concluded that gas–liquid equilibrium prevailed at the surface if the liquid was relatively clean and the surface freshly formed (Cullen and Davidson, 1956; Scriven and Pigford, 1958). If surface active agents are present, this picture may change. A discussion may be found in the book by Sherwood, Pigford and Wilke (1975). It is not expected that surface resistance is a problem in the experiments discussed here.

15.3   Laminar Jet

15.3.1   Background

The laminar jet apparatus was devised to provide a means of contacting liquid and gas with very short exposure times. For fast reactions this is useful in that pseudo first order reaction may be assumed even at atmospheric pressure.

The earliest reference found for this apparatus seems to be Matsuyama (1950, 1953) in Kyoto, Japan, but these references seem to be available in Japanese only. Laminar jet work was also done at the University of Delaware, Newark for the PhD dissertations of Eipper (1955) and Manogue (1957) as referenced by Scriven and Pigford (1958, 1959) also based at the University of Delaware. No papers have been found from Eipper or Manogue. Further early work on laminar jets was done in Cambridge (Cullen and Davidson, 1957).

15.3.2   Principle and Experimental Layout

The term laminar jet alludes to a liquid jet that runs through a gas space from one point to another thus creating a well-defined cylindrical interface. The jet is, as the name suggests, laminar such that the mathematics of diffusion associated with the use of the penetration theory fully describes the transport in the liquid phase. The jet's surface is smooth and the jet is supposed to be rod-like without tapering, such that there is a constant velocity profile both radially and axially.

The jet runs vertically downwards. It is formed at a nozzle that may be a hole in a thin plate, see Figure 15.3.

Scriven and Pigford (1958, 1959) studied both fluid dynamics of the laminar jet and investigated the hypothesis of surface resistance or the lack of it. They made the nozzle out of a 3.5 mm thick stainless steel plate where the hole was turned such that the upper half represented a quadrant of an ellipse while the lower half was straight and had a diameter

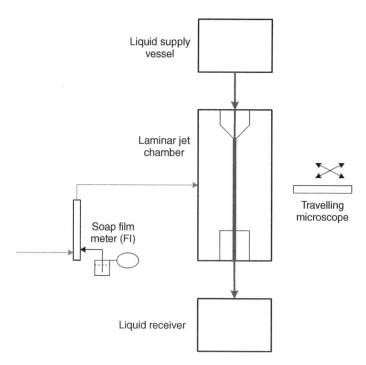

Figure 15.3   *The principles of a laminar jet apparatus.*

of 1.5 mm. Another 1950s work on the laminar jet was done by Raimondi and Toor (1959) in Pittsburgh where surface resistance was also studied. Sharma and Danckwerts (1963) present another variation using a square edged orifice where they ground down a glass capillary until there was hardly anything left. They obtained satisfactory results this way while Scriven and Pigford and Raimondi and Toor discarded this approach. Clarke (1964) investigated alternative nozzles. He concluded that a hole cut in a thin sheet was the best. A bell shaped nozzle where the liquid flowed through some channel of length was suspected of giving a retarded outer liquid layer, and it was observed that the jet contracted. Haimour and Sandall (1984a) at the University of Santa Barbara more recently used laminar jets with a square edged hole 0.51 mm diameter made from a 0.08 mm thick stainless steel plate. Even more recently, at the University of Regina in Canada Aboudheir *et al.* (2004) studied the use of nozzles with round holes. It is only known that they used the same dimensions as Haimour and Sandall. No detailed information about the Regina hole shape is given.

Typical dimensions are a plate thickness of 0.1 mm or less and a hole diameter around 0.5 mm. The hole can be made in a number of ways. In the literature 'cutting', 'drilling' and 'turning' has been referred to (Clarke, 1964; Haimour and Sandall, 1984a,b; Scriven and Pigford, 1958). Such discs can also be bought with holes made by laser from Lenox Laser (Glen Arm, MD, USA) as done by Ying and Eimer (2012a). It is customary to make provisions for ability to alter the jet length, 1–15 cm are mentioned in the literature. Typical liquid exposure times are in the region of 0.001–0.030 seconds.

The liquid is supplied to the nozzle typically from a constant level overflow vessel that keeps the liquid pressure constant to ensure a steady flow as shown in Figure 15.1.

The liquid plunges into a circular channel slightly bigger in diameter than the jet but no bigger than that the jet and immediately adheres to the channel walls such that no gas is pulled down with it. Clarke (1964) mentions 20–30% bigger diameter than the nozzle hole as suitable. This is in line with statements from other sources. The liquid then runs out to a receiver via a level control arrangement. Liquid is collected periodically to accurately determine the liquid flow rate. Some versions of the laminar jet are fitted with adjustment screws to allow the jet to be aligned with the reception hole but it is also possible to make and operate the unit without this feature.

The gas space around the jet is enclosed. It is filled with pure gas of the type to be investigated. As the gas is absorbed and possibly reacts with the absorbent, fresh gas is supplied via a flow meter. The gas reservoir is kept at atmospheric pressure. There are options for how to do this.

The liquid to be investigated is prepared up front, contained in a reservoir vessel from where it is pumped to the overflow vessel at a rate a little greater than that used for the jet. The excess liquid flows back to the reservoir. For simple operation the liquid vessels are, like the gas reservoir, pressure controlled by communication with the atmosphere. Typically 15–20 l of solution are prepared. Preparation includes stripping the liquid of gas by use of vacuum and possibly heat (Figure 15.4).

15.3.3 Mathematics and Practicalities

The mathematics are straightforward once the operating conditions are ensured to keep the experiment within the confines of the pseudo first order reaction regime when there is chemical reaction to be quantified. If there is only physical absorption to study, diffusion

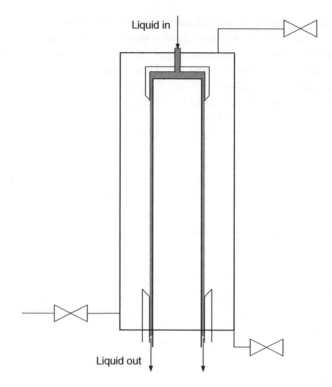

Figure 15.4   *The principles of a wetted wall apparatus.*

coefficients may be determined. The mathematics is accounted for by Equations 10.9a and 12.17 for physical and chemical absorption, respectively.

15.3.4   Past Users

Some consider it to be the most flexible of the laboratory rigs for kinetic measurements. There have been many users of a laminar jet apparatus in the past. The group at Cambridge, UK with Davidson and Danckwerts is one of the better known. This group did extensive work on absorption with chemical reaction in the period 1950–1980 with Danckwerts being one of the leading people in this field. The group at University of Delaware was very active in the same period with Scriven and Pigford producing some often referenced papers. Later, Astarita completed his PhD at Delaware in 1960, and he was later associated with both this university and the University in Napoli in Italy. His research spans the late 1950s to roughly 1990 but the involvement with the laminar jet belongs to the early years of this period. Orville Sandall's group at the University of California Santa Barbara is another notable effort. Their work took place in the period 1980–2000. More recently a laminar jet apparatus was built at University of Regina (Aboudheir *et al.*, 2004), and very recently at Telemark University College (Ying and Eimer, 2012a). These groups are specially mentioned because they have published widely in the field of gas treating. There are many more groups that have used a laminar jet over the years but these are the ones that have published a lot in the gas treating field. There is also the 1984 work by Babu, Narsimhan and Phillips (1984) for studies of SO_2 absorption.

15.4   Wetted Wall

15.4.1   Background

The wetted wall apparatus was devised to provide a means of contacting liquid and gas with controlled exposure times. A distinction may be made between long and short wetted wall columns. Short ones are used if the desire is to use the measurements for estimating mass transfer coefficients and deduce reaction kinetics and diffusion coefficients. This chapter discusses short columns. Long columns tend to experience a rippled liquid surface, a feature that may define what a long wetted wall column is.

It is difficult to say when this apparatus was first used, but it is at least described by Chilton and Colburn (1934), by Chambers and Sherwood (1937), and for CO_2 related work by van Krevelen and van Hooren (1948). Several designs of wetted wall columns have seen the light over the years. Danckwerts gives a good review in his book (1970). Liquid may flow downwards both inside and outside a tube, assuming a pipe is used. Outside flow is to be preferred since it makes the desirable observation of the liquid film easier. Without this observation it is not really possible to be absolutely certain that the wall is properly wetted and the surface ripple free. The biggest difference between designs is probably how the liquid is introduced. A distinction between short and long wetted walls is necessary. Long wetted walls will have rippling of the liquid surface that thus becomes undefined. The present discussion is around short wetted wall columns where the surface is well defined, at least intended to be, and the data can be interpreted in terms of diffusion theory. Lynn, Straatemeier and Kramers (1955a,b) did work with both types. They describe wetted wall columns of 12–22 cm length as 'long'. In their work with long wetted wall columns they used a surface active agent to prevent ripples. The results could still be explained by the penetration theory while assuming no surface mass transfer resistance.

The term wetted wall alludes to the liquid running over a wall wetted by the liquid. It is placed in a gas space where the wetted wall creates a well-defined interface. The liquid film running over the wetted wall is laminar such that the mathematics of diffusion as used in the penetration theory fully describes the transport in the liquid phase. The liquid surface is ideally smooth (Danckwerts, 1970) but work has been done in the past where it is not, in which case the data interpretation becomes different. Surface rippling is particularly a problem with long wetted wall columns and for high liquid flow rates. For the kind of work implied in this chapter, rippling should be avoided. There are entry and exit effects for the liquid that should be addressed but there are many publications where this seems to be ignored. The entry effects are associated with how the liquid is introduced, a point at which the liquid may be at a different velocity than its terminal velocity achieved as it flows downwards. Danckwerts (1970) states that this error may be addressed by good design and quotes an error of 1% accomplished in one case. More serious are exit effects that are caused by build-up of a rigid film arising from accumulation of surface active materials where the liquid exits. Rippling of the surface is another error and this enhances mass transfer to the extent that the data may not be interpreted in the usual way with the use of the penetration theory. Surface active additives have sometimes been used to help create a smooth surface flow (Lynn, Straatemeier and Kramers, 1955b; Nysing and Kramers, 1958).

The liquid flow, onto, on and off the wetted wall has been in focus in a number of past works. Roberts and Danckwerts (1962) used grooves for removing the liquid. That is to say that the liquid affected by rigid film build-up at the exit only affect the much smaller surface in the grooves rather than the full the circumference of the wetted wall pipe. The grooves reduce the contact area for the problem part of the surface. By this they managed to eliminate the exit problems.

The liquid is introduced via a nozzle over the top of the wetted wall. It is common that the liquid is fed on the inside of the tube and cascades over the top edge. Just empty tubes have also been used. According to Danckwerts (1970) controlling inlet flow problems would, however, dictate another solution. Typical liquid exposure times are in the region of $0.1-1$ second, and typical interfacial areas are $10-100\,cm^2$ (Astarita, Savage and Bisio, 1983). The length of a short wetted wall has been as low as 1 cm and could be as long as $10-12\,cm$. Using longer walls would give ripples unless surfactants were used.

The overall arrangement of the wetted wall apparatus is very much like that of the laminar jet apparatus. The liquid is supplied to the top nozzle arrangement typically from a constant level overflow vessel that keeps the liquid pressure constant to ensure a steady flow. The collected liquid then runs out to a receiver via a level control arrangement. Liquid is collected periodically to accurately determine the liquid flow rate. See Figure 15.1.

The gas space around the wetted wall is enclosed to form a closed gas chamber. As the gas is absorbed and possibly reacts with the absorbent, fresh gas is supplied via a flow meter. The gas reservoir is kept at atmospheric pressure. There are options for how to do this. Gas flow-through may also be arranged.

The liquid to be investigated is prepared up front, contained in a reservoir vessel from where it is pumped to the overflow vessel at a rate a little greater than used to form the liquid film. The excess flows back to the reservoir. For simple operation the liquid vessels are, like the gas reservoir, pressure controlled by communication with the atmosphere. Typically $15-20\,l$ of solution are prepared. Preparation includes stripping the liquid of gas by use of vacuum and often heat as well.

15.4.2   Mathematics and Practicalities

The mathematics are straightforward once operating conditions are ensured to keep the experiment within the confines of the pseudo first order reaction regime when there is chemical reaction to be quantified. If there is only physical absorption to study, diffusion coefficients may be determined.

The mathematics to treat the data is the same as for the laminar jet but the liquid element's exposure time is much higher. Due to this, action must often be taken to reduce the partial pressure of the reacting gas. It could be achieved by vacuum, or nitrogen may be used to dilute the gas. In the latter case the gas side mass transfer resistance must probably be accounted for.

15.4.3   Past Users

In the field of gas treating, the group at Cambridge with Danckwerts made use of a wetted wall column (Danckwerts, 1955). There was a group using it at MIT in the 1960s at least

(Brian, Vivian and Matiatos, 1967). It is presently used by Rochelle's group at University of Texas at Austin (Pacheco, Kaganoi and Rochelle, 2000; Dang and Rochelle, 2003). Meng-Hui Li and co-workers in Taiwan is another wetted wall user with a number of publications on gas treating (Sun, Yong and Li, 2005; Yong and Li, 2009). There are several other users too, most of them in other fields than gas treating.

15.5 Single Sphere

15.5.1 Background

The single sphere apparatus has also been referred as the Wetted Sphere and One-Sphere. (There is also an apparatus with more than one sphere where the liquid is mixed to an extent between spheres, hence 'single' sphere here.) It was devised as an improvement on the wetted wall in the sense that it all but eliminates the entrance and exit effects experienced with a basic wetted wall column. The vertical section is also short and this is advantageous with respect to avoiding ripples on the surface. The single sphere provides a means of contacting liquid to gas with short exposure times. Contact times 0.1–1 second are typical (Astarita, Savage and Bisio, 1983). This is the same range as for a wetted wall column. This apparatus was first made and used by Lynn, Straatemeier and Kramers (1955c) in Delft, Netherlands and by Cullen and Davidson (1956; 1957) in Cambridge, UK.

15.5.2 Principle and Experimental Layout

The term single sphere alludes to the liquid running over a sphere placed in an enclosed gas space. The liquid runs from top to bottom thus creating a well-defined interface. The liquid film that flows over the single sphere is laminar such that the mathematics of diffusion as applied in the penetration theory fully describe the transport in the liquid phase. The liquid surface is smooth. Because of the spherical shape of the film, it is stretched and contracted as it flows down the sphere. The mathematics become a little more complex than for a laminar jet, but it is manageable (Wild and Potter, 1968).

The liquid is introduced via a nozzle over the top of the sphere. One option is to make this as a hole where a rod is placed in the centre to make an annular space where the liquid flows down and spreads itself around the sphere. This rod is then used to suspend the sphere. Versions of the sphere where the liquid is distributed without the top rod are also known. There is also a rod at the bottom where the liquid is collected and runs down this rod to a small diameter tube surrounding it. This rod is a possible alternative for holding the sphere in place. The liquid is supplied to the top nozzle arrangement typically from a constant level overflow vessel that keeps the liquid pressure constant to ensure a steady flow as shown in Figure 15.1 with further details shown in Figure 15.5.

The overall arrangement of the single sphere apparatus is very much like the laminar jet and wetted wall apparatuses. Here, the liquid runs over the sphere that typically has a diameter 35–50 mm, although Lynn, Straatemeier and Kramers (1955c) used spheres from 10 to 29.5 mm. The sphere material could be polished steel or Hastelloy, but Cullen and Davidson (1956) used table tennis balls (!) with a steel rod through (equipment does not need to be expensive). Liquid flows are in the order of 1 cm^3/s. The liquid collected at the

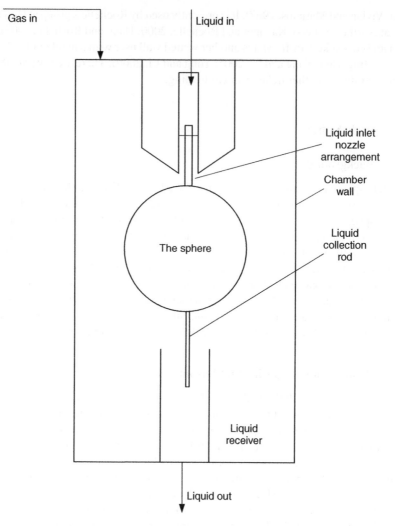

Figure 15.5   *The principles of a single sphere apparatus.*

bottom of the sphere then runs out to a receiver via a level control arrangement. Liquid is collected periodically to accurately determine the liquid flow rate.

The gas space around the single sphere is enclosed. As the gas is absorbed and possibly reacts with the absorbent, fresh gas is supplied via a flow meter. The gas reservoir is kept at atmospheric pressure. There are options for how to do this.

The liquid to be investigated is prepared up front, contained in a reservoir vessel from where it is pumped to the overflow vessel at a rate a little greater than needed for wetting the sphere. The excess flows back to the reservoir. For simple operation the liquid vessels are, like the gas reservoir, pressure controlled by communication with the atmosphere. It is convenient to prepare 15–20 l of solution. Parallel runs are made, and typically the flow will be varied within the batch.

15.5.3   Mathematics and Practicalities

The mathematics is a little more complicated than for the laminar jet and wetted wall experiments. However, it can be dealt with (Wild and Potter, 1968) and it is straightforward once operating conditions are set to ensure the experiment is kept within the confines of the pseudo first order reaction regime when there is chemical reaction to be quantified. If there is only physical absorption to study, diffusion coefficients may be determined.

15.5.4   Past Users

There have been many users of a single sphere laminar jet apparatus in the past. The group in Delft (Lynn, Straatemeier and Kramers, 1955c) has already been mentioned. Davidson and Danckwerts' group in Cambridge used it for CO_2 absorption measurements in the period from the late 1950s to around 1970. Orville Sandall's group at University of California Santa Barbara produced another notable effort. Their work took place in the period 1980–2000.

15.6   Stirred Cell

15.6.1   Background

The stirred cell has also been around for a long time and it has been used by many over the years. It is difficult to trace its origins in this field. Danckwerts' in his book on gas–liquid reactions (1970) does not discuss the stirred cell in his experimental chapter. He and co-workers have, however, used it in the 1960s (Danckwerts and McNeil, 1967). Early work of stirred cells was different from the present use where it is arranged to enable assumption of a diffusion controlled liquid layer. The technique seems to have gathered popularity from around 1980 with the work of van Swaaij's group in Twente (Versteeg, Blauwhoff and van Swaaij, 1983). Experiments may be done with a much lower use of chemicals than the liquid flow-through equipment like laminar jet, wetted wall and single sphere. Less than half a litre is typically used. The general flowsheet is also simpler than for the other methods. See Figure 15.2.

15.6.2   Principle and Experimental Layout

A stirred cell in this context is simply a reactor with a stirrer that ensures that both the gas and the liquid phase are well stirred such that they may be assumed to be homogeneous. The variation shown in Figure 15.6 shows two stirrers in the gas and one in the liquid mounted on the same shaft. Another variety is separate drives for the gas and liquid side stirrers. Blades could any reasonable shape or number. The surface itself should be kept undisturbed such that there is mass transfer by diffusion only. A stirred cell is illustrated in Figure 15.6. To avoid vortex formation it is equipped with baffles. Stirrer speeds from 60 to 140 rpm have been reported in the literature. A satisfactory speed will depend on the shape of the stirrers themselves and must be experimentally established from case to case. Stirrers may be driven by shafts going through the ends of the cell or by coupling to external magnet drives.

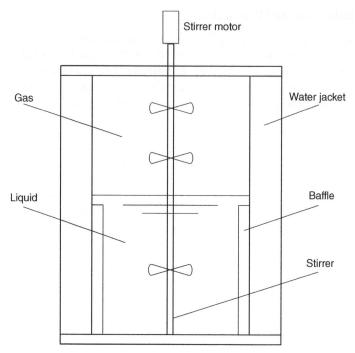

Figure 15.6 *The principle of a typical stirred cell. Separate axles and drivers may be used for the stirrers in the gas and liquid.*

There are a number of ways a stirred cell may be operated. It may be used with fluid(s) flowing through, but it is more common not to do this. The cell may be operated with constant pressure and measurement of gas make-up, or alternatively it may be closed and the rates determined by observing the fall in gas pressure.

Contact times are long, and the pressure of the gas may need to be reduced to enable operation under pseudo first order reaction conditions. The simplest way to do this is by operation under vacuum, but it may also be done by diluting the gas. If the gas is diluted, it will be necessary to do extra measurements to establish the gas phase mass transfer resistance.

15.6.3   Mathematics and Practicalities

Stirred cells in this context may be operated as batch for both phases, or it could be operated isobaric by adding gas as it is consumed, or there may be a flow through of gas. The mathematics in treating the observations is different in these cases. Contact times are by nature much, much longer here than for the previously described equipment.

Given batch operation with respect to both phases, a mass balance with respect to the transferred species may be set up for the gas phase:

$$\text{Accumulated} = \text{In} - \text{out}$$

There is no reaction in the gas phase such that this term is irrelevant, nor is there any transfer into the gas such that 'in = 0'. If the gas phase is indicated by G and the liquid phase by L, the balance becomes:

$$\frac{d}{dt}\left[\frac{V_G p_i}{RT}\right] = -k_L^0 EAC_i^{Li} \tag{15.1}$$

Here V is volume, p is pressure, R is the gas constant, T is temperature, k_L^0 is the liquid side mass transfer coefficient, A is interfacial area and C is the concentration of the physically dissolved component i in the liquid. Superscript ^{Li} on C refers to the liquid side of the interface as before. It is assumed that any back pressure of component 'i' from the solution may be neglected. Using Henry's law relation:

$$p_i = H_i C_i^{Li} \tag{15.2}$$

in Equation 15.1, this may be integrated to obtain:

$$k_L^0 E = \frac{V_G H_i}{ART}\frac{\ln p_2 - \ln p_1}{t_2 - t_1} \tag{15.3}$$

Everything on the right hand side is either known or measured. To get from the left hand side to reaction kinetics, relations in Chapters 10 and 12 must be used.

15.6.4   Past Users

The stirred cell was used as one of many methods used by the research groups at the University of Cambridge (Danckwerts and McNeil, 1967; Alper and Danckwerts, 1976; Laddha and Danckwerts, 1981). They covered a number of systems including ammonia, monoethanolamine (MEA), K_2CO_3 and arsenite as catalyst.

The Twente University group lead by van Swaaij and later Versteeg is perhaps the most extensive user of the stirred cell over a period from the early 1980s to the present (Versteeg, Blauwhoff and van Swaaij, 1983; Blauwhoff, Versteeg and van Swaaij, 1984).

There are others, but it is beyond the present scope to give a complete review of stirred cell usage in gas–liquid reactions.

15.7   Stopped Flow

15.7.1   Background

The stopped flow technique is an old established technique amongst chemists. However, it did not appear on the scene of CO_2 absorption into amines until the early 1980s (Barth *et al.*, 1981). This technique offers alternative possibilities and seems to have gathered popularity in the last 20 years or so. It is commercially available. The increased popularity can probably be attributed to better detectors and computer/computing technologies. Example vendors are Hi-Tech Scientific and BioLogic SAS.

15.7.2   Principle and Experimental Layout

In this apparatus two reactant solutions are prepared and filled in two supply syringes. From there the solutions are pumped through a mixer to obtain a well-mixed solution that subsequently moves on to fill a reactor where the flow is stopped, hence the name of the

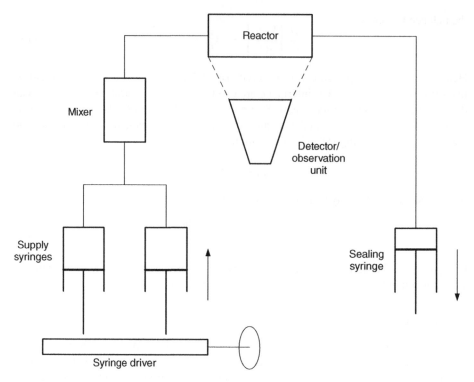

Figure 15.7   *A schematic diagram of a stopped flow apparatus.*

technique. At the other end there is sealing syringe that is instrumental in isolating the reactants in the reactor. There is also a detector that observes the progress of the reaction taking place in the reactor. This detector is the clue to making this work. After all, unless the reaction progression is mapped, there will be no kinetic data coming out (Figure 15.7).

A good version of such a set-up is described by Kierzkowska-Pawlak and co-workers (Kierzkowska-Pawlak, Siemieniec and Chacuk, 2013; Siemieniec, Kierzkowska-Pawlak and Chacuk, 2013). In their version they measured the electrical conductance of the solution using A/C current and a Fast-Fourier-Transform (FFT) to detect and log the reaction progression. They successfully measured reaction kinetics between CO_2 and diethanolamine (DEA) and also ethylethanolamine.

The prepared solutions are a solvent solution of the amine to be investigated and another solution where CO_2 has been dissolved in the solvent. Most reported experiments have used water as the solvent, Rayer, Henni and Li in Regina used the technique also with nonaqueous or hybrid solutions (2013).

On the practical side suitable solutions of the amine (or equivalent) reactant are made up. The CO_2 will be fed as an aqueous solution (or a solvent solution as the case might be), the concentration of which is determined from the partial pressure of CO_2 above it combined with knowledge of its Henry's coefficient. Obviously care must be taken to ensure that the CO_2 solution is in equilibrium with the gas above. CO_2 could be fed to this solution under pressure. This would mean that care must be taken with the mixture and the pressure in the cell to avoid spontaneous stripping of CO_2.

15.7.3 Mathematics and Practicalities

A practical difference with such a set-up compared with the foregoing techniques described is that CO_2 is likely to be consumed until some form of practical equilibrium is met. This has to be accounted for in the interpretation of the observations. The previously described techniques will have a continuous addition of CO_2 to the solution. In the stopped flow apparatus there will be a pseudo first order situation with the amine concentration remaining constant. Considering the amount of CO_2 fed to the cell, the temperature increase from the reaction is negligible (less than 0.5 K).

The extent of the reaction will be observed as a function of time over a period of seconds. To obtain the reaction kinetics constant the reaction rate equation is fitted to the observed time profile. This is straightforward.

The amine concentration and temperature may be varied from run to run such that a comprehensive kinetic investigation may be carried out.

15.7.4 Past Users

The first users of the stopped flow technique in this context were Barth *et al.* (1981) and Barth, Tondre and Delpeuch (1983). They mixed the CO_2 and the amine solution in a small vessel and observed the rate of progress of reaction by UV spectrometer while using an indicator in the solution to produce the colour change observed. Indicators used were thymol blue and m-cresol purple. Past users of the stopped flow technique are summarised

Table 15.1 *Stopped flow detectors used at various establishments for CO_2-amines research.*

Establishment	Detector	Temperature (°C)	Amines include
Technical University Lodz[a]	A/C conductivity FFT	15–60	DEA, EMEA, MDEA
University of Regina[b]	Conductance	20–40	DEMEA, MEA, DEA, MDEA and more
University of Newcastle, NSW, Australia[c]	1H-NMR	30	MEA
University of Kuwait[d]	Electrical conductivity	5–35	MMEA, and so on
King's College, London, UK[e]	Electrical conductivity	25	DEA, MEA, TEA, MDEA
Universite de Nancy, France[f]	Colour change of pH indicator	25	TEA, MDEA, DEA

Corresponding persons and references:
[a]Kierzkowska-Pawlak (Kierzkowska-Pawlak, *et al.*, 2013; Kierzkowska-Pawlak and Chacuk, 2010; Siemieniec *et al.*, 2013).
[b]Henni (Henni, 2008; Li, *et al.*, 2007; Rayer *et al.*, 2013).
[c]Maeder (McCann *et al.*, 2009).
[d]Ali (Ali, *et al.*, 2000; Ali, 2005).
[e]Crooks (Crooks and Donnellan, 1988).
[f]Tondre (Barth *et al.*, 1981; Barth, *et al.*, 1983).

in Table 15.1. The University of Kuwait was also early in adopting this technique (Ali, 2000, 2005; Alper and Danckwerts, 1990). The University of Regina used it for studies of DEMEA, EDA and EEA (Li, Henni and Tontiwachwuthikul, 2007; Henni, 2008), and continue to use it (Kadiwala, Rayer and Henni, 2012; Rayer, Henni and Li, 2013). There is also the Australian work at CSIRO and the University of Newcastle NSW (McCann *et al.*, 2009), and finally as already discussed previously, the Technical University of Lodz in Poland.

15.8   Other Mass Transfer Methods Less Used

15.8.1   Rapid Mixing

Hikita *et al.* (1977) used a rapid mixing method to determine the reaction rate between CO_2 and MEA, DEA and TEA. Their apparatus consisted of a very small mixing chamber where CO_2 and the amine solution were injected before proceeding to an observation chamber where the rate of reaction was determined by measuring the temperature difference along the tube that was 10 mm inner diameter and 500 mm long. Linear velocity in the tube was 5.5 m/s with a Reynolds number given as 17 800.

15.8.2   Rotating Drum

The way it works resembles a wetted wall column. A rotating drum with its axis parallel to the liquid level is partially submerged and rotated dragging a liquid film as its surface leaves the liquid. This liquid film is then exposed to the gas studied. Rate of absorption is determined by measuring make-up gas to keep the pressure constant. It was used by Danckwerts and Kennedy in 1954 and later by Oishi *et al.* in 1965 (Danckwerts, 1970). This equipment suffers from the same end effects as a wetted wall column and the lack of later use demonstrates that this was not a particular success, although publishable measurements were made.

15.8.3   Moving Band

This is another apparatus that works resembling the function of a wetted wall column. The technique was used by Govindan and Quinn in 1964 (Danckwerts, 1970). Here, a band moves in a loop where it is partially submerged in liquid. It moves through a gas space with gas of interest with a liquid film adhering to the band. Again rate of absorption may be observed by measuring the rate of make-up gas. Danckwerts' own comment in his book was that this technique was more elaborate to rig up than the laminar jet without giving any real advantage.

15.8.4   Kinetic Measurement Techniques Summarised

There are five techniques that seem to be the candidates today if you were to build a rig in your laboratory. They have been discussed previously and are summarised in Table 15.2. It has to be realised that very reactive substances with high kinetic reaction constants will necessitate working with very short contact times, or to be slowed down by lowering the

Table 15.2 An overview of the main experimental techniques[a].

Parameters	Laminar jet	Watted wall	Single sphere	Stirred cell	Stopped flow
Operational mode	Continuous	Continuous	Continuous	Contin, batch	Batch
Pressure (kPa)	Atmospheric	Atmospheric	Atmospheric	Atmospheric, vacuum	Atmospheric
Temperature, °C	Depends on construction				
Typical gas volume (l)	A few	A few	A few	Order of 1	ml
Typical liquid volume (l)	5–20	5–20	5–20	0.5	ml
Contact time (s)	0.001–0.020	0.1–1	0.1–0.5	Seconds-minutes	Seconds
Contact area (cm^2)	1–8	10–100	20–80	40–110	None
Liquid side MT coefficient (m/s)	0.00016–0.0016	0.00004–0.00016	0.00005–0.0001	0.000016–0.0001	–
Diffusivities present?	Yes	Yes	Yes	No	No
Reaction kinetics?	Yes	Yes	Yes	Yes	Yes
Gas solubilities?	No	No	No	Yes	No

[a]Sources include, Danckwerts (1970), Astarita et al. (1983) and Kierzkowska-Pawlak (2013).

CO_2 partial pressure as is done with the stirred cell. Regarding the stopped flow technique it is a question of finding a detector with a very short response time.

15.9   Other Techniques in Gas–Liquid Mass Transfer

In the course of 80 years or so of gas treating a number of laboratory units have seen the light. They could still be used with success, but the event of modern computers and their number crushing power has shifted the emphasis on what is desirable. The trend is clearly to do specific measurements and do 'complicated estimates'. In the past experiments were done such that they were directly scalable to design columns. Sounds good, but the disadvantage was reduced ability to reuse expensive experimental data.

Small scale packed columns have been used extensively. The principle of similarities that is a key to the old style unit operations approach had to be used with all the difficulties and uncertainties involved.

Wetted wall columns came early. These were in the main of the long variety, and ripples were allowed. The aim was to establish mass transfer coefficients or height of transfer units.

Discs on string and spheres on string were introduced and used to measure mass transfer that was used directly. There was no possibility of deriving information of the chemical rates.

The use of stirred cells to simulate columns section by section by simulating surface renewal rates was introduced by Danckwerts and McNeil (1967).

15.10   Equilibrium Measurements

In gas treating theoretical work, both physical and 'chemical' solubilities are used. The dissolved quantities are very different in the two cases and the measurements must be approached by different techniques in practice. In the discussion next, it is not specifically stated that the temperature is controlled, but that is to be assumed. As before, the temperature may be controlled by either jackets or by submersion in either an air chamber or in water.

15.10.1   Physical Solubilities

Since the 1980s physical solubilities have largely been measured by an equilibrium cell as illustrated in Figure 15.8 (Haimour and Sandall, 1984a,b).

It is based on a constant pressure where the dissolved gas is quantified by measuring the volume difference between the start where the gas is introduced and the end where equilibrium has been reached. Pressure is kept constant by manually adjusting a U-tube manometer. Usually the pressure is kept in balance with the atmosphere but a small difference could in principle be used as long as it was constant. The equilibrium cell is evacuated then filled with pre-saturated pure gas of interest, next the liquid is injected and the volume recorded. The liquid is stirred and equilibrium is usually reached within an hour or so. The gas volume is again recorded after equilibrium is reached. A variation is to evacuate with the liquid in place to saturate the gas space before the gas is introduced.

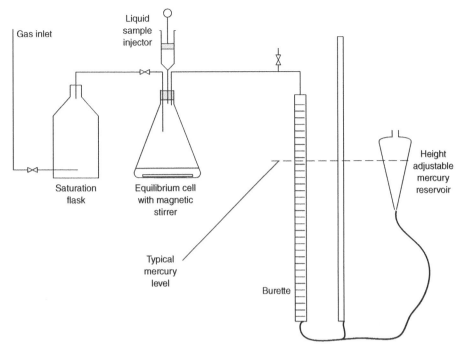

Figure 15.8 *The principle of the traditional equilibrium cell for measuring physical solubilities. The equipment is normally submerged in a water bath (not shown) for temperature control. The figure is based on work by Haimour and Sandall (1984a,b).*

Recently this technique has been modified by Ying, Eimer and Yi (2012b). In this version the volume of gas dissolved is measured by a mercury drop moving in a calibrated tube at constant pressure. See Figure 15.9. An advantage is the reduced mercury inventory and the elimination of the need to keep on adjusting the U-tube levels.

There are also other ways of doing it. At Twente Oyevaar, Morssinkhof and Westerterp (1989) used a closed glass vessel with a magnetic stirrer but the solubility was determined via measuring the pressure before gas started to dissolve and after equilibrium was reached. Littel, Versteeg and van Swaaij (1992), also at Twente, used one of their normal stirred cells where the solubility was again determined via pressure difference.

15.10.2   Chemical Solubilities

When dealing with chemical solvents, the solvent's capacity for gas is much bigger. Using the preceding technique would not be possible or convenient. It is resorted to a technique extensively used for measuring vapour–liquid equilibria for distillation problems. The apparatus involved is illustrated in Figure 15.10. Here, an amount of liquid is injected into an evacuated cell where the gas or gas mixture of choice is also introduced. The system is closed off and the gas is circulated via a gas pump until equilibrium is reached. This could take 1–2 h. Both gas and liquid are then sampled and analysed. An apparatus like this has been used by Mather's group in the University of Alberta, Edmonton in the

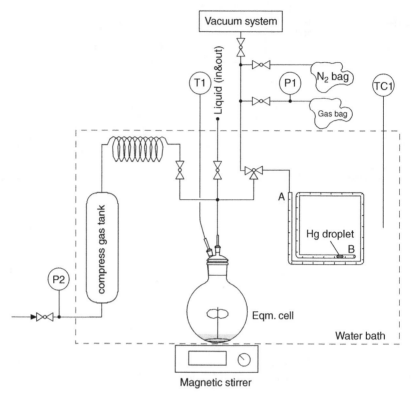

Figure 15.9   *The modified equilibrium cell made by Ying (2013). (Reproduced with permission from Ying, (2013). Copyright 2012, American Chemical Society.)*

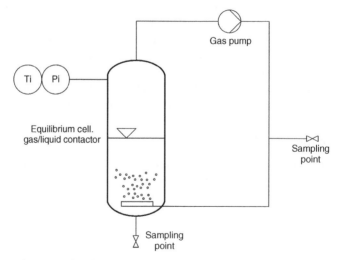

Figure 15.10   *The principle of an equilibrium cell where samples of both phases are extracted for analysis once equilibrium is achieved. Extra gas buffer volume may be added.*

period 1970–2000 (Jou, Mather and Ott, 1995). It is customary to pre-load the liquid with the gas investigated and to make a gas mixture resembling the expected equilibrium. This is done to save time and to exercise some control of which equilibrium point that will be obtained. If for example, we want to know the equilibrium pressure of CO_2 over a solution at a loading around 0.4 mol acid gas per mol of chemical, then it makes sense to pre-load the solution to that level and let the gas composition change. Due to the difference in the number of moles in the two phases, the liquid phase will not change much even if the gas phase does.

15.11   Data Interpretation and Sub-Models

When deriving the chemical kinetics of acid gases' reaction with the active chemical in the absorbent solution, other physical data or properties of the solution are used as part of the mathematical treatment. Obviously the same data should also be used when the derived data are employed for analysing the mass transfer kinetics, and indeed columns, as sub-models become implicit parts of the model. It is important to stick to a consistent set of data. This set may well be built from different sources, but is important that the same data are used all the time. Conversely, when using derived data from a correlation, it is prudent to check what supporting data they used in its derivation.

For chemical kinetics derived from, for example a laminar jet, the use of physical solubility of the gas and its diffusivity in the liquid is involved. The diffusivity is usually derived via measurements based on measuring with N_2O and using the N_2O analogy.

Using different sub-models when using data compared to those used when they are derived may introduce unnecessary error. It may be argued that there are uncertainties involved anyway but it is better to be consistent and avoid systematic errors for no reason.

References

Aboudheir, A., Tontiwachtikul, P., Chakma, A. and Idem, R. (2004) Novel design for the nozzle of a laminar jet absorber. *Ind. Eng. Chem. Res.*, **43**, 2568–2574.

Ali, S.H. (2005) Kinetics of the reaction of carbon dioxide with blends of amines in aqueous media using the stopped-flow technique. *Int. J. Chem. Kinet.*, **37**, 391–405.

Ali, S.H., Merchant, S.Q. and Fahim, M.A. (2000) Kinetic study of reactive absorption of some primary amines with carbon dioxide in ethanol solution. *Sep. Purif. Technol.*, **18**, 163–175.

Alper, E. and Danckwerts, P.V. (1976) Laboratory scale-model of a complete packed column. *Chem. Eng. Sci.*, **31**, 599–608.

Alper, E. (1990) Reaction mechanism and kinetics of aqueous solutions of 2-amino-2-methyl-1- propanol and carbon dioxide. *Ind. Eng. Chem. Res.*, **29**, 1725–1728.

Astarita, G., Savage, D.W. and Bisio, A. (1983) *Gas Treating with Chemical Solvents*, John Wiley & Sons, Inc.

Babu, D.R., Narsimhan, G. and Phillips, C.R. (1984) Absorption of sulfur dioxide in calcium hydroxide solutions. *Ind. Eng. Chem. Fundam.*, **23**, 370–373.

Barth, D., Tondre, C. and Delpeuch, J.-J. (1983) Stopped-flow determination of carbon dioxide-diethanolamine reaction mechanism: kinetics of carbamate formation. *Int. J. Chem. Kinet.*, **15**, 1147–1160.

Barth, D., Tondre, C., Lapal, G. and Delpeuch, J.-J. (1981) Kinetic study of carbon dioxide reaction with tertiary amines in aqueous solution. *J. Phys. Chem.*, **85**, 3660–3667.

Blauwhoff, P.M.M., Versteeg, G.F. and van Swaaij, W.P.M. (1984) A study on the reaction between CO_2 and alkanolamines in aqueous solutions. *Chem. Eng. Sci.*, **39**, 207–225.

Brian, P.L.T., Vivian, J.E. and Matiatos, D.C. (1967) Interfacial turbulence during the absorption of carbon dioxide into monoethanolamine. *AIChE J.*, **13**, 28–36.

Chambers, F.S. and Sherwood, T. (1937) Absorption of nitrogen dioxide by aqueous solutions. *Ind. Eng. Chem.*, **29**, 1415–1422.

Chilton, T.H. and Colburn, A.P. (1934) Mass transfer (absorption) coefficients from data on heat transfer and fluid friction. *Ind. Eng. Chem.*, **26**, 1183–1187.

Clarke, J.K.A. (1964) Kinetics of absorption of carbon dioxide in monoethanolamine solutions at short contact times. *Ind. Eng. Chem. Fundam.*, **3**, 239–245.

Crooks, J.E. and Donnellan, J.P. (1988) Kinetics of the formation of N,N-dialkylcarbamate from diethanolamine and carbon dioxide in anhydrous ethanol. *J. Chem. Soc., Perkin Trans. 2*, 191–194.

Cullen, E.J. and Davidson, J.F. (1956) The effect of surface active agents on the rate of absorption of carbon dioxide by water. *Chem. Eng. Sci.*, **6**, 49–56.

Cullen, E.J. and Davidson, J.F. (1957) Absorption of gases in liquid jets. *Trans. Faraday Soc.*, **53**, 113–120.

Danckwerts, P.V. (1955) Gas absorption accompanied by chemical reaction. *A.I.Ch.E. J.*, **1**, 456–463.

Danckwerts, P.V. (1970) *Gas-Liquid Reactions*, McGraw-Hill..

Danckwerts, P.V. and McNeil, K.M. (1967) The absorption of carbon dioxide into aqueous amine solutions and the effects of catalysis. *Trans. Inst. Chem. Eng.*, **45**, T32–T48.

Dang, H. and Rochelle, G.T. (2003) CO_2 absorption rate and solubility in monoethanolamine/piperazine/water. *Sep. Sci. Technol.*, **38**, 337–357.

Davidson, J.F. and Cullen, E.J. (1957) The determination of diffusion coefficients for sparingly soluble gases in liquids. *Trans. Inst. Chem. Eng.*, **35**, 51–60.

Eipper, J.E. (1955) PhD thesis. University of Delaware, Newark.

Haimour, N. and Sandall, O.C. (1984a) Molecular diffusivity of hydrogen sulfide in water. *J. Chem. Eng. Data*, **29**, 20–22.

Haimour, N. and Sandall, O.C. (1984b) Absorption of carbon dioxide into aqueous methyldiethanolamine. *Chem. Eng. Sci.*, **39**, 1791–1796.

Henni, A. (2008) Reply to "Comments on "Reaction kinetics of CO_2 in aqueous ethylenediamine, ethyl ethanolamine, and diethyl monoethanolamine solutions in the temperature range of 298-313 K using the stopped-flow technique. *Ind. Eng. Chem. Res.*, **47**, 991–992.

Hikita, H., Asai, S., Ishikawa, H. and Honda, M. (1977) The kinetics of reactions of carbon dioxide with monoethanolamine, diethanolamine and triethanolamine by a rapid mixing method. *Chem. Eng. J.*, **13**, 7–12.

Jou, F.-Y., Mather, A.E. and Otto, F.D. (1995) The solubility of CO_2 in a 30 mass percent monoethanolamine solution. *Can. J. Chem. Eng.*, **73**, 140–147.

Kadiwala, S., Rayer, A.V. and Henni, A. (2012) Kinetics of carbon dioxide (CO_2) with ethylenediamine, 3-amino-1-propanol in methanol and ethanol, and with 1-dimethlamino-2-propanol and 3-adimethylamino-1-propanol in water using stopped-flow technique. *Chem. Eng. J.*, **179**, 262–271.

Kierzkowska-Pawlak, H. and Chacuk, A. (2010) Kinetics of carbon dioxide absorption into aqueous MDEA solutions. *Ecol. Chem. Eng. S*, **17**, 463–474.

Kierzkowska-Pawlak, H., Siemieniec, M. and Chacuk, A. (2013) Investigation of CO_2 and ethylethanolamine reaction kinetics in aqueous solutions using the stopped-flow technique. *Chem. Pap.*, **67**, 1123–1129.

van Krevelen, D.W. and van Hooren, C.J. (1948) Kinetics of gas-liquid reactions, part II. Application of general theory to experimental data. *Recl. Trav. Chim. Pays-Bas*, **67**, 587–599.

Laddha, S.S. and Danckwerts, P.V. (1981) Reaction of CO_2 with ethanolamines: kinetics from gas-absorption. *Chem. Eng. Sci.*, **36**, 479–482.

Li, J., Henni, A. and Tontiwachwuthikul, P. (2007) Reaction kinetics of CO_2 in aqueous ethylenediamine, ethyl ethanolamine, and diethyl monoethanolamine solutions in the temperature range of 298-313 K using the stopped-flow technique. *Ind. Eng. Chem. Res.*, **46**, 4426–4434.

Littel, R.J., Versteeg, G.F. and van Swaaij, W.P.M. (1992) Solubility and diffusivity data for the absorption of COS, CO_2, and N_2O in amine solutions. *J. Chem. Eng. Data*, **37**, 49–55.

Lynn, S., Straatemeier, J.R. and Kramers, H. (1955a) Absorption studies in the light of penetration theory. I. Long wetted-wall columns. *Chem. Eng. Sci.*, **4**, 49–57.

Lynn, S., Straatemeier, J.R. and Kramers, H. (1955b) Absorption studies in the light of penetration theory. II. Absorption by short wetted-wall columns. *Chem. Eng. Sci.*, **4**, 58–62.

Lynn, S., Straatemeier, J.R. and Kramers, H. (1955c) Absorption studies in the light of penetration theory. III. Absorption by wetted spheres, singly and in columns. *Chem. Eng. Sci.*, **4**, 63–67.

Manogue, W.H. (1957) PhD dissertation. University of Delaware, Newark.

Matsuyama, Y. (1950) *Chem. Eng. Jpn.*, **14**, 245 (Not obtained. Referenced to credit origin).

Matsuyama, Y. (1953) *Mem. Fac. Eng. Kyoto Univ.*, **15**, 142. (Not obtained. Referenced to credit origin).

McCann, N., Phan, D., Wang, X., Conway, W., Burns, R., Attala, M., Puxty, G. and Maeder, M. (2009) Kinetics and mechanism of carbamate formation from CO_2 (aq), carbonate species, and monoethanolamine in aqueous solution. *J. Phys. Chem. A*, **115**, 5022–5029.

Nysing, R.A.T.O. and Kramers, H. (1958) Absorption of CO_2 in carbonate bicarbonate buffer solutions in a wetted wall column. *Chem. Eng. Sci.*, **8**, 81–89.

Oyevaar, M.H., Morssinkhof, R.W.J. and Westerterp, K.R. (1989) Density, viscosity, solubility and diffusivity of CO_2 and N_2O in solutions of diethanolamine in aqueous ethylene glycol at 298 K. *J. Chem. Eng. Data*, **34**, 77–82.

Pacheco, M.A., Kaganoi, S. and Rochelle, G.T. (2000) CO_2 absorption into aqueous mixtures of diglycolamine and methyldiethanolamine. *Chem. Eng. Sci.*, **55**, 5125–5140.

Raimondi, P. and Toor, H.L. (1959) Interfacial resistance in gas absorption. *AIChE J.*, **5**, 86–92.

Rayer, A.V., Henni, A. and Li, J. (2013) Reaction kinetics of 2-((2-aminoethyl) amino) ethanol in aqueous and non-aqueous solutions using the stopped flow technique. *Can. J. Chem. Eng.*, **91**, 490–498.

Roberts, D. and Danckwerts, P.V. (1962) Kinetics of CO2 absorption in alkaline solutions – I. Transient absorption rates and catalysis by arsenite. *Chem. Eng. Sci.*, **17**, 961–969.

Scriven, L.E. and Pigford, R.L. (1958) On the phase equilibrium at the gas-liquid interface during absorption. *AIChE J.*, **4**, 439–444.

Scriven, L.E. and Pigford, R.L. (1959) Fluid dynamics and diffusion calculations for laminar liquid jets. *AIChE J.*, **5**, 397–402.

Sharma, M.M. and Danckwerts, P.V. (1963) Fast reactions of CO2 in alkaline solutions – (a) Carbonate buffers with arsenite, formaldehyde and hypochlorite as catalysts (b) Aqueous monoisopropanolamine (1-amino-2-propanol) solutions. *Chem. Eng. Sci.*, **18**, 729–735.

Sherwood, T.K., Pigford, R.L. and Wilke, C.R. (1975) *Mass Transfer*, McGraw-Hill..

Siemieniec, M., Kierzkowska-Pawlak, H. and Chacuk, A. (2013) Reaction kinetics of CO_2 in aqueous diethanolamine solutions using the stopped-flow technique. *Ecol. Chem. Eng. S*, **19**, 55–66.

Sun, W.-C., Yong, C.-B. and Li, M.-H. (2005) Kinetics of the absorption of carbon dioxide into mixed aqueous solutions of 2-amino-2-methyl-1-propanol and piperazine. *Chem. Eng. Sci.*, **60**, 503–516.

Versteeg, G.F., Blauwhoff, P.M.M. and van Swaaij, W.P.M. (1983) The effect of diffusivity on gas-liquid mass transfer in stirred cells. Experiments at atmospheric and elevated pressures. *Chem. Eng. Sci.*, **42**, 1103–1119.

Wild, J.D. and Potter, O.E. (1968) Gas absorption with chemical reaction into liquid flowing on a sphere. *Inst. Chem. Eng. Symp. Ser.*, **28**, 30–38.

Ying, J. (2013) Mass transfer kinetics of carbon dioxide into concentrated aqueous solutions of monoethanolamine. PhD dissertation. Telemark University College.

Ying, J. and Eimer, D.A. (2012a) Measurements and correlations of diffusivities of nitrous oxide and carbon dioxide in monoethanolamine + water by laminar liquid jet. *Ind. Eng. Chem. Res.*, **51**, 16517–16524.

Ying, J., Eimer, D.A. and Yi, W. (2012b) Measurements and correlation of physical solubility of carbon dioxide in (monoethanolamine + water) by a modified technique. *Ind. Eng. Chem. Res.*, **51**, 6958–6966.

Yong, C.-B. and Li, M.-H. (2009) Kinetics of the absorption of carbon dioxide into mixed solutions of piperazine and triethanolamine. *J. Chem. Eng. Jpn.*, **42**, 29–38.

16

Absorption Equilibria

16.1 Introduction

Information on absorption equilibria are needed to perform mass transfer estimates as the difference between partial pressure in the gas and the equilibrium pressure from the solution constitutes the mass transfer driving force. In real processes the back pressure from the solution must be taken into account.

A picture of the multiple equilibria striving to settle in the system is illustrated in Figure 16.1. Gas being absorbed is first physically absorbed before it enters into one or more chemical reactions. In parallel to this there are a number of ionic oriented equilibria in the solutions that are assumed to always be established as ionic reactions are considered to be instantaneous. There are also reactions that shift the absorbed species from one form to another, for example CO_2 is shifted between carbamate and bicarbonate when primary and secondary amines are used. Typical absorption isotherms are shown in Figure 16.2.

Relations may be written assuming ideal gas and ideal liquid, meaning that fugacity and activity coefficients are neglected. Given a few adjustable parameters it is possible to make representative mathematical models on such an idealised basis. From a theoretical point of view it is, however, more satisfactory to account for the non-idealities. One argument in support of this is these models' capability of extrapolating data, particularly in the dilute solution regime. It is not necessarily so that the more complicated models give better results.

The isotherms shown in Figure 16.2 are typical for model range coverage. The partial pressures cover several decades, and the solution loadings cover 2–3. Published results and models show little scatter on such a plot. However, when model and measurements are inspected more closely in the region of industrial interest, it may well be that large deviations are found. If such deviation is systematic over a region, the explanation is more likely to be model shortcomings than experimental scatter. The essence of this is that no model should be accepted at face value without some quality check before it is used for process design. In these days of readymade commercially available computer programs it is easy to forget the old programming wisdom 'garbage in, garbage out'.

Gas Treating: Absorption Theory and Practice, First Edition. Dag Eimer.
© 2014 John Wiley & Sons, Ltd. Published 2014 by John Wiley & Sons, Ltd.

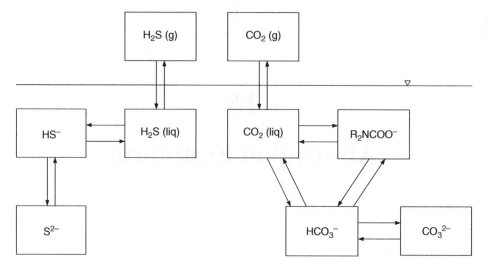

Figure 16.1 *Overview of equilibria for H_2S and CO_2 one by one.*

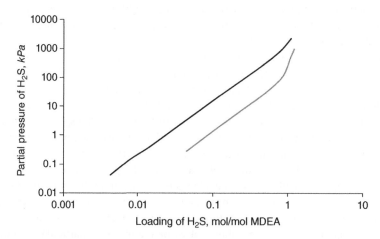

Figure 16.2 *A typical set of equilibrium isotherms. The system is H_2S-MDEA-water. The upper curve is for 100°C and the lower for 40°C. The data are taken from Jou, Mather and Otto (1982).*

16.2 Fundamental Relations

A fundamental principle in thermodynamics is that the path followed to get to a state is irrelevant to the properties of that state. On this background the reaction mechanisms involved in the absorption of acid gases do not influence the resulting equilibrium. Thus any reaction equation may be used to define the equilibrium relations given that they represent real relations and that all the species involved are represented.

Danckwerts and McNeil (1967) introduced a reaction concept in the 1960s that is still in use when analysing absorption equilibria. Kent and Eisenberg (1976) built on this when they made their much used and quoted equilibrium model. In the ensuing reaction concept this has been extended to deal with a mixture of up to four amines and both CO_2 and H_2S. Furthermore it is shown how the Vapour-Liquid Equilibria (VLEs) for the non-acid-gas-species can be included in the scheme. The reactions are:

Protonation of the amine:

$$R^aR^bR^cNH^+ \leftrightarrow H^+ + R^aR^bR^cN \tag{16.1}$$

$$R^dR^eR^fNH^+ \leftrightarrow H^+ + R^dR^eR^fN \tag{16.2}$$

$$R^gR^hR^iNH^+ \leftrightarrow H^+ + R^gR^hR^iN \tag{16.3}$$

$$R^jR^kR^lNH^+ \leftrightarrow H^+ + R^jR^kR^lN \tag{16.4}$$

The Rs denote any functional group in organic chemistry. The letter superscripts are used to enable distinguishing between functional groups. They are, however, not necessarily different. Here one or more of the 'Rs' may also represent a hydrogen atom such that also primary and secondary amines are represented by this notation. An extension to more amines is straightforward.

The next group of equations are the so-called carbamate reversion reactions. Tertiary amines do not form carbamates. The third 'R' on each amine is replaced by an H when writing these equations since only two would be relevant for a carbamate ion. In the case of a tertiary amine the associated equilibrium constant may simply be assigned a very high number to account for the reaction being all the way over to the right hand side of the equation. And the H on the amine should of course be substituted for the appropriate 'R' form in the equations previously.

$$R^aR^bNCOO^- + H_2O \leftrightarrow R^aR^bNH + HCO_3^- \tag{16.5}$$

$$R^dR^eNCOO^- + H_2O \leftrightarrow R^dR^eNH + HCO_3^- \tag{16.6}$$

$$R^gR^hNCOO^- + H_2O \leftrightarrow R^gR^hNH + HCO_3^- \tag{16.7}$$

$$R^jR^kNCOO^- + H_2O \leftrightarrow R^jR^kNH + HCO_3^- \tag{16.8}$$

Then there are the acid dissociation reactions for H_2O, CO_2 and H_2S:

$$H_2O + CO_2 \leftrightarrow H^+ + HCO_3^- \tag{16.9}$$

$$H_2O \leftrightarrow H^+ + OH^- \tag{16.10}$$

$$HCO_3^- \leftrightarrow H^+ + CO_3^{2-} \tag{16.11}$$

$$H_2S \leftrightarrow H^+ + HS^- \tag{16.12}$$

$$HS^- \leftrightarrow H^+ + S^{2-} \tag{16.13}$$

Finally there are the physical solubilities of CO_2 and H_2S expressed by simple relations involving partition coefficients referred to as H_i, which is the same as the Henry's coefficients in the limiting case when the partial pressures approach zero.

$$p_{H_2S} = H_{H_2S} \cdot [H_2S] \tag{16.14}$$

$$p_{CO_2} = H_{CO_2} \cdot [CO_2] \tag{16.15}$$

The equilibrium model may, or may not, include the water volatility. The water absorption equilibrium must be included if the heat effect associated with it is to be accounted for. It may (for convenience) be written using a partition coefficient H without implying that Henry's law is representative. That assumption may not be made here.

$$p_{H_2O} = H_{H_2O} \cdot [H_2O] \qquad (16.16)$$

In principle, the various amines in the solution are also volatile but for normal analysis of most absorption columns this may be neglected. The amine volatilities should be included in the desorption column estimates since the temperature is much higher here. Their partial pressures may also be expressed in the same form as before:

$$p_{RaRbRcN} = H_{RaRbRcN} \cdot [R^aR^bR^cN] \qquad (16.17)$$

$$p_{RdReRfN} = H_{RdReRfN} \cdot [R^dR^eR^fN] \qquad (16.18)$$

$$p_{RgRhRiN} = H_{RgRhRiN} \cdot [R^gR^hR^iN] \qquad (16.19)$$

$$p_{RjRkRlN} = H_{RjRkRlN} \cdot [R^jR^kR^lN] \qquad (16.20)$$

For the partition coefficients H_i it is clear that they are concentration dependent. Fugacity and/or activity coefficients should be used. However, if models are fitted for limited ranges of interest for the analysis at hand, this could be dispensed with. Early models like that of Kent and Eisenberg did this but their model did not represent the volatility of anything but the acid gases. It is possible to combine models, for example by using the Kent and Eisenberg model for the acid gases and use something else for water while neglecting the amine volatilities. Clearly it becomes important to understand the choices made and realise the implications of the approximations involved. This also goes for the use of commercial simulation packages.

For each of the chemical Equations 16.1–16.13 there is an equilibrium relation. These 13 relations may be written as follows (order as previously).

$$K_1 = \frac{[R^aR^bR^cN][H^+]}{[R^aR^bR^cNH^+]} \qquad (16.21)$$

$$K_2 = \frac{[R^dR^eR^fN][H^+]}{[R^dR^eR^fNH^+]} \qquad (16.22)$$

$$K_3 = \frac{[R^gR^hR^iN][H^+]}{[R^gR^hR^iNH^+]} \qquad (16.23)$$

$$K_4 = \frac{[R^jR^kR^lN][H^+]}{[R^jR^kR^lNH^+]} \qquad (16.24)$$

$$K_5 = \frac{[R^aR^bNH][HCO_3^-]}{[R^aR^bNCOO^-][H_2O]} \qquad (16.25)$$

$$K_6 = \frac{[R^dR^eNH][HCO_3^-]}{[R^dR^eNCOO^-][H_2O]} \qquad (16.26)$$

$$K_7 = \frac{[R^gR^hNH]\left[HCO_3^-\right]}{[R^gR^hNCOO^-][H_2O]} \tag{16.27}$$

$$K_8 = \frac{[R^jR^kNH]\left[HCO_3^-\right]}{[R^jR^kNCOO^-][H_2O]} \tag{16.28}$$

$$K_9 = \frac{[H^+]\left[HCO_3^-\right]}{[H_2O][CO_2]} \tag{16.29}$$

$$K_{10} = \frac{[H^+][OH^-]}{[H_2O]}, \text{but normally we use}: K_W = [H^+][OH^-] \tag{16.30}$$

$$K_{11} = \frac{[H^+]\left[CO_3^{2-}\right]}{\left[HCO_3^-\right]} \tag{16.31}$$

$$K_{12} = \frac{[H^+][HS^-]}{[H_2S]} \tag{16.32}$$

$$K_{13} = \frac{[H^+][S^{2-}]}{[HS^-]} \tag{16.33}$$

These equilibrium relations have been listed based on the concentrations of the species. There is nothing wrong with that, but it means that the equilibrium constants are concentration dependent. Early equilibrium models were made on this basis. When newer equilibrium models are made, activities are normally used such that variations in the equilibrium constants are limited to the temperature influence. As will become clear in the following, activities are referred to some reference state that can be different for the various species involved. It is, for example, customary to use pure water as its reference state and possibly for the other solvents while ions are referred to their aqueous infinite dilution state. Another approach is to refer also the 'other solvents' (i.e. the amines) to their aqueous infinite dilution state. Further discussion of this subject may be found in the thermodynamics literature, for example the book by Kontogeorgis and Folas (2010) but also in the more general literature (Smith and van Ness, 1975).

16.3   Literature Data Reported

Absorption equilibria have been reported by numerous authors over the last 80 years or so. Some have been more prolific with papers than others. There are a number of single papers from laboratories and they will clearly, rightly or wrongly, be viewed more critically than papers coming from laboratories that have been active over a longer period with several papers published. The group at University of Alberta Edmonton stands out by a continuous stream of papers over a 30 year period.

Table 16.1 *Acid gas absorption equilibria published.*

References	Amine(s) used	Only CO_2	Only H_2S	Mix CO_2 and H_2S
Isaacs, Otto and Mather (1977)	Sulfinol (diisopropanolamine (DIPA) and Sulfolan + water)	X	X	–
Jou, Mather and Otto (1982)	MDEA	X	X	X
Weiland, Chakravarty and Mather (1993)	MEA, DEA, MDEA, DGA	X	X	X
Jou, Otto and Mather (1995a)	MEA	X	–	–
Jou, Otto and Mather (1995b)	TEA	–	–	X
Ma'mun, Jakobsen and Svendsen (2006)	AEEA	X	–	–
Huttenhuis, Agrawal, Hogendoorn and Versteeg (2007)	DEA	x	–	–
Huttenhuis, Agrawal, Hogendoorn and Versteeg (2007)	MDEA	X	X	–
Faramarzi, Kontogeorgis, Thomsen and Stenby (2009)	MEA, MDEA	X	–	–
Dong, Chen and Gao (2010)	MPA	X	–	–

A list of papers published includes those given in Table 16.1. This list does not attempt to be all inclusive. The emphasis is to embrace the most common systems by quoting the most recent papers that will allow the interested reader to establish a more comprehensive data collection for the system of interest by following up references quoted in these papers.

It is beyond the scope of this text to engage in critical reviews of data published. This is, however, a much needed task as it is clear that published measurements are of a variable quality as demonstrated by the spread of results. If the quality of data is critical to the task at hand, it is necessary to evaluate the literature in sufficient detail. Alternatively the newest paper summarising the literature may be used or at least taken as a starting point. Appropriate allowance should be made in any process design to account for the uncertainty in data.

16.4   Danckwerts–McNeil

Danckwerts and McNeil (1967) laid the fundament for the approach to representation of absorption equilibria outlined earlier but are not much quoted for their equilibrium model as such. Their model is really a set of equilibrium constants given for measurements at 18°C. According to Kent and Eisenberg (1976) the model showed big differences between the predicted gas pressures and measured values. This is only to be expected as their paper does not allude to any fitting of parameters by regression of model versus experimental values. The equilibrium constant given for the carbamate reversion reaction in 2 M monoethanolamine (MEA) is taken from McNeil's PhD thesis with no details given in the paper.

16.5 Kent–Eisenberg

Kent and Eisenberg (1976) worked for Exxon Research and Engineering and took an industrialist's approach to the problem. They needed a model that could be used for computer simulations and decided to fit a number of equilibrium constants by doing regression analysis using literature values. Their work included H_2S and CO_2 equilibria for aqueous solutions of MEA and Diethanolamine (DEA). No mixtures of amines were considered but the presence of both gases simultaneously was accounted for.

In their model they fit equilibrium data for the reaction:

$$R^aR^bR^cNH^+ \leftrightarrow H^+ + R^aR^bR^cN \tag{16.1}$$

to data for H_2S absorption equilibrium pressures. The amine ionisation constant thus becomes a 'free parameter' for regression analysis. This they do first for MEA data and then for DEA data. In both cases they fit it to data temperature by temperature and in the end the equilibrium constant is observed to follow an Arrhenius relationship. From a theoretical point of view the use of the amine ionisation constant as a free parameter may certainly be questioned. However, they score a success in representing data. Their equilibrium constant for the ionisation of the amine must obviously not be used as a description of this ionisation.

Next they use the carbamate reversion equilibrium constant to fit absorption equilibrium data for CO_2. Again it is done first for MEA and then for DEA, and the temperature relationship is developed as before. In this way they use data from the H_2S regression in the fitting of parameters for CO_2, and they are in the end successfully representing data for cases when both CO_2 and H_2S are present. Their paper accounts well for the literature used and their numerical work in fitting the coefficients.

The Kent and Eisenberg model is really the first recognised practical absorption equilibrium model and has been widely quoted and used. It is included in at least two flowsheet simulators as one option. In view of its empirical nature with no use of activities the Kent and Eisenberg model gives surprisingly good results.

This model has been extended later to cover more data. Hu and Chakma (1990) and Kritpiphat and Tontiwachwuthikul (1996) worked with amino methyl propanol (AMP). Chakma and Meisen (1990) extended the work with CO_2 and DEA. Patil, Malik and Jobson (2006) worked with CO_2 and H_2S in Methyldiethanolamine (MDEA).

16.6 Deshmukh–Mather

Deshmukh and Mather (1981) at the University of Alberta, Edmonton developed their equilibrium model with the work of Kent and Eisenberg as a starting point in the sense that they used the same chemical equations. However, they pointed out that the solution was not ideal and that activity coefficients had to be introduced. They also treated the gas phase as non-ideal and computed fugacities using the Peng–Robinson equation of state. The activity coefficients of the ionic species 'k' were calculated using an extended Debye–Hückel (Guggenheim version) equation (as given by Deshmukh and Mather):

$$\ln\left(\gamma_k\right) = \frac{-Az_k^2 I^{0.5}}{1 + b_k I^{0.5}} + 2\sum \beta_{kj} m_j \tag{16.34}$$

A and b_k are constants, 'I' is the ionic strength of the solution, z_k is the valency of component k, β_{kj} is an interaction parameter that may be fitted to data and m_j is the molality of species j. Deshmukh and Mather fitted the binary interaction parameters β_{kj} to ternary data for the systems CO_2-water-MEA and H_2S-water-MEA.

In addition to the chemical equations and the equilibrium relations that follow from earlier, there were a charge balance and mass balances to account for CO_2 and H_2S. The method was a challenge with respect to computation time but that was at that time, that is 1981. Their development was hampered by limited data for some of the reactions but they claimed good results in representing acid gas absorption in MEA.

The Deshmukh–Mather model is essentially an extension of the Kent–Eisenberg model where they account for non-idealities. It has often been quoted, but it seems to have found limited use in simulation packages with ProTreat being the only one using it that this author has come across. The model was extended to mixed amines by Chackravarty (1985). Parameters for this equation to represent H_2S and CO_2 in MEA, DEA, MDEA and diglycolamine (DGA) were later worked out by Weiland, Chackravarty and Mather (1993).

16.7   Electrolyte NRTL (Austgen–Bishnoi–Chen–Rochelle)

The model was first published by Austgen *et al.* (1989) based on work done at the University of Texas Austin. The extension of the NRTL model to electrolyte systems was described by Chen and Evans (1986). Their model is thermodynamically consistent.

In the 1989 version the choice of standard reference states for all solvents (water and amines) is the pure liquid at the system temperature and its vapour pressure. The standard state for the ionic species is the ideal, infinitely dilute aqueous solution at the system temperature and pressure. The Henry's coefficients are also referred to the infinitely dilute aqueous solution. Bishnoi and Rochelle (2000) published further work using the same model. They kept the reference state for the solutes but calculated the activity coefficients first the same way as solvents before 'normalising' them by dividing by the activity coefficient calculated for infinitely dilute solution of water. The NRTL model for CO_2 absorption has found its way into the commercial flowsheeting packages.

16.8   Li–Mather

Li and Mather (1994) published a model that was able to predict CO_2 solubilities in aqueous mixtures of amines. Parameters for MEA-MDEA were given. The model is activity coefficient based. It makes use of excess Gibbs free energy to account for the non-idealities and this excess function is made up of contributions from short-range forces and long range forces (Debye–Hückel). Their version of the Debye–Hückel equation is due to Clegg and Pitzer rather than the Guggenheim version used by Deshmukh and Mather.

Their choice of standard reference states for all solvents (water and amines) is the pure liquid at the system temperature and its vapour pressure. The standard state of the ionic species is the ideal, infinitely dilute aqueous solution at the system temperature and pressure. The Henry's coefficients are also referred to the infinitely dilute aqueous solution.

Binary and ternary interaction coefficients are fitted.

This model has achieved a limited follow-up in the literature (perhaps this is due to shortcomings in their paper that made reproduction by others difficult: Y.-G. Li, Personal Communication with John Arild Svendsen, Norsk Hydro (with the author as co-worker, 2005). However, it has found its way into both the ASPEN-HYSYS and ASPEN PLUS flowsheeting programs as options.

16.9   Extended UNIQUAC

UNIQUAC was introduced as an alternative to NRTL in the 1970s. Its advantage was the use of two rather than three parameters for describing equilibria. The introduction of UNIFAC, which is a method/system capable of predicting parameters for UNIQUAC if they are not known, soon followed. In the 1990s an extension to UNIQUAC that allowed representation of electrolytic systems was introduced.

The extended UNIQUAC has been successfully used to represent absorption equilibria for CO_2, H_2S and alkanolamines (Kaewsichan *et al.*, 2001; Faramarzi *et al.*, 2009). Even the carbonate systems used in, for example the Benfield process has been represented (Tang, Spoek and Gross, 2009).

One strength of the UNIQUAC/UNIFAC method is that its application to new solvent systems, including mixtures, is easier since parameters may be estimated without the need for experiments. This should at least be applicable to a mixture of liquids. If fitting parameters to acid gas solubilities in a new absorbent, this would be useful. No literature has so far been found suggesting that the acid gas absorption equilibria may be predicted *a priori*.

16.10   EoS – SAFT

Equations of state (EoS) have been used to correlate and predict VLEs for a long time. Equations like Peng–Robinson and Soave–Redlich–Kwong have been very successful for hydrocarbon systems and in the field of cryogenics. However, these equations have not been used for the systems normally met in gas treating except to describe gas phase non-idealities. Considerable work has gone towards developing an EoS that could be more generally used. The breakthrough came with the attention to the use of the Helmholtz free energy and equations derived from this have been successfully used to represent equilibria between acid gases and alkanolamines (Huttenhuis *et al.*, 2007).

In a different type of approach to the work on Helmholtz free energy referred to previously, another type of EoS was developed. Based on perturbation theory a new model known as SAFT (Statistical Associating Fluid Theory) was developed. This is well summarised by Kontogeorgis and Folas (2010).

SAFT, or a variant of it, has been successfully applied to represent absorption equilibria of CO_2 in alkanolamines (Button and Gubbins, 1999; Rozmus, de Hemptinne and Mougin, 2011).

16.11   Other Models

The models described are not the only ones published to deal with acid gas absorption into alkanolamines. The reason for giving special attention to these models is simply that they are perceived to be either referred to by many or that they have achieved success as part of flowsheeting programmes that the reader may easily come across.

Other models have been published by:

- Li and Shen (1993)
- Kamps, Balaban, Jödecke (2001).

References

Austgen, D.M., Rochelle, G.T., Peng, X. and Chen, C.-C. (1989) Model for vapor-liquid equilibria for aqueous acid gas-alkanolamine systems using the electrolyte-NRTL equation. *Ind. Eng. Chem. Res.*, **28**, 1060–1073.

Bishnoi, S. and Rochelle, G.T. (2000) Physical and chemical solubility of carbon dioxide in aqueous Methyldiethanolamine. *Fluid Phase Equilib.*, **168**, 241–258.

Button, J.K. and Gubbins, K.E. (1999) SAFT prediction of vapour-liquid equilibria of mixtures containing carbon dioxide and aqueous monoethanolamine and Diethanolamine. *Fluid Phase Equilib.*, **158–160**, 175–181.

Chakma, A. and Meisen, A. (1990) Improved Kent–Eisenberg model for predicting CO_2 solubilities in aqueous Diethanolamine (DEA) solutions. *Gas Sep. Purif.*, **4**, 37–40.

Chakravarty, T., Phukan, U.K. and Weiland, R.H. (1985) Reaction of acid gases with mixtures of amines. *Chem. Eng. Prog.*, **81** (4), 32–36.

Chen, C.-C. and Evans, L.B. (1986) A local composition model for the excess Gibbs energy of aqueous electrolyte systems. *AIChE J.*, **32**, 444–454.

Danckwerts, P.V. and McNeil, K.M. (1967) The absorption of carbon dioxide into aqueous amine solutions and the effects of catalysis. *Trans. Inst. Chem. Eng.*, **45**, T32–T49.

Deshmukh, R.D. and Mather, A.E. (1981) A mathematical model for equilibrium solubility of hydrogen sulphide and carbon dioxide in aqueous alkanolamine solutions. *Chem. Eng. Sci.*, **36**, 355–362.

Dong, L., Chen, J. and Gao, G. (2010) Solubility of carbon dioxide in aqueous solutions of 3-amino-1-propanol. *J. Chem. Eng. Data*, **55**, 1030–1034.

Faramarzi, L., Kontogeorgis, G.M., Thomsen, K. and Stenby, E.H. (2009) Extended UNIQUAC model for thermodynamic modelling of CO_2 absorption in aqueous alkanolamine solutions. *Fluid Phase Equilib.*, **282**, 121–132.

Hu, W. and Chakma, A. (1990) Modelling of equilibrium solubility for CO_2 and H_2S in aqueous amino methyl propanol (AMP) solutions. *Chem. Eng. Commun.*, **94**, 53–61.

Huttenhuis, P.J.G., Agrawal, N.J., Hogendoorn, J.A. and Versteeg, G.F. (2007) Gas solubility of H_2S and CO_2 in aqueous solutions of *N*-methyldiethanolamine. *J. Pet. Sci. Eng.*, **55**, 122–134.

Isaacs, E.E., Otto, F.D. and Mather, A.E. (1977) Solubility of hydrogen sulphide and carbon dioxide in a Sulfinol solution. *J. Chem. Eng. Data*, **22**, 317–319.

Jou, F.-Y., Mather, A.E. and Otto, F.D. (1982) Solubility of H_2S and CO_2 in aqueous methyldiethanolamine. *Ind. Eng. Chem. Process Des. Dev.*, **21**, 539–544.

Jou, F.-Y., Otto, F.D. and Mather, A.E. (1995a) The solubility of CO_2 in a 30 mass percent monoethanolamine solution. *Can. J. Chem. Eng.*, **73**, 140–147.

Jou, F.-Y., Otto, F.D. and Mather, A.E. (1995b) Solubility of hydrogen sulphide and carbon dioxide in aqueous solutions of triethanolamine. *J. Chem. Eng. Data*, **41**, 1181–1183.

Kaewsichan, L., Al-Bofersen, O., Yesavage, V.F. and Selim, M.S. (2001) Predictions of the solubility of acid gases in monoethanolamine (MEA) and Methyldiethanolamine (MDEA) solutions using the electrolyte-UNIQUAC model. *Fluid Phase Equilib.*, **183–184**, 159–171.

Kamps, A.P.-S., Balaban, A., Jödecke, M., Kuranov, G., Smirnova, N.A. and Maurer G. (2001) Solubility of single gases carbon dioxide and hydrogen sulphide in aqueous solutions of N-methyldiethanolamine at temperatures from 313 to 393 K and pressures up to 7.6 MPa: new experimental data and model extension. *Ind. Eng. Chem. Res.*, **40**, 696–706.

Kent, R.L. and Eisenberg, B. (1976) Better data for amine treating. *Hydrocarbon Process.*, **55**, 87–90.

Kontogeorgis, G.M. and Folas, G.K. (2010) *Thermodynamic Models for Industrial Applications*, John Wiley & Sons, Ltd..

Kritpiphat, W. and Tontiwachwuthikul, P. (1996) New modified Kent-Eisenberg model for predicting carbon dioxide solubility in aqueous 2-amino-2-methyl-1-propanol (AMP) solutions. *Chem. Eng. Commun*, **144**, 73–83.

Li, Y.-G. and Mather, A.E. (1994) Correlation and prediction of the solubility of carbon dioxide in a mixed alkanolamine solution. *Ind. Eng. Chem. Res.*, **33**, 2006–2015.

Li, M.H. and Shen, K.-P. (1993) Solubility of hydrogen sulphide in aqueous mixtures of monoethanolamine with N-methyldiethanolamine. *J. Chem. Eng. Data*, **38**, 105–108.

Ma'mun, S., Jakobsen, J.P. and Svendsen, H.F. (2006) Experimental and modelling study of the solubility of carbon dioxide in aqueous 30 mass % 2-(2-aminoethyl)amino)ethanol solution. *Ind. Eng. Chem. Res.*, **45**, 2505–2512.

Patil, P., Malik, Z. and Jobson, M. (2006) Prediction of CO_2 and H_2S solubility in aqueous MDEA solutions using an extended Kent and Eisenberg model. *Inst. Chem. Eng. Symp. Ser.*, **152**, 498–510.

Rozmus, J., de Hemptinne, J.-C. and Mougin, P. (2011) Application of GC-PPC-SAFT EoS to amine mixtures with a predictive approach. *Fluid Phase Equilib.*, **303**, 15–30.

Smith, J.M. and van Ness, H.C. (1975) *Introduction to Chemical Engineering Thermodynamics*, McGraw-Hill.

Tang, X., Spoek, R. and Gross, J. (2009) Modeling the phase equilibria of CO_2 and H_2S in aqueous electrolyte systems at elevated pressure. *Energy Procedia*, **1**, 1807–1814.

Weiland, R.H., Chakravarty, T. and Mather, A.E. (1993) Solubility of carbon dioxide and hydrogen sulphide in aqueous alkanolamines. *Ind. Eng. Chem. Res.*, **32**, 1419–1430.

17

Desorption

17.1 Introduction

What exactly is desorption? It is often described as the opposite of absorption but it is not quite that simple. True enough, for some purposes and situations it may be analysed on this basis but life is a little more complicated. In absorption the pressure is generally 'made up' from the presence of an inert gas while in desorption, at least in gas treating as discussed, there is not really any inert gas as such (although water vapour also serves as a diluent in addition to provide energy). In desorption the pressure is made up of the partial pressures of all species present. From thermodynamics it follows that the pressure set through some control valve will determine the temperature of the solution being subject to stripping. This, of course implies that there is a source of heat since the temperature will be high and the desorption process is endothermic. The likeness to distillation is obvious. In a column, be it trayed or packed, there will be an interface between gas and liquid where mass transfer can take place. If there is a pool of bubbling liquid, however, the incipient bubble formation is hard as the surface forces are inversely proportional to the bubble radius. On the other end of the scale, there is desorption by flashing, which is when an absorbent comes out from the absorber at pressure and this pressure is let down over a valve (or any other pressure reducing contraption for that matter). If the solutions become sufficiently supersaturated, gas will spontaneously flash off as bubbles. In a process this can be exploited for full or partial desorption. A vessel with two outlets is then provided such that the gas is given a chance of escape from the liquid. This is the setting for discussing desorption in more detail.

First a look to past literature to see what information is available. When Danckwerts (1970) published his book in 1970 his chapter on desorption cited only two references and the treatment was by and large that desorption was the opposite of absorption. Textbooks at the time did not give special attention to desorption. Danckwerts was in good company and I am sure that he and other authors were knowledgeable regarding the differences. It was just that desorption had (and still has) received much less attention than absorption in terms of research but there are now many publications also on desorption although to a

Gas Treating: Absorption Theory and Practice, First Edition. Dag Eimer.
© 2014 John Wiley & Sons, Ltd. Published 2014 by John Wiley & Sons, Ltd.

much lesser extent in-depth studies looking at the mechanisms and detailed rate data like for absorption. The more extreme conditions used for desorption make relevant units more expensive to install in a laboratory, and this situation has something to do with the lack of studies of details. Another aspect is that the reaction kinetics becomes very fast due to the much higher temperature, and this makes measurements more difficult. When analysing the process, it would be easy to assume that the desorption reaction was instantaneous and proceed on that basis. A number of experimental desorption investigations have, however, been made at temperatures of 80°C and lower, and recently there is the work of Jamal, Meisen and Jim Lim (2006a,b) working up to 110°C.

The field of desorption at large was reviewed by Shah and Sharma (1976). The theory of desorption with chemical reaction was extensively treated by Astarita and Savage in two articles (1980a,b) with further discussion of desorption from carbonate solutions (Savage, Astarita and Joshi, 1980) as a practical example. Some of this discussion of desorption mass transfer rates may also be found in the book by Astarita, Savage and Bisio (1983). They are mostly concerned with the size of the desorption driving force and how that influences the mass transfer enhancement due to chemical reaction in the regime where desorption is controlled by mass transfer limitations.

There are many efforts focusing on flowsheet variations and optimisation out of the many efforts that form part of the development work within desorption. Significant savings in operational costs have been identified by doing this. The aim has in general been to reduce the reboil needs of the desorption column. A number of such publications are listed in Table 17.1. There are further discussions in the patent literature but that has been left out here. The so-called activated methyldiethanolamine (MDEA) process would constitute such a patent, and that is more than 30 years old by now. Generally there is no need for stripping acid gas down to extremely low loadings to meet treated gas specifications when most of this gas is absorbed from higher partial pressures. A so-called Split-flow (SF) process with two absorbent feed points using lean and 'semi-lean' absorbent is a way to save energy. This is further discussed in Chapter 20.

Other efforts have focused on mapping the energy needs to reboil the solution at the bottom of the desorption column. Such publications are listed in Table 17.2.

Table 17.1 *Publications on desorption section flowsheet variations.*

WHO	Absorbent	Novelty	Energy advantage
Aroonwilas (2004)	MEA	Split-flow (SF)	30% ++
Jassim and Rochelle (2006)	MEA	Vapour recompression, (VR) and multipressure, (MP)	MP: 3–11%
			VR: 3–8%
Oyenekam and Rochelle (2006)	MEA, PZ	MP and vacuum stripping (VS)	Not given
Oyenekam and Rochelle (2007)	MEA, MEA/PZ, K2CO3/PZ, MDEA/PZ	'Double matrix', MP-SF, flashing feed stripper, internal heat exchange stripper	Matrix configuration is best, but only 10% lower than base

Table 17.2   *Publications on reboil needs.*

WHO	Absorbent	Technique	Energy
Sakwattanopong, Aroonwilas and Veawab (2005)	MEA, DEA, MDEA, AMP, MEA-MDEA, MEA-AMP, DEA-MDEA	Column with reboiler	Optimum reboil duty varying from 1 500 to 5 000 kJ/kg CO_2
Bougie and Iliuta (2010)	AHPD, AMPD, AEPD, MEA, PZ, AMP	Lab-rig: VLE cell with heating	Not explicitly established. Regeneration efficiency best for first listed absorbent, and so on
Tobiesen, Juliussen and Svendsen (2008)	MEA	Column with reboiler	3 700–11 200 kJ/kg

AHPD = 2-amino-2-hydroxymethyl-1,3-propandiol, AMPD = 2-amino-2-methyl-1,3-propandiol and AEPD = 2-amino-2-ethyl-1,3-propandiol.

There are also a number of people that have investigated the rate of desorption for mass transfer in a variety of apparatuses. Their goals have varied from screening studies of new absorbents to more fundamental investigations of mass transfer and reaction kinetics. A summary of such work is listed in Table 17.3. The table includes only works that were flagged as desorption oriented studies.

Desorption work has been described by a number of authors. Of the above Singh and Versteeg (2008) used a stirred flask to measure CO_2 evolved from a variety of absorbents to investigate the effect of side groups on the amines α-carbon. The effect of amine molecule structure was also studied by Bonenfant, Mimeault and Hausler (2003). Zhang *et al.* (2008) used a stirred cell to measure mass transfer rates in the range 80–130°C from monoethanolamine (MEA), diethanolamine (DEA), MDEA and diethylenetriamine (DETA). In addition there are more focused studies by Xu *et al.* (1995) who made rate measurements in a packed column using the system methyldiethanolamine-piperazine (MDEA-PZ). Then there is the paper by Yeh, Pennline and Resnik (2001) who claimed regeneration was easier with amino methyl propanol (AMP) compared to MDEA but data given in support was sketchy. Jamal, Meisen and Jim Lim (2006a,b) used a novel, hemispherical apparatus to investigate the desorption rates of CO_2 from solutions of MEA, DEA, AMP and MDEA as well as combinations thereof. Kierzkowska-Pawlak and Chacuk (2011) used a stirred cell that was also a calorimeter to analyse desorption rates. There is also the work by Cadours *et al.* (1997), but although 'desorption' is in the title, it is more relevant for the discussion of H_2S capture.

Desorption differs in the combination of chemistry with mass transfer from absorption. Here the component(s) to be desorbed must first be (re)created through a chemical reaction that is endothermic. The release of gas is not necessarily limited by diffusional mass transfer like in absorption. If the pressure is quickly reduced, there will quite possibly be enough super-saturation for gas to flash off spontaneously as bubbles. This would be the case in a typical flash vessel in this process whereas the desorption column would be mass transfer

Table 17.3 *Publications on desorption mass transfer and reaction rates for screening purposes. That is, no fundamental properties were measured.*

WHO	Absorbent	Technique	Temperature	Ratings
Bonenfant, Mimeault and Hausler (2003)	MEA, TEA, AEE, triethylamine, pyridine, pyrrolidine, AEE, DNH2, AEPDNH2	Boiling flask	89–101°C	Loadings given as function of time of boiling. (up to 3 minute)
Bougie and Iliuta (2010)	AMP, AEPD, AMPD, AHPD, PZ	Lab-rig: VLE cell with heating	353–393 K	Gives only CO_2 reduction as function of time
Choi, Seo and Jang (2009)	MEA-AMP	Desorption column	363–383 K	Favour blends
Singh and Versteeg (2008)	MEA, 3-amino-1-propanol, 4-amino-1-butanol, 5-amino-1-pentanol, and so on	Boiling flask	Boiling	Investigated structure relationships to desorption
Zhang, Shi and Wei (2008)	AMP	Reflux boiling	358–403 K	Regeneration effluent given as 86–98%

TEA = triethanolamine, AEE = 2-(2-aminoethylamino)ethanol and VLE = vapour-liquid equilibria.

oriented. At normal desorption temperatures, like 110°C and above, the chemical reactions would be fast and bordering on being instantaneous. However, if an acid gas loaded solution was left at lower temperatures, say below 70°C, the reaction rates would be finite and need to be accounted for in the way discussed in earlier chapters. Thuy and Weiland (1976) speculate on the difference between absorption and desorption based on room temperature CO_2 desorption from water. There may be a mass transfer resistance at the surface that is not observed during absorption. Furthermore, the nature of the surface could influence the rate of desorption by making bubble formation easier with a rough surface being preferred. However, they were not able to substantiate this. The observation that mass transfer coefficients are lower for desorption was not shared by Hamborg, Kersten and Versteeg (2010) for physical absorption systems and the same group subsequently came to the same conclusion for chemically reactive systems (2012). There is also the similarity between desorption and distillation. In both cases the pressure in the vapour phase is determined by the equilibrium partial pressures over the liquids.

17.2   Chemistry of Desorption

The reaction mechanism involved when CO_2 is absorbed has been the subject of considerable research and discussion. The matter of zwitterion or Crooks–Donnellan's theory has not been resolved (see Chapter 4 for discussion and documentation). For desorption nobody seems to have discussed how the reverse reaction proceeds. The mathematics of mass transfer will of course be influenced by this. Similarly for H_2S, little attention has been given to its desorption in the literature.

The reaction mechanisms proposed leave options for desorption pathways. Since there is no authoritative view on mechanisms, we are left to explore the field by hypothesising and analysing the consequences.

17.2.1   Zwitterion Based Analysis

If we simply reverse Equations 4.18 and 4.19, we could write:

$$\left(BH^+\right)\left(R_1R_2NCOO^-\right) = R_1R_2N^+HCOO^- + B \tag{17.1}$$

and:

$$R_1R_2N^+HCOO^- = R_1R_2NH + CO_2 \tag{17.2}$$

To get from the carbamate (adduct) to CO_2 there are two reactions in series. Either one could be a rate determining step, or the kinetics could be of the same order such that the mathematics would need to reflect consecutive reactions. The base B will mainly be the amine but it should be remembered that both OH^- and H_2O are also bases in this reaction scheme. Furthermore one or both reactions may, or may not, be instantaneous or approach that situation at desorption temperatures.

17.2.2   Crooks–Donnellan

In the case of the Crooks–Donnellan mechanism we reverse the Equations 4.24–4.27:

$$\left(CO_2\right)\left(R_1R_2NCOO^-\right)\left(H_3O^+\right) = CO_2 + \left(R_1R_2NH\right)\left(H_2O\right) \tag{17.3}$$

and

$$\left(CO_2\right)\left(R_1R_2NCOO^-\right)\left(R_1R_2NH_2^+\right) = CO_2 + \left(R_1R_2NH\right)\left(R_1R_2NH\right) \tag{17.4}$$

There are thus two 'sources of carbamate' to provide full reversion. The second step is merely a reversion of the 'complexes'. An interesting observation is that with the Crooks–Donnellan mechanism CO_2 is a product of the first step as opposed to the zwitterion mechanism where CO_2 does not reappear until after the second step. Although the CO_2 appears after the first step, the kinetics of the second step is still of interest because the rate of disappearance of the intermediate(s) will influence the rate of the first step since the reactions must be viewed as reversible.

17.2.3   Alternative Mechanisms

It is conceivable that neither of the described mechanisms is correct. Neither has been formulated to describe the desorption reaction, and all testing has been associated with absorption kinetics except for some NMR studies looking at intermediate species.

One obvious potential alternative path is via carbamate reversion, that is where:

$$R_1R_2NCOO^- + H_2O = HCO_3^- + R_1R_2NH \tag{17.5}$$

and then:

$$HCO_3^- = H_2O + CO_2 \qquad (17.6)$$

Under absorption conditions this carbamate reversion is reputedly slow although no directly measured kinetic data has been found. There is only indirect evidence by CO_2 absorption being slow above loadings of 0.5 mol/mol. However, when the temperature is increased from 40 to 110°C, the kinetic constant would be expected to increase by a factor of 2^7 (=128) based on the old rule of thumb that the kinetic constant is doubled every 10 K increase. In this case the reaction may have become fast enough to be reckoned with anyway.

When tertiary amines are used, there is no carbamate involved. As the chemistry is perceived, this should make possible desorption routes more simple since there is one type of compound that can be ruled out.

17.2.4   For Tertiary Amines

Here the CO_2 is bound to the protonated amine in the form of a bicarbonate and to a lesser extent as carbonate. It is reasonable to expect that this complex will dissociate in one step:

$$\left(R_1R_2R_3NH^+\right)\left(HCO_3^-\right) = R_1R_2R_3N + CO_2 + H_2O \qquad (17.7)$$

In mixed absorbents where both tertiary amines and either primary or secondary amines are present, the picture is even more complicated in that all these reactions may be in operation during desorption.

17.2.5   H₂S Desorption

H_2S will be bound in the same way whether the amine is primary, secondary or tertiary: $(R_1R_2R_3NH^+)(HS^-)$. It is again reasonable to assume a one-step dissociation step leading to desorption:

$$\left(R_1R_2NH_2^+\right)(HS^-) = R_1R_2NH + H_2S \qquad (17.8)$$

Or:

$$\left(R_1R_2R_3NH^+\right)(HS^-) = R_1R_2R_3N + H_2S \qquad (17.9)$$

17.3   Kinetics of Reaction

Only one study has been found that has measured kinetic constants by a fundamental approach at temperatures approaching those of desorption (Jamal, Meisen and Jim Lim, 2006a,b). Their work was based on the use of a hemisphere that is based on the same principle as a single sphere apparatus in that the interfacial area is known and the liquid flows over a spherically curved surface. In this case only the upper half of the sphere was used with the shape 'below equator' being redone for efficient drainage. The desorption process was studied at temperatures ranging from 60 to 110°C. Loadings for CO_2 ranged from 0.02 to 0.7 mol/mol, and moistened nitrogen was used as stripping gas. They studied the absorbents MEA, DEA, MDEA and AMP. It is clear from this work that extrapolation of kinetic reaction constants derived from absorption data at temperatures of 60°C and below deviate significantly from this work.

Studies such as the one reported by Jamal, Meisen and Jim Lim (2006a,b) would be the 'gold standard' that would allow really rigorous analysis to be made. However, such data may not be available without conducting experiments with the chemicals targeted for use. Without such data it may be necessary to base the analysis on extrapolations of data that *are* available. There is also the possibility of just assuming such fast reaction kinetics that the process for practical purposes becomes instantaneous in which case the process becomes diffusion limited anyway. Reaction paths could still be relevant to mass balances.

Apart from the work discussed there does not seem to be many published data on reaction kinetics under desorption conditions. Rates of CO_2 desorption have been studied by Dugas and Rochelle (2009) for a mixture of MEA and PZ in the range 40–60°C. Kierzkowska-Pawlak and Chacuk (2011) have studied CO_2 desorption from aqueous MDEA solutions at 20–60°C. There is also work done by the research group in Twente (Hamborg, Kersten and Versteeg, 2010; Hamborg and Versteeg, 2012) at temperatures from 25 to 60°C. According to Jamal, Meisen and Jim Lim (2006a) there is considerable uncertainty in extrapolating these data to standard desorption conditions. There may, however, be no alternative in a specific case, better to extrapolate than to guess. Assumptions should always be evaluated.

The desorption reactions sketched out in Section 17.2 may in fact be so fast that the reaction mechanism as such are irrelevant for mass transfer estimates. Once the reactions are fast enough, the mass transfer rates will tend towards mass transfer with infinitely fast reaction where the only influence of the reaction is how far the species will need to diffuse before conversion. When the reaction is this fast, its path and kinetics are no longer needed as the transfer process becomes diffusion controlled as described in Chapter 12. As the reaction:

$$\text{Adduct} \rightarrow \text{amine} + \text{acid gas} \tag{17.10}$$

becomes instantaneous, the adduct, where the CO_2 or H_2S is 'stored', should spontaneously split according to the reaction equation, an example is the desorption of H_2S according to Equations 17.8 and 17.9. This would, however, lead to immediate flashing of gas bubbles. As discussed next in Section 17.4, bubbling desorption will not take place unless the equilibrium pressure of H_2S over the solution is at least equal to the total system pressure. In the desorption column this is not the case. The explanation for this has to be that the chemical reaction is in a range where it is reversible as well as nearly instantaneous. Hence, the reaction term for production of the acid gas 'i' is:

$$r_i = k_{desorption}[adduct] - k_{absorption}[C_i][amine] \tag{17.11}$$

and the second term is significant. The desorption of CO_2 could be seen the same way, but here there are alternative paths as evident from Reactions 17.1–17.7.

17.4 Bubbling Desorption

The previous discussion has been concerned with desorption in quiescent liquids. That means that no bubbles are produced in the bulk of the liquid. It is known from opening, for example, bottles of carbonated water that bubbles will spontaneously form when the bottle is depressurised on opening. As will be further discussed in Chapter 20, the same phenomenon is exploited in natural gas treating when the rich absorbent from the absorber is subjected to pressure let-down in a so-called 'flash vessel'. A suitable pressure for this vessel is chosen and dissolved gases desorb by bubbling out of the liquid. This kind of mass transfer is faster by a few orders of magnitude compared with the reaction-diffusion mechanism discussed so far.

Desorption in the bubbling regime has received only limited attention although a number of workers have had a go over the years. Such work can be traced back to 1950, but we shall start in the 1970s. Thuy and Weiland (1976) experimented with an apparatus based on spheres-on-a-string and mapped the release of CO_2 from water. Bubbling will only occur when the equilibrium pressure of CO_2 over the liquid becomes higher than the total pressure. However, they confirmed the result of the earlier work that bubbling could be avoided even if the CO_2 pressure was as high as 3.75 times the total pressure in which case the solution would be supersaturated. Such supersaturation is only possible if the pressure is let down very slowly. They also observed that the bubbling desorption became 30–40% faster when they roughened the surface of the spheres. Presumably this provided nucleation sites akin to those found on modern day special boiler tube surfaces. This work was extended from room temperature measurements (Weiland, Thuy and Liveris, 1977). Here up to 90°C was tested in a stirred cell reactor where the object was to study the effect of solvent vapour pressure on the transition from bubbling to quiescent desorption. They found, however, that there was no effect of the solvent's vapour pressure. Hikita and Konishi (1984, 1985) and Hikita, Asai and Konishi (1988, 1987) followed up on this research angle. They used a stirred cell to study stripping of super-saturated CO_2 from water. The bubbling regime was divided into a low and high bubbling regime. It is in the latter that mass transfer rates are significantly improved by increasing Reynolds number as the stirring is stepped up. The work was later extended by looking at the effect of electrolytes. Their presence improved mass transfer further, allegedly by preventing coalescence of bubbles. A further contribution in this work was measuring drag coefficients for single bubbles rising. They also looked at the combination of absorption and bubbling desorption in the system 'inert gas' -SO_2-CO_2-$NaHCO_3$ (aq) in two papers.

Kierzkowska-Pawlak and Chacuk (2010) looked at CO_2 desorption from physical solvents at 20–50°C in what is essentially a stirred cell. In this work desorption as a function of driving force was studied. When the driving force increased beyond a certain value, the relation between mass transfer and driving force was no longer linear. For the lower driving forces the desorption was by diffusionally controlled mass transfer whereas bubbles started to form as the driving force increased. They defined a size referred to as *supersaturation*, which was:

$$\sigma = \frac{C_{CO_2}^{Lb} - C_{CO_2}^{L*}}{C_{CO_2}^{L*}} \tag{17.12}$$

In their work this was used as a parameter to correlate mass transfer against system and process variables in a model derived from nucleation theory. However, the model was specific to the test equipment used.

Similar fundamental work with amine solutions has not been found. This discussion could, however, be applicable given that the concentrations of CO_2(aq) and H_2S(aq) can be reliably estimated.

Flash vessels are in general not dimensioned according to the results and correlations discussed earlier. They are more likely to be dimensioned based on rules-of-thumb for a minimum residence time for the liquid. These rules-of-thumb are not easy to find in the literature. A guidance for allowable gas velocity in the flash tank may be derived from Branan (1994). If the gas density is around $3\,kg/m^3$, the gas velocity in a horizontal flash tank should probably be below 0.3 m/s, for a vertical tank it could 25% higher. Wagner and Judd (2006) suggest the liquid hold-up time should be 5 min if the gas is dry and carry no hydrocarbons, that is <2% C_2+. If the gas causes pick-up of hydrocarbons they suggest 20 min for liquid hold-up with skimming off arrangement to deal with the hydrocarbons. These guidance figures are empirical. The crux of the matter is allowing gas bubbles sufficient time to disengage from the liquid. A corollary to this is ensuring a good liquid path to avoid any liquid 'bypassing the liquid hold-up volume'. Arranging the liquid flow in such a way that the gas bubbles have the shortest possible way to travel to the liquid's surface is also good practice.

17.5   Desorption Process Analysis and Modelling

The basic theory covering desorption with chemical reaction was introduced by Astarita and Savage (1980a,b). Section 12.4.4 discusses this work and the implications. Essentially CO_2 and/or H_2S are transported from the bulk of the liquid to the gas–liquid interface both as free gas dissolved in the liquid and as chemically bound. The concentration of the chemically bound form of the gases is very much greater and will dominate the transport picture. Hence, the acid gases have a faster journey towards the interface than they have on their way into the bulk during absorption. It should also be clear that all reactions and transport processes are speeded up by the higher temperature, especially the reactions.

On a more macroscopic scale desorption analysis was also included in the 1985 publication by Blauwhoff and co-workers (1985) at the University of Twente. This part of their work is essentially textbook material but they do point out and discuss the problem of a pinch point arising when, for example MEA is stripped to very low CO_2 loadings. See Figure 17.1. They refer to the region of leaner loadings below the pinch point loading as a 'stripping limited' regime as compared to a region above that was referred to as 'heat limited' regime. The essence behind these terms is that in the stripping limited region the amount of water vapour condensing to provide desorption heat is negligible compared to the total amount of water vapour present to act as a vapour phase diluent to lower the partial pressure of CO_2. The operating line is approximately linear in this region. In the heat limited region, however, the water condensation is significant in the sense that the operating line is not straight.

A number of people have reported dedicated process models to deal with the desorption process over the years (Oyenekam and Rochelle, 2006; Tobiesen, Juliussen and Svendsen, 2008; Weiland, Rawal and Rice, 1982). The models as such are not published but the results are of interest. Weiland and co-workers checked their results against a pilot plant with good results, but their report was mainly about establishing design data. Oyenekam and Rochelle analysed alternative process configuration with a view to performing desorption with a modified process. Tobiesen, Svendsen and Hoff (2005) also made a model and verified it against pilot plant data. Their main contribution is perhaps to point out, on the basis of their model, that there is a significant energy saving to be achieved if the lean

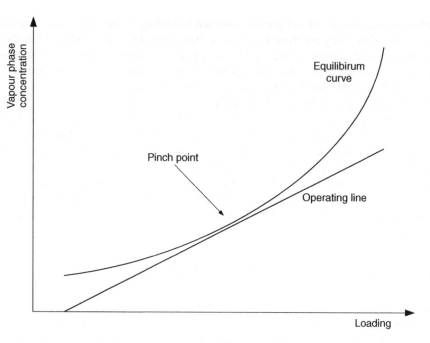

Figure 17.1   *Desorption column equilibrium curve and operating line.*

absorbent is less well regenerated. They looked at 30% (wt) MEA solutions and concluded that lean CO_2 loadings of 0.20–0.24 were to be preferred over 0.15–0.18 (all loadings are in mol CO_2 per mol MEA), which seemed to be the regeneration range discussed 20 years ago. In recent years these higher lean loadings have certainly been in focus for flue gas treating proposals. It is surprising that this insight had not materialised long before but other absorbent solutions had in the main ousted MEA for treating natural gas and in synthetic gas trains. There is also the work of Hamborg *et al.* (2011) that points to ways of enhancing the desorption process.

Modern processes in high pressure applications tend to make use of flash regeneration as much as possible augmented by some of the solutions being subjected to a complete desorption column with reboil.

A specific point relating to the need for desorption of H_2S is made by Huffmaster (1997). He points out that to achieve 4 ppm (roughly 0.25 grains per 100 ft^3) of H_2S in the treated gas, the lean absorbent to the absorber must have an equilibrium H_2S pressure 1/3–1/4 lower than the partial pressure of H_2S in the treated gas.

17.6   Unconventional Approaches to Desorption

The use of rotating packed beds and variations on those must be classed as unconventional but only as far as the choice of equipment goes. Such equipment was discussed in Chapter 9 and is flagged here only as a reminder. However, the short residence times involved is believed to enable operation at higher temperatures and pressures without leading to more degradation of the chemicals involved. This would lead to savings in the CO_2 compressor if the CO_2 produced was to be fed into enhanced oil recovery (EOR) or CO_2 capture and storage (CCS) schemes.

There have been attempts at finding additives to the solution in an attempt to reduce the energy input needed to provide 'dilution water vapour'. Hamborg and co-workers (2011) in Twente experimented with methanol and ethanol as additives to the solvent. Methanol had a measureable effect on the stripping of CO_2 from aqueous MDEA solutions whereas there seemed to be a much smaller effect from ethanol. Several concentrations of alcohol added were tested and even less than 10% seemed useful. Tobiesen and Svendsen (2006) in Trondheim attempted the use of hydrocarbons. They were looking towards having an add-on hydrocarbon loop around the desorber. The process worked but they concluded that energy for reboil increased.

A very different approach was taken by Feng *et al.* (2010) at the University of Queensland. They experimented with adding organic acids to the rich absorbent solution to ease the stripping of CO_2. Acids tried included suberic, phthalic and oxalic. It was foreseen that there would be a separate loop for these acids where they would be added to the rich absorbent and precipitated from the lean solution as crystals. A concept like this was also suggested by Eimer *et al.* (2005) without any details.

There was also work published in the 1980s by Quinn and Pez (1990) and Quinn, Pez and Appleby (1994) where amine salts were used in what could be referred to as a 'melting point swing' process. The concept was that the salt melt would absorb CO_2 while it would desorb when it was solidified. Since the melting point was around 60°C, waste heat could be used to drive the process. Unfortunately the cycle capacity of the identified salts was too low to be of practical use.

Very recently there has been work published on phase change solvents (Liebenthal *et al.*, 2013). The full results and potential of this line of investigation is still to come.

It is interesting to observe that there are exciting developments in the pipeline for a process that has been around for 80 years and would be considered mature by most.

References

Aroonwilas, A. (2004) Evaluation of split-flow scheme for CO_2 absorption process using mechanistic mass-transfer and hydrodynamic model. Proceedings from GHGT-7, the 7th International Conference on Greenhouse Gas Control Technologies.

Astarita, G. and Savage, D.W. (1980a) Theory of chemical desorption. *Chem. Eng. Sci.*, **35**, 649–656.

Astarita, G. and Savage, D.W. (1980b) Gas absorption and desorption with reversible instantaneous chemical reaction. *Chem. Eng. Sci.*, **35**, 1755–1764.

Astarita, G., Savage, D.W. and Bisio, A. (1983) *Gas Treating With Chemical Solvents*, John Wiley & Sons, Inc., New York.

Blauwhoff, P.M.M., Kamphuis, B., van Swaaij, W.P.M. and Westerterp, K.R. (1985) Absorber design in sour natural gas treatment plants: impact of process variables on operation and economics. *Chem. Eng. Process.*, **19**, 1–25.

Bonenfant, D., Mimeault, M. and Hausler, R. (2003) Determination of the structural features of distinct amines important for the absorption of CO_2 and regeneration in aqueous solution. *Ind. Eng. Chem. Res.*, **42**, 3179–3184.

Bougie, F. and Iliuta, M.C. (2010) Analysis of regeneration of sterically hindered alkanolamines aqueous solutions with and without activator. *Chem. Eng. Sci.*, **65**, 4746–4750.

Branan, C.R. (ed) (1994) *Rules of Thumb for Chemical Engineers*, Gulf Publishing.

Cadours, R., Bouallou, C., Gaunand, A. and Richon, D. (1997) Kinetics of co2 desorption from highly concentrated and co2-loaded methyldiethanolamine aqueous solutions in the range 312–383 K. *Ind. Eng. Chem. Res.*, **36**, 5384–5391.

Choi, W.-J., Seo, J.-B., Jang, S.-Y., Jung, J.H. and Oh, K.-J. (2009) Removal characteristics of CO_2 using aqueous MEA/AMP solutions in the absorption and regeneration process. *J. Environ. Sci.*, **21**, 907–913.

Danckwerts, P.V. (1970) *Gas-Liquid Reactions*, McGraw-Hill.

Dugas, R. and Rochelle, G.T. (2009) Absorption and desorption rates of carbon dioxide with monoethanolamine and piperazine. *Energy Procedia*, **1**, 1163–1169.

Eimer, D.M., Sjøvoll, N., Eldrup, R., Heyn, O., Juliussen, M. and McLarney, O. Swang (2005) Creative chemical approaches for carbon dioxide removal from flue gas, in *Carbon Dioxide Capture for Storage in Deep Geologic Formations – Results from the Carbon Capture Project*, Capture and Separation of Carbon Dioxide from Combustion Sources, vol. **1** (ed D.C. Thomas), Elsevier, pp. 189–200.

Feng, B., Du, M., Dennis, T.J., Anthony, K. and Perumal, M.J. (2010) Reduction of energy requirement of CO_2 desorption by adding acid into CO_2-loaded solvent. *Energy Fuels*, **24**, 213–219.

Hamborg, E.S., Derks, P.W.J., van Eik, E.P. and Versteeg, G.F. (2011) Carbon dioxide removal by alkanolamines in aqueous organic solvents. A method for enhancing the desorption process. *Energy Procedia*, **4**, 187–194.

Hamborg, E.S., Kersten, S.R.A. and Versteeg, G.F. (2010) Absorption and desorption mass transfer rates in non-reactive systems. *Chem. Eng. J.*, **161**, 191–195.

Hamborg, E.S. and Versteeg, G.F. (2012) Absorption and desorption mass transfer rates in chemically enhanced reactive systems. Part I: chemical enhancement factors. *Chem. Eng. J.*, **198–199**, 555–560.

Hikita, H., Asai, S. and Konishi, Y. (1987) Absorption of sulphur dioxide into aqueous sodium bicarbonate solutions accompanied by bubbling-desorption of carbon dioxide, in *Proceedings of the 2nd International Chemical Reaction Engineering Conference*, vol. **2** (eds D.S. Kulkani and S. Masjelkar), John Wiley & Sons, Inc., New York, pp. 30–44.

Hikita, H., Asai, S. and Konishi, Y. (1988) A criterion for gas supersaturation in simultaneous absorption and desorption with an instantaneous irreversible reaction. *Chem. Eng. Sci.*, **43**, 2907–2909.

Hikita, H. and Konishi, Y. (1984) Desorption of carbon dioxide from supersaturated water in an agitated vessel. *AIChE J.*, **30**, 945–951.

Hikita, H. and Konishi, Y. (1985) Desorption of carbon dioxide from aqueous electrolyte solutions supersaturated with carbon dioxide in an agitated vessel. *AIChE J.*, **31**, 697–699.

Huffmaster, M.A. (1997) Stripping requirements for selective treating with Sulfinol and amine systems. Laurance Reid Gas Conditioning Conference, Norman, OK.

Jamal, A., Meisen, A. and Jim Lim, C. (2006a) Kinetics of carbon dioxide absorption and desorption in aqueous alkanolamine solutions using a novel hemispherical contactor – 1. Experimental apparatus and mathematical modelling. *Chem. Eng. Sci.*, **61**, 6571–6589.

Jamal, A., Meisen, A. and Jim Lim, C. (2006b) Kinetics of carbon dioxide absorption and desorption in aqueous alkanolamine solutions using a novel hemispherical contactor – 2. Experimental results and parameter estimation. *Chem. Eng. Sci.*, **61**, 6590–6603.

Jassim, M.S. and Rochelle, G.T. (2006) Innovative absorber/stripper configurations for CO_2 capture by aqueous monoethanolamine. *Ind. Eng. Chem. Res.*, **45**, 2465–2472.

Kierzkowska-Pawlak, H. and Chacuk, A. (2010) Carbon dioxide desorption from saturated organic solvents. *Chem. Eng. Technol.*, **33**, 74–81.

Kierzkowska-Pawlak, H. and Chacuk, A. (2011) Kinetics of CO_2 desorption form aqueous N-methyldiethanolamine solutions. *Chem. Eng. J.*, **168**, 367–375.

Liebenthal, U., Pinto, D.D.D., Monteiro, J.G.M.-S., Monteiro, H.F. and Svendsen, A. Kather (2013) Overall process analysis and optimisation for CO_2 capture from coal fired power plants based on phase change solvents forming two liquid phases. *Energy Procedia*, **37**, 1844–1854.

Oyenekam, B.A. and Rochelle, G.T. (2006) Energy performance of stripper configurations for CO_2 capture by aqueous amines. *Ind. Eng. Chem. Res.*, **45**, 2457–2464.

Oyenekam, B.A. and Rochelle, G.T. (2007) Alternative stripper configurations for CO_2 capture by aqueous amines. *AIChE J.*, **53**, 3144–3154.

Quinn, R., G.P. Pez, (1990) Use of salt hydrates as reversible absorbents of acid gases. US Patent 4,973,456.

Quinn, R., G.P. Pez, J.B. Appleby, (1994) Process for reversibly absorbing acid gases from gaseous mixtures. US Patent 5,338,521.

Sakwattanopong, R., Aroonwilas, A. and Veawab, A. (2005) Behaviour of reboiler heat duty for CO_2 capture plants using regenerable single and blended alkanolamines. *Ind. Eng. Chem. Res.*, **44**, 4465–4473.

Savage, D.W., Astarita, G. and Joshi, S. (1980) Chemical absorption and desorption of carbon dioxide from hot carbonate solutions. *Chem. Eng. Sci.*, **35**, 1513–1522.

Shah, Y.T. and Sharma, M.M. (1976) Desorption with and without chemical reaction. *Trans. Inst. Chem. Eng.*, **54**, 1–41.

Singh, P. and Versteeg, G.F. (2008) Structure and activity relationships for CO_2 regeneration from aqueous amine-based absorbents. *Proc. Saf. Environ. Prot.*, **86**, 347–359.

Thuy, L.T. and Weiland, R.H. (1976) Mechanisms of gas desorption from aqueous solution. *Ind. Eng. Chem. Fundam.*, **15**, 286–293.

Tobiesen, F.A., Juliussen, O. and Svendsen, H.F. (2008) Experimental validation of a rigorous desorber model for CO_2 post-combustion capture. *Chem Eng Sci*, **63**, 2641–2656.

Tobiesen, F.A. and Svendsen, H.F. (2006) Study of a modified amine-based regeneration unit. *Ind. Eng. Chem. Res.*, **45**, 2489–2496.

Tobiesen, F.A., Svendsen, H.F. and Hoff, K.A. (2005) Desorber energy consumption amine based absorption plants. *Int. J. Green Energy*, **2**, 201–205.

Wagner, R., B. Judd, (2006) Fundamentals – gas sweetening. Laurance Reid Gas Conditioning Conference, Norman, OK.

Weiland, R.H., Thuy, L.T. and Liveris, A.N. (1977) Transition from bubbling to quiescent desorption of dissolved gases. *Ind. Eng. Chem. Fundam.*, **16**, 332–335.

Weiland, R.H., Rawal, M. and Rice, R.G. (1982) Stripping of carbon dioxide from monoethanolamine solutions in a packed column. *AIChE J.*, **28**, 963–973.

Xu, G.-W., Zhang, C.-F., Qin, S.-J. and Zhu, B.-C. (1995) Desorption of CO_2 from MDEA and activated MDEA solutions. *Ind. Eng. Chem. Res.*, **34**, 874–880.

Yeh, J.T., Pennline, H.W. and Resnik, K.P. (2001) Study of CO_2 absorption and desorption in a packed column. *Energy Fuel*, **15**, 274–278.

Zhang, P., Shi, Y., Wei, J., Shao, W. and Ye, Q. (2008) Regeneration of 2-amino-2-methyl-1-propanol used for carbon dioxide absorption. *J. Environ. Sci.*, **20**, 39–44.

18

Heat Exchangers

18.1 Introduction

The literature features numerous articles on absorbents, process configuration, kinetics and absorption equilibria, but there is little material to be found on the heat exchangers in the process. These units are much less special to this process compared to the columns and special units like the reclaimers, but they are nevertheless indispensable parts of the process. Failure to get the heat exchange system functional may lead to a malfunctioning process failing to achieve treated gas specifications and a penalty in the form of extra needs for energy.

Technical design of these heat exchangers can be dealt with by use of information in the general heat transfer literature. However, there are a few points that are worth discussing. This chapter attempts to summarise what little dedicated literature there is on this subject and also discusses special considerations that might be given when choosing equipment.

18.2 Reboiler

18.2.1 Introduction

The reboiler's function is to provide stripping steam for the desorption column. This is achieved by boiling the lean absorbent solution coming off the desorber bottom.

A secondary result of the reboiler is the provision of an extra equilibrium stage for desorption although the stage efficiency will not be 100%. In the process of boiling off water vapour, a significant amount of CO_2 is also boiled off.

18.2.2 Heat Media

Any heating medium can, in principle, be used. For natural gas treating heat is often provided by burning gas, and the hot exhaust may flow through fire tubes submerged in the

Gas Treating: Absorption Theory and Practice, First Edition. Dag Eimer.
© 2014 John Wiley & Sons, Ltd. Published 2014 by John Wiley & Sons, Ltd.

reboiler. With this kind of operation there is a severe risk of hot spots on the exchange surface, and these are detrimental to the upkeep of solution integrity. Increased degradation of the solution may result. An alternative is to use a heating oil that may in turn be heated by burning gas. If steam is available, this is a good choice for driving the reboiler. The temperature driving force is then easy to control, there will be no hot spots and condensing steam has a much higher heat transfer coefficient such that less heat exchanger surface will be needed compared with the use of hot exhaust gas or heating oil. However, steam systems are often unavailable at gas treating plants.

18.2.3   Kettle Reboiler Design

Kettle reboilers seem to dominate when choosing a design for this reboil application. There are alternatives, but the kettle design is the one that is receiving attention in the literature. The layout of a kettle reboiler is illustrated in Figures 18.1 and 18.2.

The kettle reboiler typically has an end box with a divider where the heating medium enters on one side and leaves from the other. The heating medium is taken through the boiling absorbent in U-tubes in what may be called the boiling section. There is a third part at the other end formed by a baffle where liquid flows over with the baffle acting as a weir. The liquid overflow is withdrawn and recycled to the absorption process. This end sump also allows the liquid to be degassed before pumping. Its size must reflect this. If a bubble is to be allowed to rise, the downward liquid velocity must be kept lower than the velocity of rise for the bubble (Figure 18.3).

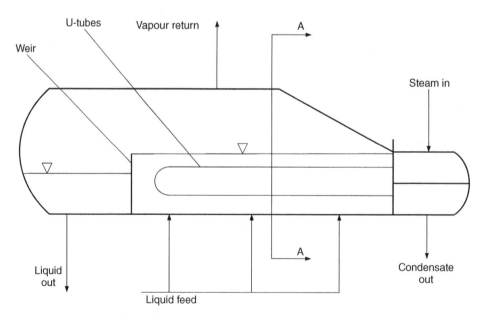

Figure 18.1   *Sketch of a kettle reboiler depicting its internal weir, U-tube heat exchanger pipes and steam box with divider as well indications of nozzle positions.*

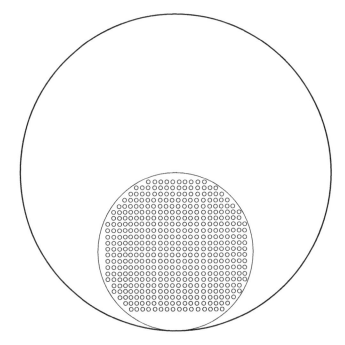

Figure 18.2　*A kettle reboiler. End view at section AA in Figure 18.1.*

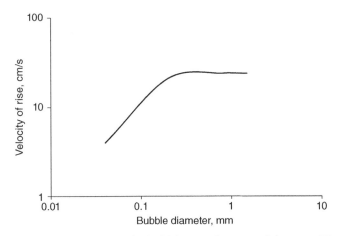

Figure 18.3　*Terminal velocity of rise for bubbles as a function of diameter. The plateau is in the range 22–26 cm/s. The figure is after Sherwood, Pigford and Wilke (1975).*

Lean liquid off the column bottom enters at the bottom of the reboiler along its boiling section. In a big reboiler there should be more than one inlet nozzle to provide a better distribution of liquid and help to avoid stationary 'pockets'. There should likewise be more than one vapour draw-off nozzle over the top.

18.2.4   Reboiler Specifics

Although the literature on the reboiler in this application is scarce, a few pointers may be found.

It is recommended to keep a minimum of 15 cm liquid layer height above the bundle. Below the bundle there should be allowed space to let the incoming liquid distribute itself under the bundle such that stagnant zones are avoided.

The boiling liquid should essentially be in the pool boiling regime. If the boiling becomes too vigorous, the tubes may be blanketed by gas thus lowering the rate of heat transfer. Tong and Tang (1997) discuss aspects of this phenomenon referred to as Taylor bubbles.

It is also recommended to keep the stripping of CO_2 down to avoid too much gas in the boiling liquid. This must be taken care of by ensuring that sufficient CO_2 is desorbed already in the desorption column itself. If there is a lot of gas stripping in the reboiler, removing tubes from the bundle should be considered, in order to create 'channels' for gas escape as described in Figure 18.4.

The Gas Processors Supplier Association (GPSA) Engineering and Data Book (Chapter 9) gives $14.2-20.5\,kW/m^2$ ($4500-6500\,Btu/h{\cdot}ft^2$) as a typical range for overall boiling heat flux. This is based on steam as heat medium and amine solutions boiling. Peyghambarzadeh, Jamialahmadi and Fazel (2009) has made an experimental laboratory study of nucleate pool boiling with horizontal tubes using aqueous monoethanolamine (MEA) and diethanolamine (DEA), but there was no CO_2 nor H_2S involved (2009). The results are thus not directly translatable.

For large plants parallel units must be used. There may be as many as 10 such units needed for a really large flue gas CO_2 recovery plant. Making the reboilers more efficient would thus easily remove whole units from the flowsheet.

18.2.5   Alternatives to Kettle Reboiler

In principle any kind of reboiler arrangement could be used to serve the desorption column. However, there are a number of considerations to be made. The GPSA Engineering and

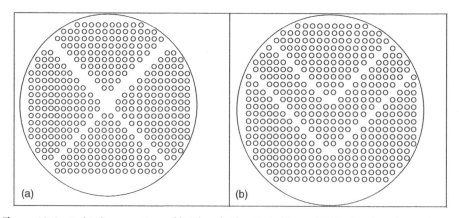

Figure 18.4   (a,b) Cross-section of kettle reboiler (A-A, Figure 18.1) showing alternative gas release channels made by removing tubes.

Data Book (Chapter 9) gives a check chart to guide the choice of reboiler. If the reboiler in amine applications is considered as low to moderately fowling at its typical pressure of 1.5–2 bar, then they list

- Kettle
- Vertical thermosyphon
- Horizontal thermosyphon.

as alternative reboiler selections. This application is not particularly fouling if the solution is properly maintained, that is, kept clean by filtering. Another consideration is foaming as the amine solutions are characterised as 'foaming'. This means that shear forces applied should be limited in order not to induce foam formation.

The thermosyphons are interesting options as they could probably be made at a lower cost than the kettle that has a rather large shell compared to the bundle.

Plate-and-frame heat exchangers (PFHEs) have also been suggested. It is known that they are in service as evaporators.

There are challenges associated with deviations from the trodden path, and it is beyond the scope of this text to argue the case for the alternatives.

18.3   Desorber Overhead Condenser

18.3.1   Introduction

The function of the desorber overhead condenser is to recover water and amine from the desorbed CO_2 gas. The water (with maybe some amine) recovered is used as reflux in a small section in the top of the column itself to further help with the recovery. The condenser is thus an integrated part of the recovery and reflux system.

In flue gas treating, and other applications where the CO_2 must be dried, a low dew point out of the condenser will also be beneficial to the downstream gas drying system.

18.3.2   The Reflux System

The reflux system shown in Figure 18.5 consists of a small column section where the up-going gas is contacted with the down-flowing reflux liquid, then the condenser, which is followed by a separator or reflux vessel. The liquid reflux is pumped from this vessel back into the top of the column. The gas desorbed leaves over the top of the reflux vessel.

Because of the size of the units and the need to pump cooling water to a high level (or pressure), the condenser, reflux vessel and reflux pump are normally mounted at ground level. The gas would in any case need a duct to bring it down if it was to be piped away.

18.3.3   The Condenser Design

The condenser for this application is little described in the literature. It is essentially a condenser where there is a lot of inert gas present and this must be taken into account when making the thermal design. A shell and tube based design is the base case.

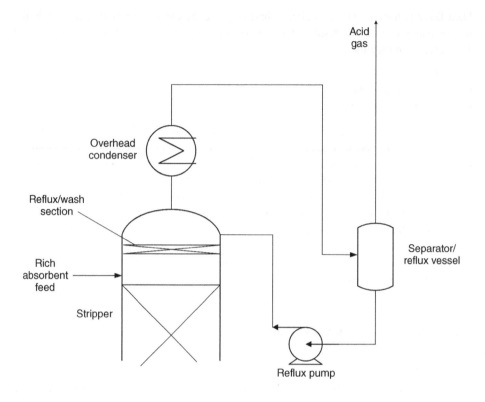

Figure 18.5   *The desorption column overhead reflux system.*

One feature of this condenser is the presence of condensed water that can be made quite acid by the presence of CO_2. As this is inevitable, the choice of construction materials must take this into account.

18.3.4   Alternatives

Plate- and frame heat exchangers have been suggested as a cheaper alternative. They have been cleared for condenser use although not specifically for this application.

18.4   Economiser or Lean/Rich Heat Exchanger

18.4.1   Introduction

The regenerated absorbent solution, referred to as lean amine, is returned to the absorber. It is, however, at a high temperature when it leaves the reboiler and its heat is recovered by pre-heating the rich amine coming from the absorber before this stream is piped to the top of the desorber.

18.4.2   Design Considerations

In a flue gas CO_2 capture plant based on 400 MW combined cycle gas turbine (CCGT) power plant the liquid circulation is likely to be around 2200 m^3/h, and the rich amine is heated from around 50 to 110°C. This gives a thermal load around 147 MW. What does this imply?

Traditionally, shell and tube heat exchangers have been used for this service. It is not realistic to build a single shell heat exchanger for the heat load as defined previously. Shell and tube heat exchangers have a specific heat exchanger area as described in Figure 18.6. Clearly the area density depends on both the outer tube diameter and the pitch. Pitch is defined as distance between tube centres. The minimum pitch is 1.25 times the tube diameter (Coulson and Richardson, 1977).

A square pitch is often preferred because it makes it easier to do mechanical cleaning of the tubes. Tube diameters of 25 mm (outer) are most common. The longer the tubes, the more robust they need to be to avoid vibration problems. Baffles will of course help with respect to this. It is also a question of commercially available tube lengths. Real heat exchanger designs involve making compromises between pressure drop and heat transfer performance, and in a liquid–liquid exchanger like the economiser the designer would look for a good match of heat transfer coefficients for the two sides.

Coulson and Richardson (1977) suggest that heat transfer coefficients would be in the range 570–1700 W/m^2·K for organic solvents and 4500–11 300 W/m^2·K for water. Since the fluids in the economiser are aqueous solutions of organic solvents, it may be argued that

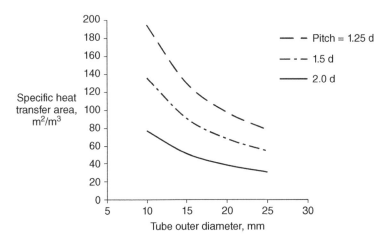

Figure 18.6 *Specific heat transfer area for shell and tube heat exchangers as function of outer tube diameter (d) and pitch (specified as a number times the diameter). The figure is based on a square pitch. A triangular pitch gives roughly a 30% larger area.*

Table 18.1   Heat transfer area needed for a 147 MW economiser as function of temperature driving force.

ΔT (K)	Area, A (m²)	Energy recovery (%)
5	23 520	93
10	11 760	86
15	7 840	79

Energy recovery is based on lean absorbent from reboiler being 120°C and rich from absorber being 50°C.

heat transfer coefficients are somewhere in between, say 2500 W/m²·K. Using this value for both sides and neglecting the thermal resistance of the pipe wall, the overall heat transfer coefficient would be 1250 W/m²·K. This is a number that would need to be checked by making proper design estimates.

Using the argued heat transfer coefficient, the economiser will have a relation between area and temperature driving force.

$$A \cdot \Delta T = 117000 \text{ m}^2 \cdot \text{K} \tag{18.1}$$

With roughly the same fluid flow on both sides of the heat exchanger and assuming no CO_2 desorption, the approach temperature and the average temperature driving force would be roughly the same. The trade-off between energy recovery, as expressed by temperature difference and heat exchanger area, will then be as described in Table 18.1.

The heat transfer area needed is very big in a heat exchanger context. Shell and tube heat exchangers have commonly been used for this service in the gas industry. Heat exchanger tubes are normally maximum 6 m in length, and 25 mm tubes are most commonly used. Given a pitch of 1.25×25 mm, specific heat transfer area would then be in the order of 80 m²/m³ (Figure 18.6). If we use a length to diameter ratio for a shell and tube exchanger (exclusive of heads) of 3 to 1, the volume of the active exchanger will be 18.8 m³. The area that could then be incorporated in one exchanger is accordingly $80 \times 18.8 = 1500$ m². Clearly even with a 10 K driving force there is a need for eight such exchangers. A more common size of shell and tube exchangers is 500–750 m². If 500 m² exchangers were used, there would be a need for 24. Mounting this many exchangers in series and/or in parallel as needed will involve significant piping and manifolding. If a better heat recovery is wanted, the number of heat exchangers goes up. It doubles if we go from 10 to 5 K as driving force.

Plate and frame heat exchangers have been discussed as an alternative choice (Choi *et al.*, 2005; Reddy, Johnson and Gilmartin, 2008). The gasketed variety has been around since the 1920s and is well established as a product. The gaskets have traditionally limited its use to maximum 25 bar, but technological development is going on perpetually. Plate and frame exchangers may easily have a surface density up to 200 m²/m³, and heat transfer numbers achieved is twice that of shell and tubes. Furthermore, the price per unit area is significantly lower than that of shell and tubes (Hesselgreaves, 2001; Reay, 1999). Use of plate and frame heat exchangers has certainly been advocated in the context of CO_2 capture from flue gases. These heat exchangers have already been used in natural gas treating plants (Harbison and Handwerk, 1987; MacKenzie *et al.*, 1987). MacKenzie and co-workers point out that plate exchangers offer more surface per unit volume and that plates should be in 316 SS and they recommend EPDM gaskets stating that these gaskets should overcome past problems of leaking gaskets in this service. They also advocate single pass and high velocities of the liquid to help preventing solids settling on the surfaces. On the other hand they also recommend a 0.14 bar (2 psi) pressure drop on the rich side as maximum to keep acid gas from desorbing. The latter is an issue but it can also be handled by keeping the pressure high enough to prevent this happening. The warm end temperature is a key factor in this equation.

A very interesting development by Zhou, Wu and Tu (2008) is launched in their study of compact heat exchangers where it was concluded in favour of a PTFE based plate-fin heat exchanger. It seems to be on the market in China and it is maybe one for the future.

The size of the problem is considerably smaller for most natural gas treating plants. However, there are many very big plants built also in this area even if most plants (by number) are smaller ones.

18.5   Amine Cooler

The challenge is much the same as for the economiser, but the heat load would be much smaller since the cooling is from 60–70°C down to 40–50°C, that is less than 1/3 of the economiser. The cold side would feature cooling water but alternatively a coolant cooled by air or water may be used. It could, as an alternative, be an air cooler.

18.6   Water Wash Circulation Cooler

The problem matter is the same as for the amine cooler but the heat load would be smaller still.

18.7   Heat Exchanger Alternatives

As the size of chemical process trains have increased, there has also been a lot of attention paid to increasing the capacity of heat exchangers and making them more efficient. From the concept of process intensification it emerged that it was better to have many parallel narrow channels for fluid flow compared to fewer channels with a larger cross section. This comes from the consideration that smaller channels provide more area per unit volume than the bigger ones and, furthermore, the pressure drop influence of channel diameter is less than its beneficial influence on area. The books by Reay (1999) and Hesselgreaves (2001) review available compact heat exchangers.

There are a number of design ideas around. The one that has made its way into the oil and gas industry is the Heatric 'Printed Circuit' heat exchanger (PCHE). There is a 'sales picture' from Heatric of one PCHE replacing three shell and tube exchangers, and the PCHE is smaller than even one of the shell and tubes. The fact that the PCHE has made it into the oil and gas industry suggests that PCHEs should also be considered for the absorption-desorption process where they would considerably reduce the number of units needed. Their price per area is probably still higher than that for the shell and tube exchangers, but the installed cost may be something else (Eimer and Eldrup, 2013).

PFHE has already been alluded to. They have been analysed for use as economiser and are shown to come out with a lower cost than the standard shell and tubes.

Interesting work has also been done on thin film polymer heat exchangers. These could be very interesting for the absorption–desorption process in flue gas treating in view of the benign pressures and temperatures involved. The heat resistance in the polymer compared to stainless steel should not be a show-stopper (Ramshaw, 1993; Reay, Ramshaw and Harvey, 2008).

References

Choi, G.N., Chu, R., Degen, B., Wen, H., Richen, P.L. and Chinn, D. (2005) CO_2 removal from power plant flue gas – cost efficient design and integration study, in *Carbon Dioxide Capture for Storage in Deep Geological Formations – Results from the CO_2 Capture Project*, vol. **1** (ed D.C. Thomas), Elsevier, pp. 99–376.

Coulson, J.M. and Richardson, J.F. (1977) *Chemical Engineering*, 3rd edn, vol. **1**, Pergamon Press.

Eimer, D. and Eldrup, N.H. (2013 Chapter in the book:) in *Process Intensification Technologies for Green Chemistry: Engineering Solutions for Sustainable Chemical Processing* (eds K.V.K. Boodhoo and A. Harvey), Wiley-Blackwell.

Harbison, J.L., G.E. Handwerk, (1987) Selective removal of H_2S utilizing generic MDEA. Paper H, Laurance Reid Gas Conditioning Conference, Norman, OK.

Hesselgreaves, J.E. (2001) *Compact Heat Exchangers*, Pergamon Press.

MacKenzie, D.H., Prambil, F.C., Daniels, C.A. and Bullin, J.A. (1987) Design and operation of a selective sweetening plant using MDEA. *Energy Prog.*, **7**, 31–36.

Peyghambarzeh, S.M., Jamialahmadi, M., Fazel, S.A.A. and Azizi, S. (2009) Experimental and theoretical study of pool boiling heat transfer to amine solutions). *Barz. J. Chem. Eng.*, **26**, 33–43.

Ramshaw, C. (1993) The opportunities for exploiting centrifugal fields. *Heat Recovery Syst. CHP*, **13**, 493–513.

Reay, D. (1999) *Compact Heat Exchangers*, CADDET Analysis Series, vol. **25**, Centre for the Analysis and Dissemination of Demonstrated Energy Technologies.

Reay, D., Ramshaw, C. and Harvey, A. (2008) *Process Intensification*, Elsevier Butterworth-Heinemann.

Reddy, S., D. Johnson, J. Gilmartin, (2008) Fluor's econamine FG Plus[SM] technology for CO_2 capture at coal-fired power plants. Power Plant Air Pollutant Control "Mega" Symposium, Baltimore, MD, August 25–28.

Sherwood, T.K., Pigford, R.L. and Wilke, C.R. (1975) *Mass Transfer*, McGraw-Hill.

Tong, L.S. and Tang, Y.S. (1997) *Boiling Heat Transfer and Two-Phase Flow*, Taylor & Francis Group, Washington, DC.

Zhou, G.-Y., Wu, E. and Tu, S.-T. (2008) Techno-economic study on compact heat exchangers. *Int. J. Energy Res.*, **32**, 1119–1127.

Further Reading

GPSA (1987) *Engineering Data Book*, 10th edn, vol. **1 & 2**, GPA Global.

Maddox, R.N. (1977) *Gas and Liquid Sweetening*, Campbell Petroleum Series, 2nd edn, Campbell Petroleum Corpration, Norma, OK.

Perry, R.H. and Green, D. (eds) (1984) *Perry's Chemical Engineers' Handbook*, 6th edn, McGraw-Hill.

19

Solution Management

19.1 Introduction

This chapter is about the less glamorous features of the absorption–desorption process. Nevertheless, the process will not operate properly unless these aspects of the process are taken care of in an adequate manner. The need to keep the absorbent solution clean to achieve trouble free operation is stressed in many conference papers and in the occasional journal article. It is, however, difficult to translate this into classroom teaching as it is not possible to go into a laboratory and replicate operational problems, and academic staff often don't have hands-on experience of practical plant operations. Essentially, operational stability is about attention to detail. It is often said that the devil is disguised in the details.

Information on this subject is found in bits and pieces in the literature. There is always the book by Kohl and Nielsen (1997). More recently the subject has been reviewed by Cummings, Smith and Nelson (2007) and their paper gives a lot of information on most aspects of solution management, although their angle is that of a service provider to the natural gas industry. A review of contaminants in amine gas treating service was given by Alvis and Jenkins (2004). They give extensive lists of contaminants for various amine solutions and recommendations for limits. An interesting review of amine solution colours and the compounds that cause the colours is given by Parnell (2000).

Keeping the solution clean is not an exact science. There is no absolute specification of allowable content of impurities. Various plants and different operators approach this problem in different ways. A list of pointers is, however, given in the review by Alvis and Jenkins (2004).

If the solution becomes contaminated, foaming is a likely outcome depending on plant load. Foaming will in turn cause difficulty with separation of phases, entrainment and increased pressure drop and possible flooding of columns. The treated gas will quickly go off-spec. When this happens, production time will be lost as the gas may not enter the transport system.

Gas Treating: Absorption Theory and Practice, First Edition. Dag Eimer.
© 2014 John Wiley & Sons, Ltd. Published 2014 by John Wiley & Sons, Ltd.

It is reclaiming in amine absorption systems that are getting most attention in gas treating meetings but glycols are also subjected to reclamation.

This chapter will discuss various measures introduced to ensure as trouble free operation as possible. The simplest way is what may be termed 'bleed and feed'. This implies that a portion of the absorbent inventory is regularly bled and replaced with fresh absorbent as make-up. No attempt is made to recover any fresh unspoilt amine according to this method. It is expensive, and it gives a poor image with regard to environmental aspect. The latter issue may in some situations lead to loss of profitable business opportunities and at worst may lead to being shut down. There are many reasons to take this issue seriously.

19.2   Contaminant Problem

The nature of contamination will differ between absorbent systems and from application to application. How to deal with this must obviously be adapted to the case at hand. For example, monoethanolamine (MEA) forms HEEU (a urea) that is usually dealt with in a thermal reclaimer. A similar approach will deal with diamines and oxazolidones, which have in recent years come into focus as more in-depth understanding of these absorption systems has become available. If components more acidic than the standard acid gases come in with the gas, there will be so-called heat stable salts formed (HSSs). They could be formate, acetate, propionate, glyconate, oxalate, chloride, thiocyanate, thiosulfate and sulfates. They will build up in the process if left unattended. The cation for all these salts will be a protonated amine that will in this way be prevented from serving to combine with CO_2 or H_2S and reduces the absorbing capacity of the plant. It may be appropriate to point out that also CO_2 and H_2S form salts with the protonated amine but these salts are labile and decomposes when subjected to heat in the desorber.

HSS may in turn cause corrosion problems. If this is accepted up front, more expensive construction materials may be chosen. Left without any action taken, iron sulfide is likely to be present in the solution. This again may trigger more degradation reactions.

19.3   Feed Gas Pretreatment

In natural gas treating the feed gas will vary significantly from plant to plant. The gas may carry liquid droplets. These may be hydrocarbon condensate or other materials that will contaminate the absorbent solution. Another possibility is that the feed gas is a so-called 'retrograde' gas, that is, a gas where condensate will form if the pressure drops. The pressure drop associated with an absorption column is enough to bring this about.

It is common practice to install a scrubber upstream of the absorber to catch droplets and particles from the feed gas. The gas velocity in such a scrubber must be low enough to allow the drops or particles to settle. Settling could be counter-current to gas flow, or it may follow a downward trajectory in a horizontal scrubber. In either case it is necessary to allow sufficient time to let the droplets settle. Terminal velocity of liquid droplets falling through a gas is very dependent on the droplet size. A cut-off size must be decided upon. It is also good practice to install a demister in the gas outlet no matter the flow configuration.

A retrograde gas represents a bigger challenge. In such a case the absorber should be designed to have as small a pressure drop as possible. Ideally the gas should be pre-heated to avoid condensation, but this is not always practical. It could, of course, also be expensive. Thermal insulation of the absorber should be considered. Failure to avoid hydrocarbon condensate in the absorbent will cause excessive load on the other equipment installed to keep the solution clean. Operational problems are a likely outcome.

Within a synthesis gas (syngas) train feed gas contamination is usually not a big problem, but an upstream filter may be needed if an upstream catalyst is prone to dust.

When treating flue gases from power plants or equivalent for CO_2 abatement, pretreatment is a real necessity because the feed gas will most likely contain acidic components like SO_x and NO_x. These are acidic enough to react irreversibly with alkanolamines and form HSS. In this case the gas may also need to be cooled before entering the absorption tower. This cooling may be done by direct contact with cooling water (the Direct Contact Cooler or DCC). Such a cooling process will also be a water scrubbing process that may remove water soluble contaminants. Any contaminants caught by the water, whether they are particles or absorbed acid gases, will need to be removed from the circulating water to prevent build-up. Water treatment could be filtering, a bleed or both. The cooler is where the heat extracted from the gas is eventually dumped to cooling water or air depending on the local choice (Figure 19.1).

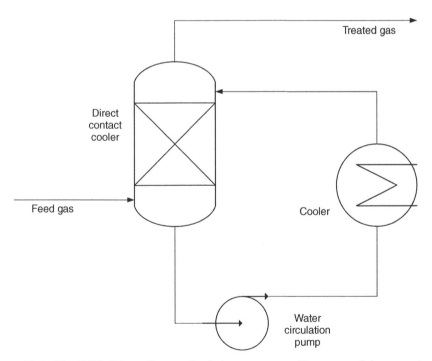

Figure 19.1 *The DCC (Direct Contact Cooler) arrangement. Treatment of the water in the circulation loop not shown.*

Pretreatment is in essence a necessity. Care must be shown when defining what kind of pretreatment that is necessary. Clearly, it is imperative to investigate the feed gas to ensure that all potential contaminants are identified. Only then may appropriate steps be taken to handle the problem. Cost effective opportunities in the process train considered should always be looked for.

19.4   Rich Absorbent Flash

In syngas treating a flash unit is desirable on the rich absorbent stream to recover hydrogen that may be compressed and returned to the process (flowsheet possibilities are shown in Figures 20.1, 20.3 and 20.4). The higher the volumetric flow rate of absorbent, the more hydrogen would be absorbed, and the more justification for adding a hydrogen recovery scheme. In syngas treating physical solvents are often used, and the associated flow rates are generally higher as the acid gas solubilities per unit volume are lower than for the chemical absorbents.

The picture is more complicated when it comes to natural gas treating. When the pressure is high, like above 50 bar say, there will be a significant co-absorption of hydrocarbons, particularly if the gas is rich in C_2+ (Wagner and Judd, 2006). In those cases it is customary to add a flash unit and use the flash gas as local fuel gas. It may also be possible to recycle this flash gas to the suction side of a gas compressor. A slip stream of lean absorbent may be used to wash CO_2 and/or H_2S out of the fuel as deemed necessary. One guidance given is to add a flash stage if $C_2+ >8\%$. Pressure considerations for the flash unit are a bit more complex. The flash tank pressure could be anything from 0.5 to 5 barg, but for any plant this may be different and that also goes for the 8% C_2+ rule. All such decisions are based on economics and local needs. If the flash tank pressure is set too low, typically less than 1 bar above the desorber pressure, a pump will be needed to transport the rich amine to the desorption column. Although a low flash pressure is beneficial for gas stripping, an extra pump adds an undesirable cost element and complexity to the plant.

The flashing of hydrocarbons from the absorbent solution reduces the risk of foam formation. The addition of a flash vessel is thus also good for the operation of the process.

In flue gas treating a rich liquid flash vessel is really a non-issue as the absorber pressure is generally too low to make it interesting.

Any general guidelines like those discussed here should be challenged on a case by case basis. A challenge could initially be evaluated based on simple back-of-the-envelope estimates. If the challenge looks interesting, estimates could be firmed up.

19.5   Filter

Ideally the whole absorbent stream should be filtered. However, this would mean a costly installation, especially for big plants with high circulation rates. Normally a slip stream of 10–20% of the absorbent stream is filtered. This is said to be adequate. Sometimes one might hear stories of such arrangements not working properly, however. There have also been cases where the filtering has been bypassed because 'they would be changing filters all the time'. A plant in question had severe operational problems. Another example is from a plant in Forestburg, Canada where all of the liquid stream was mechanically filtered but only 5% was cleaned with an active carbon filter (MacKenzie *et al.*, 1987).

It is normal practice to install the filter system on the cooled lean absorbent returning to the absorber but there are also cases where the rich liquid could or should be targeted.

19.5.1   Active Carbon Filter

The purpose of the active carbon adsorbent ('filter') is to remove dissolved hydrocarbons and surface active contaminants that would cause foaming if left to accumulate in the solution.

Active carbon is a general purpose adsorbent that can be used to remove a variety of dissolved substances from the absorbent solution by adsorption. This process is also rate limited. The substance is transferred through the liquid phase by convection and diffusion. As adsorption is often associated with the removal of substances present in low concentrations, the process is relatively slow. The adsorber will at any one time typically have an equilibrium section where the adsorbent is saturated with the contaminant. This is followed by a mass transfer section, followed by a section yet to be put into use. The mass transfer section will move downstream as time goes on. When the 'unused section' section disappears, there will be a breakthrough of the substance otherwise removed, and the adsorbent will need to be regenerated or changed. The adsorbed substance's concentration profile and adsorption bed are illustrated in Figure 19.2.

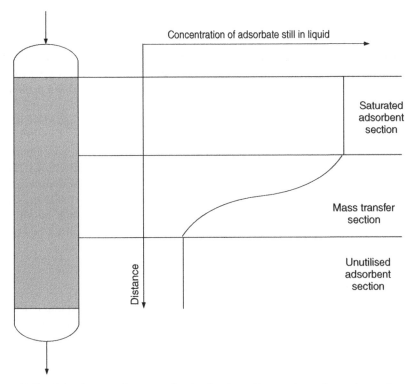

Figure 19.2 *Adsorbent column and adsorbent sections when the column is partially saturated.*

The surface available to adsorption is essentially available in the pores of the adsorbent where most of the surface is present. The pores are liquid filled and so small that there is no convection. Hence the mass transfer process is diffusion controlled and it is a slow mass transfer process. Mass transfer data, or pore data, must be obtained for each product and system. These may be obtained, but these processes are mostly designed based on experience figures, rules-of-thumb if you like. The adsorbent vendor would normally assist with adsorber process design.

Wagner and Judd (2006) give the adsorption liquid load guidance as 0.14–0.28 cm/s liquid velocity based on empty tower. This defines the tower cross-sectional area or diameter. Once the capacity for the substance to be removed is obtained from the vendor, the length of tower required for the desired operational period may be estimated. The length will be dominated by the capacity consideration rather than the mass transfer zone. Sheilan *et al.* (2008) give 15 min of hold-up as adequate and typical adsorption capacities as 15–35 kg contaminant per 100 kg of active carbon. This capacity depends on the type and concentration of the contaminant captured.

The relevant adsorption capacity, or experienced capacity, will be available from vendors with experience of supplying the oil and gas industry. The alternative is to do tests for oneself but in reality this would need to be supplied with a real absorbent solution that would not be available until after the plant was built. Most operators would seek advice. The filter system is after all just a minor part of the plant and hence there is little incentive for saving a small percentage on the cost of this unit. Good operation of the plant would be the priority. Harbison and Handwerk (1987) describe a gas treating plant in Wyoming where the active carbon bed was changed after 20 months of service from start-up. In the interim it had been steamed for 24 hours eight times. It was a full stream treatment. Other ways of regeneration are, in principle, possible but the most likely alternative is to replace the carbon with fresh product and dispose of the spent product. Alternatives here could involve incineration or return to the provider given that product stewardship is offered (at an expense).

The size, as in length, of the carbon bed is not critical. If it is smaller than ideal, it simply means more frequent replacements.

What would the diameter of such a carbon adsorber if a slip stream constituting 10% of the absorbent circulation of the plant indicated in Sections 8.7.3 and 11.7.1? The total absorbent flow is 0.0415 m^3/s. A slip stream of 10% is 0.00415 m^3/s. The mid-range superficial velocity in the carbon bed suggested above would be 0.20 cm/s = 0.0020 m/s. In this case the carbon bed cross-sectional area would be (0.00415 m^3/s)/(0.0020 m/s) = 2.1 m^2. This implies a diameter of 1.6 m. Treating the whole stream would require a bed of 5.1 m diameter.

19.5.2   Mechanical Filter

A mechanical filter is added to remove particulates from the absorbent solution. It is common to use a 10 μm filter, but there are many accounts of filtering problems being solved by choosing a finer filter.

Normal practice is to install a mechanical filter upstream of the active carbon filter to avoid clogging of the carbon bed. There is also normally a filter downstream of the carbon filter to prevent carbon particles from the bed from entering the rest of the process where it might contribute to foaming problems.

If fine filters are needed, there may be a case for a pre-filter using a coarser filter. This may prolong filter life. A fine filter will of course stop also the coarser particles that will in turn act as an extra filter and there is a strong possibility that these finer particles will clog up the passages in this porous layer.

There are plants about that feature mechanical filtering of both the rich and the lean absorbent, even full stream filtration of the rich liquid (Harbison and Handwerk, 1987). In this case there was full stream filtering of both streams. Following plant start-up the rich stream filter cartridges were changed every 2–3 weeks while the lean stream filter was changed twice per year.

Cartridge filters are commonly used. They will have to be replaced at regular intervals at some expense. The alternative is to have wash-back filters. These are probably more expensive to install as the piping system must allow for the back-wash, there is the need for provision of this wash water and probably its downstream clean-up.

19.6 Reclaiming

When the solution has been degraded over time, it will sooner or later be time for a bleed. This bleed potentially contains more fresh solution than the degraded variety. It is customary to attempt to reclaim as much of the fresh solution from the bleed as possible before the bleed is finally sent for waste disposal. It is also important to keep the volume of waste down as there is considerable cost associated with its disposal. How to reclaim depends on the type of absorbent used. This is due to simple features like their boiling point and possibly the chemistry involved. Some degradation may be reversed by addition of chemicals. As pointed out when describing the contaminants that are formed, they also involve the amine that is there as an absorbent chemical. It is desirable to reclaim also this part of the amine. To this end a stronger base like either NaOH or KOH or their carbonate salt is added to reclaimer liquid to free this amine. The chemical reaction involved is either:

$$\left(R_2NH_2{}^+\right)(HSS^-) + NaOH = R_2NH + Na(HSS) + H_2O$$

or

$$\left(R_2NH_2{}^+\right)(HSS^-) + Na_2CO_3 = R_2NH + Na(HSS) + NaHCO_3$$

depending on which base is chosen. Obviously, this use of a stronger but cheaper base means that more amine may be recovered (or reclaimed) from the waste stream. Not only is that a win as less replacement amine needs to be bought to make up, but it is cheaper to dispose of the waste.

19.6.1 Traditional Reclaiming

The purpose of reclaiming is to remove non-volatile compounds like HSS and high boiling point residue from the solution. By doing this systematically it is possible to operate the plant with much lower levels of HSSs, This, in turn, allows the upkeep of the high amine concentrations originally planned for and are necessary to achieve gas treating targets.

The traditional way of reclaiming the solution is to boil it in a separate reboiler mounted in parallel to the process' reboiler. It works when the active absorbent, like MEA and diglycolamine (DGA), has a high enough vapour pressure to allow it to evaporate along with water from the waste solution in the reclaimer. When the amine in question has a boiling point that is too high to boil it without the risk of thermal degradation, this type of reclaiming must be done under vacuum. Clearly it is then difficult to integrate the reclaimer with the normal reboiler and stripper. The vapour stream from the reclaimer is, when possible, routed to the desorber where it is mixed with vapour from the reboiler and thus helps to bring about the desorption.

Soda ash or equivalent is added to recover amine from the HSS. The positive ionic part of the HSSs is a protonated amine molecule. Sodium reacts more strongly with the negative ions of the HSSs and thus frees the amine molecule. This technique is also used outside of the reclaimer, particularly for the amines that cannot be reclaimed at atmospheric pressures.

Reclaiming by reboiling works very well with MEA and DGA solutions (Daughton and Veroba, 2007). The boiling point of MEA is around 170°C, and that for DGA is 221°C. Already with diethanolamine (DEA) decomposition occurs before its atmospheric boiling point is reached. The pressure has to be reduced to 50 mm Hg to bring the boiling down to 187°C. Hence vacuum has to be applied if successful reboil reclaiming is to be used. Normal boiling points of methyldiethanolamine (MDEA) and diisopropanolamine (DIPA) are both around 248°C.

The reclaimer reboiler is operated at a higher temperature than the process reboiler.

The standard configuration of a reclaimer is a kettle boiler design as illustrated in Figure 18.1. A demister is often added to the vapour outlet in a reclaimer. Its function is to prevent HSS to find its way back into the process in the form of entrained droplets.

19.6.2   Ion Exchange Reclaiming

The principle of using ion exchange to perform reclaiming has been about for a while without gaining much popularity. This is mainly due to an ion exchange process being somewhat complex thus demanding a certain level of skill from the operator. While this is a standard challenge in the chemicals industry, it introduces something new in the natural gas treating plants. This represents an operational risk. For smaller plants the benefits are insufficient to justify the operational risk and extra cost involved.

An ion exchange process was developed by Conoco around 1990 and has since been improved. It is a three stage process where cations are removed in the first column. A second column then removes what they call 'group I' anions followed by a third column that removes the rest of the anions ('group II'). The reasons for the division into two groups are their differing ability to attach to the ion exchange material and also the ease with which these can be regenerated (Keller *et al.*, 1992). Its principle is shown in Figure 19.3. The process referred to as HSSX.

An ion exchange reclaimer essentially works by anionic groups being deposited on a polymeric carrier material to catch the cations from the solution. Similarly cationic groups will attract anions on a separate carrier. Once the groups have been spent, there is no further capacity and the ion exchanger must be regenerated by washing with suitable solutions.

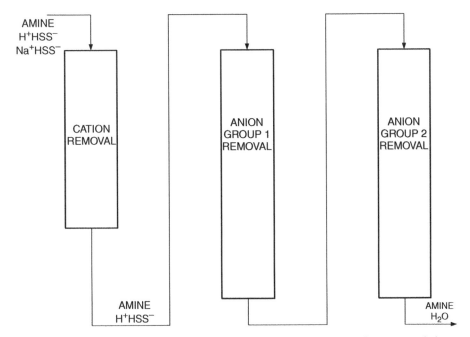

Figure 19.3 *Flowsheet of ion exchange reclaimer showing the configuration of the ion exchange columns. Regeneration is by a suitable solution, but not shown.*

These ion exchangers as presented for amine cleaning can only remove ions from the solution.

19.6.3 Electrodialysis Reclaiming

Electrodialysis is essentially an alternative way of carrying out ion exchange by using membranes and applying an electrical voltage to drive the mass transfer. The principle of the process is shown in Figure 19.4. Only ions are removed. They move in the direction of the appropriate electrode. One challenge is to make the membrane selective enough to avoid losses of significance to the waste stream. A typical electrodialysis unit will have many membranes stacked like shown but the figure only shows five membranes to avoid clutter.

A process was developed by Union Carbide (Burns and Gregory, 1995), a process that has since been further developed (Cummings, Smith and Nelson, 2007). It is based on the use of ion exchange membranes that are placed in an electric field. The electrical field ensures continuous regeneration of the membranes compared to the batch-wise regeneration of the ion exchange columns. It has met limited success so far.

19.7 Chemicals to Combat Foaming

The amine solutions have a propensity to foam, and anti-foam agents are used on a regular basis although the basic recommendation from process licensors is that its use should

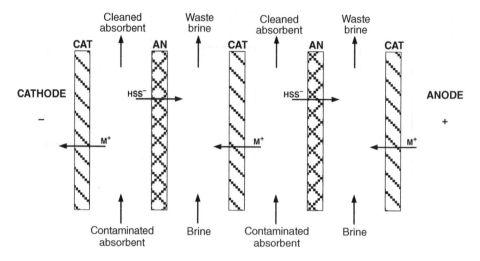

Figure 19.4   *Configuration of an electrodialysis reclaimer.*

be intermittent. The reason for amine solutions foaming is attributed to the presence of hydrocarbons absorbed from the natural gas treated and/or particles. An extensive review of foaming is given by Von Phul and Stern (2005).

Foam combating chemicals may be divided into two classes, anti-foam and defoaming chemicals (Cummings *et al.*, 2007; Von Phul and Stern, 2005). Anti-foam agents are used to prevent the formation of foam whereas defoaming agents are used to break foams that have already been formed. For amine solutions the same agent can be used for both. The word 'agent' is used intentionally to allude to the situation where these agents are proprietary recipes and the actual key chemical is a well-guarded secret. There are thousands of products available. Fortunately there is no need for the operator to screen them all as there are specialist services to be had. Finding an alternative would still involve a little trial and error.

It is generally known that where one agent fails to combat foam, another one will succeed. There is clearly no up-front guaranteed solution although past experience of operator and service provider will most often lead to the problem being solved quickly.

The foam combating agents do not remove the cause of the foaming, they attack the foam, that is, cure by removing symptoms. There are, however, foam treatment systems that works on a slip stream where the contaminants are also removed (Cummings *et al.*, 2007).

Von Phul and Stern (2005) made a survey of foaming experience by gathering information from a large number of operators. It seems that few of them had foaming problems but they still added foam combating agents on a regular basis. There were even cases where addition of foam combating agent was automatically linked to the measurement of pressure drop in the absorber.

When the concentration of foam combating agents becomes too high, they actually start to promote foam formation. If anti-foams must be used regularly as opposed to occasionally, the plant suffers from a more fundamental problem. Care must obviously be shown. In the case of silicone based anti-foams they are known to be removed by activated carbon (Von Phul and Stern, 2005). The presence of these surfactants in the solution may lead to lower mass transfer coefficients and hence less efficient absorption.

19.8   Corrosion Inhibitors

Modern plants operate with higher amine concentrations and anti-corrosion inhibitors are added. There are also inhibitors to abate oxidation by oxygen, which is particularly interesting in the newer field of flue gas treating.

19.9   Waste Handling

The absorbent bleed from an alkanolamine based treating plant is a hazardous chemical that must be disposed of accordingly. It is reckoned that it costs as much on a volume basis to get rid of the waste as it costs to buy the fresh chemical. There are companies specialising in this field. The waste may be exposed of by high temperature incineration. A cement oven is one possibility.

19.10   Solution Containment

When handling liquids in a plant of this nature, there will be a need for tanks. These tanks might be mere drums if the amount of liquid used is small as is the case with anti-foam chemicals and inhibitors.

Make-up amine may merit a tank where the amine is mixed with water before being fed to the plant, but it may equally be a case for pumping the undiluted amine into the plant steadily where it is diluted as it enters.

Process vessels like the desorber reflux drum are recommended to have a 2–5 min residence time. This is based on old rules-of-thumb from the processing industry.

There may be a need to have tank capacity to allow drainage of the working fluid from the plant.

Finally it may be required to have a containment dam around the plant in case of leakage.

19.11   Water Balance

Most absorbent solutions are aqueous and they are designed to have a specified content of water. This will eventually have an impact on the process. If there is a net loss of water, water must be added to avoid the solution becoming undesirably concentrated.

It is easy to see that a feed gas entering with little water content and leaving saturated will transport water out of the system. Gas entering as water saturated that has a significant portion of its water content removed will on the other hand bring water into the process. The process designer must pay attention to this potential problem. Operation of the plant in perfect water balance may be impractical. It may be possible to elect operation in a water positive or a water negative mode but the situation may also be forced. In either case it must be planned for removing or adding water to the process. Pure water, for example de-ionised boiler feed water (BFW), is usually necessary for use as make-up. This may seem expensive, but the treatment of waste water to be removed may be more expensive still. There is no option but to evaluate this on a case to case basis.

Quality of make-up water is important. Any salts in this water will manifest themselves as HSS in the solution forcing a higher bleed than could be achieved with salt free water make-up (Corsi *et al.*, 2002).

19.12   Cleaning the Plant Equipment

Any chemical plant will over time tend to build up scale and sludge deposits that need to be removed. Gas treating plants are no exception to this. The field has been reviewed by Canfield (2004) where mechanical and chemical methods are discussed. Problems like that tend to differ from plant to plant but it gives good insight to study the experience of others.

19.13   Final Words on Solution Management

In the last 20 years or so this area has received much more focus and a service industry has grown to provide help with various reclaiming actions and help with chemicals. It is possible to rent mobile reclaimers instead of investing in the full unit. This is particularly attractive when the plant is small, or when there is little contamination as these are situations when the reclaimer would only be used at intervals wide apart. Cummings *et al.* (2007) state that the improved reclaiming possibilities would typically save $US4 million per year in a 1 00 000 barrel-per-day refinery.

Names and ownerships will change over time and for this reason no names will be given here. These service providers are easy to find on the Internet and by studying talks given at conferences.

References

Alvis, S. and Jenkins, J. (2004) Contaminant reporting in amine gas treating service. Laurance Reid Gas Conditioning Conference, Norman, OK.

Burns, D. and Gregory, R.A. (1995) The UCARSEP process for on-line removal of non-regenerable salts from amine units. Laurance Reid Gas Conditioning Conference, Norman, OK.

Canfield, C.D. (2004) Amine system cleaning best practice. Laurance Reid Gas Conditioning Conference, Norman, OK.

Corsi, C., Betancur, R., Trovarelli, P. and Frey, C. (2002) Considerations for-design – operation of a reclaimer. Laurance Reid Gas Conditioning Conference, Norman, OK.

Cummings, A.L., Smith, G.D. and Nelson, D.K. (2007) Advances in amine reclaiming – why there's no excuse to operate a dirty amine system. Laurance Reid Gas Conditioning Conference, Norman, OK.

Daughton, D. and Veroba, B. (2007) Diglycolamine solvent quality improvement with thermal reclaiming. Laurance Reid Gas Conditioning Conference, Norman, OK.

Harbison, J.L. and Handwerk, G.E. (1987) Selective removal of H_2S utilizing generic MDEA. Paper H, Laurance Reid Gas Conditioning Conference, Norman, OK.

Keller, A.E., Kammiler, R.M., Veatch, F.C., Cummings, A.L., Thompsen, J.C. and Mecum, S.M. (1992) Heat-stable salt removal processing from amines by the HSSX process using ion exchange. Laurance Reid Gas Conditioning Conference, Norman, OK.

Kohl, A. and Nielsen, R. (1997) *Gas Purification*, 5th edn, Gulf Publishing.

MacKenzie, D.H., Prambil, F.C., Daniels, C.A. and Bullin, J.A. (1987) Design and operation of a selective sweetening plant using MDEA. *Energy Prog.*, **7**, 31–36.

Parnell, D. (2000) Colour of amine solutions. Laurance Reid Gas Conditioning Conference, Norman, OK.

Sheilan, M.H., Spooner, B.H. and van Hoorn, E. (2008) *Amine Treating and Sour Water Stripping*, 5th edn, Amine Experts, Calgary.

Von Phul, S.A. and Stern, L. (2005) Antifoam. What is it? How does it work? Why do they say to limit its use? Laurance Reid Gas Conditioning Conference, Norman, OK.

Wagner, R. and Judd, B. (2006) Fundamentals – gas sweetening. Laurance Reid Gas Conditioning Conference, February 26 – March 01, Norman, OK.

20

Absorption–Desorption Cycle

In the previous chapters specific features of the absorption process have been discussed one by one. It is time to take a holistic look at the process. The process design must in the end fulfil its objectives, which are to design a fully operational and hopefully optimized plant. There is considerable scope for optimisation. The optimisation criteria may not be the same in all situations.

Good descriptions and information on industrial practice may be found from a number of sources. These include the books of Kohl and Nielsen (1997), the Gas Processors Suppliers Association (GPSA)/Gas Processors Association (GPA) *GPSA Engineering Data Book*, 12th edition (2004), and Maddox (1977), the proceedings from the GPA's Annual Conventions, the annual proceedings from the Laurance Reid Gas Conditioning Conference and not the least the continuing 'fundamentals lectures' from the latter conference (e.g. Wagner and Judd, 2006). There is, however, a lot of proprietary information involved and there is a time lag from when this becomes knowledge in the industry to when it becomes openly available.

20.1 The Cycle and the Dimensioning Specifications

The process flowsheet has in principle been much the same as that sketched in Figure 20.1. (Variations to this flowsheet will be discussed in the next sub-section 20.2). The feed gas to be treated enters the absorption column at the bottom and flows upwards counter-currently to the absorbent that enters at the top and trickles downwards. The word trickling is used here to imply that the gas represents the continuous phase. The rich absorbent leaves the bottom of the absorber after which it is heated by the lean absorbent coming from regeneration in the desorption column and reboiler. From the heat exchange the rich absorbent continue to near the top of the desorption column from where it trickles downwards counter-currently to vapour and desorbed gas moving upwards. The vapour is generated in

Gas Treating: Absorption Theory and Practice, First Edition. Dag Eimer.
© 2014 John Wiley & Sons, Ltd. Published 2014 by John Wiley & Sons, Ltd.

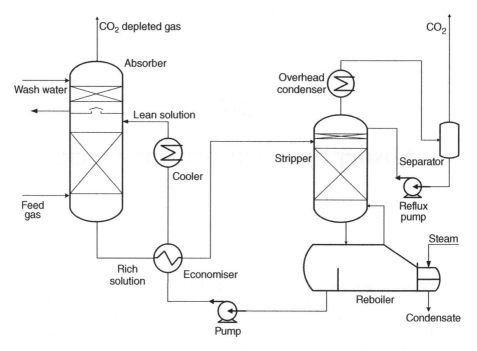

Figure 20.1 *Basic and traditional absorption–desorption cycle flowsheet.*

the reboiler. Some of it will condense on its way upwards to provide heat for the desorption of the absorbate. The vapour and desorbed absorbate is washed in a small section above the rich solution feed point where most of the amine is recovered, while CO_2 and $H_2O(g)$ will leave over the top of the desorber from where this stream will go to a condenser where the vapour will condense. The uncondensed absorbate will be separated from the condensate in a separation vessel from where the condensate will be pumped back to the top of the desorber where it is used as reflux in the small wash section in the top of the desorber. The regenerated, now lean, absorbent leaving the reboiler will be pumped back to the absorber via heat exchange with the rich absorbent before cooling by a coolant and then back to the top of the absorber.

In the very top of the absorption column, over the absorption section, there is a water wash section to recover absorbent vapours from the treated gas. This section is a feature in monoethanolamine (MEA) based plants, but may be omitted if another alkanolamine with very low vapour pressure is used.

The process and its description are simple enough. What are the dimensioning specifications? And what must be done to make it work?

In order to avoid a very abstract discussion, the absorption of CO_2 shall be used as an example, and an aqueous solution of a primary or secondary alkanolamine may be used as a sample absorbent. A 30% (wt) solution of MEA shall be used since data are widely available. (30% (wt) is roughly $5\,kmol/m^3$).

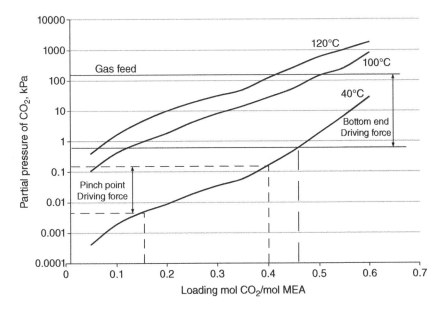

Figure 20.2 *An equilibrium diagram for CO₂ for 30% (wt) MEA in aqueous solution.*

The purpose of such a process is to meet some specification for the treated gas. A typical specification for an liquefied natural gas (LNG) train is that the CO_2 content must be removed to 50 ppm or lower which is an ambitious purification target even for an MEA plant. This has an impact on the process conditions to be achieved in the top of the absorber. At the temperature in the absorber top the equilibrium pressure of CO_2 over the lean absorbent entering must be lower than the partial pressure of CO_2 required in the treated gas. If the gas treated is at 3000 kPa, the partial pressure of CO_2 will be 0.15 kPa given 50 ppm CO_2 in the treated gas. A typical temperature would be 40°C. Using this information in the equilibrium diagram in Figure 20.2, it may be observed that the lean absorbent loading must be less than 0.39 mol CO_2/mol MEA.

There must be a reasonable driving force for the absorption also in the top of the absorber. It is elected here to set a lean loading at 0.15 mol/mol, which will give a top end driving force of roughly 0.15 kPa. The size of driving force is open for discussion, but making it smaller would require a taller column. The lean loading will in turn have a profound effect on the desorption column and the energy needed to operate it.

The specification for the gas entering will be known. The quantity and the amount of CO_2 present will determine the need for absorbent circulation rate. If the gas flow is 20 000 kmol/h, the temperature is 35°C, and the CO_2 content is 5%, then the circulation rate may be determined.

From Figure 20.2 it is seen that the rich absorbent loading could be as high as 0.65 mol CO_2/mol MEA. However, the rich loading shall be limited to 0.45 mol/mol. The rate of absorption slows down at higher loadings due to the product becoming bicarbonate rather than carbamate. The circulation flow Q m^3/h may be found from a simple CO_2 balance around the column. This is approximated by:

$$(Q \text{ m}^3/\text{h}) \, (5 \text{ k mol MEA}/\text{m}^3) \, [(0.45 - 0.15) \text{ mol } CO_2/\text{mol MEA}]$$
$$= (20\,000 \text{ k mol/h}) (0.05 - 0.02)$$
$$\text{and } Q = 400 \text{ m}^3/\text{h}$$

The approximation lies in the fact that the treated gas is 3% less than the 10 000 kmol/h entering. The accuracy is more than good enough for an initial appraisal. The absorbent flow rate and loading difference will determine the size of the regeneration part of the cycle.

The desorption column is the next consideration. The rule of thumb in gas treating is to operate the reboiler at 105–115°C. This determines the desorber pressure as the liquid will be at its bubble point in the reboiler. Since the gas stripped off will normally be released to ambient, there is little point in operating the desorber at a higher pressure than necessary. The pressure will follow from the choice of temperature in the reboiler in any case (it may be of interest to note that in flue gas applications for CO_2 abatement, the desorber pressure tends to be increased).

For absorbent regeneration to this level it may be expected to use in the order of 2 kg steam per kg CO_2 removed. This figure is widely quoted in the field of CO_2 abatement but in this field the reboiler temperature is often set to 120°C. The higher temperature tends to reduce the desorption energy needed. It is known that letting the lean absorbent loading increase to 0.20–0.25 mol/mol could reduce the energy consumption significantly. However, this optimisation is not in focus here but it is pointed out to show the possibilities as well as the constraints.

When using MEA as absorbent, it is customary to add a small water wash in the top of the absorber to recover MEA that would otherwise follow the treated gas. The reason for this measure is the vapour pressure of MEA over the absorbent. It is not high but the losses would still be significant.

Earlier (prior to 1970) designs limited the MEA concentration to 15% (wt) due to corrosion and degradation issues. Inhibitors were introduced in the 1960s, and these have allowed doubling the concentration to 30% (wt). This naturally reduces pumping power and less energy is spent heating the absorbent in the desorber.

The heat exchangers will not be discussed here, but shell and tube exchangers have been the work horse.

20.2 Alternative Cycle Variations

There are many variations on total gas pressure and content of CO_2, and H_2S for that matter, in natural gas. Not surprisingly the flowsheet as shown in Figure 20.1 has been adjusted to exploit local possibilities. It is not the aim of this text to review all possibilities practised but it is prudent to mention a couple of principles open for choice.

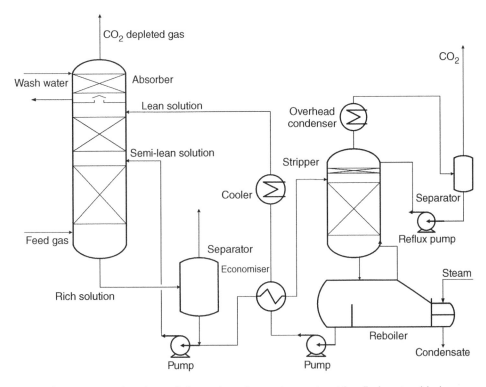

Figure 20.3   *Flowsheet of absorption–desorption cycle with a flash unit added.*

The first obvious variation to the flowsheet is to add a flash vessel to the rich absorbent stream to let some of the acid gas release prior to heating and entry to the desorber. This is illustrated in Figure 20.3 that speaks for itself. Its application would be where the partial pressures of the acid gas(es) were high enough to load up the solution to an extent where gas would release spontaneously on reduction of pressure.

Another variation that may be used with tertiary amines and physical solvents is to use flash regeneration only. This is shown in Figure 20.4. Further explanation of the flowsheet should be unnecessary. It may be the process of choice for high partial pressures of acid gas and where the specification of the treated gas is not too stringent. If the gas specification was stringent, it is possible to operate a split absorption process where, for example 10% of the absorbent is desorbed fully in a column and this leaner stream is then fed to a higher point in the absorption column. This flowsheet variation is as illustrated in Figure 20.3 but the design will be tuned towards reaching a very low acid gas content in the gas leaving the absorber.

It is also appropriate to mention the use of hot carbonate processes that are also used widely. This process, or processes, would tend to use the flowsheet shown in Figure 20.4 or a variation of this. There are, however, a number of other variations also in this field.

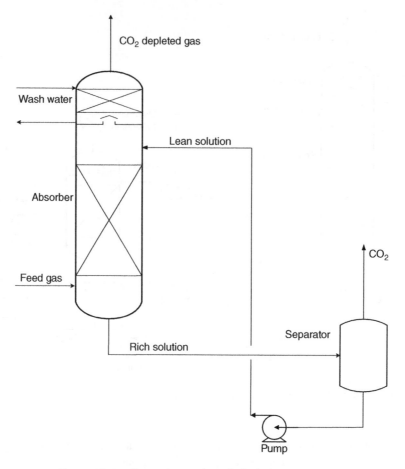

Figure 20.4   *Absorption cycle with flash desorption only.*

20.3   Other Limitations

It has already been alluded to previously that there are a number of less obvious constraints to take into account for process design than equilibrium limitations and rate of mass transfer. There are limits to loadings and absorbent concentrations due to corrosion issues. Some of these could undoubtedly be dealt with by using better materials of construction but that would increase the plant costs. The oil and gas industry has by tradition set ways of doing things as well, and these help to cement the well tried rules of thumb developed over years of practice. Related to corrosion is erosion and this is certainly a consideration as heat is added and pressure is released from the rich absorbent. The gas thus released represents a very significant volume flow and it would make the flow two-phase. This is a good recipe for erosion. Rich absorbent should preferably not undergo pressure let-down until just before the flash vessel or desorption column. Wear plates might be added to such equipment.

The dimensioning of the column diameter has not yet been discussed. It is fairly straight forward, but, for example amine solutions are classified as having a propensity for foaming. This means that the pressure drop per distance (kPa/m) should be kept a little lower than the choice that would otherwise be made.

Attention must also be paid to the composition of the feed gas. There should be a scrubber upstream of the absorber to catch any condensate. Special care must be taken if the feed gas contains hydrocarbons that might condense or to an extent dissolve in the absorbent as this represents an extra potential for foaming.

20.4 Matching Process and Treating Demands

Matching the process to the treating demands sounds like an obvious statement. However, it is worth stressing since subtle points are easy to overlook. Condensing hydrocarbons have been mentioned earlier. Avoiding cooling of the gas is one remedy for that. It may be desirable to pre-heat the gas if it is a retrograde one although this would be expensive. Such gas would condense when the pressure is reduced and there is pressure drop in a column. This might also influence the choice of column hardware.

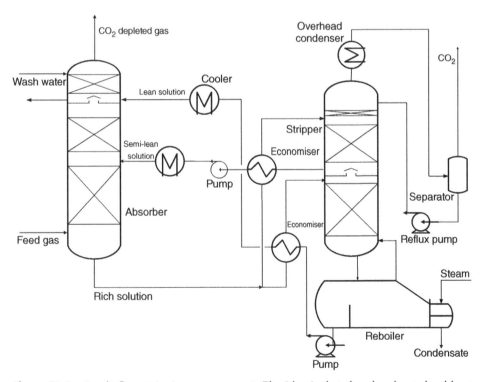

Figure 20.5 *A split-flow stripping arrangement. The idea is that the absorbent should not be more extensively regenerated than absolutely necessary. This is addressed by reducing the mass transfer driving forces in the absorber.*

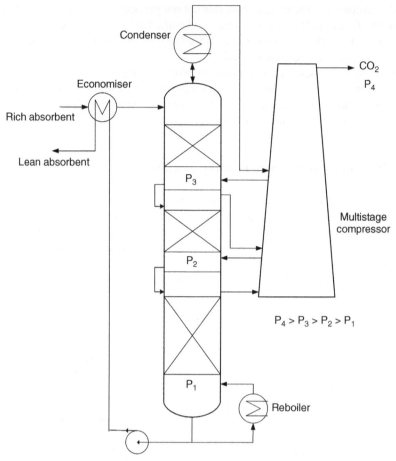

Figure 20.6 *Multipressure stripper. Since the CO_2 shall be compressed in the end, there is an energy saving side to letting the CO_2 go to the compressor at a higher pressure. Reducing the CO_2 volume flow will also lead to a smaller and cheaper compressor.*

Meeting explicit specifications is obvious.

If there is H_2S removal involved, care must be taken to make the off gas suitable for the next step in handling the sulfur. The presence of H_2S would also dictate extra attention to avoid leaks. This includes sweating liquids through seals.

There is also the need to make use of whatever energy is available locally, and that may have an impact on the plant.

20.5   Solution Management

The flowsheets shown in Figures 20.1, 20.3 and 20.4 are skeleton sketches and do not show any features for solution maintenance. This question is dealt with in Chapter 19,

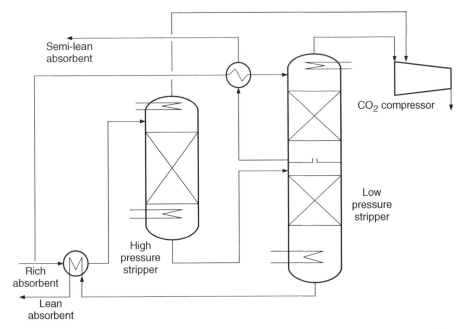

Figure 20.7   *This is the depiction of a so-called double matrix stripper. There are parallels to the split-flow process but it is noticed that part of CO$_2$ bypasses the first compressor stage. This saves energy and investment.*

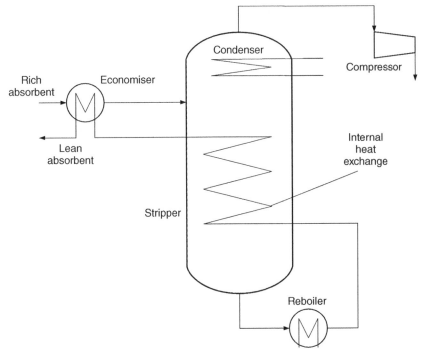

Figure 20.8   *The internal heat exchange variation. Here, the desorption heat up of the column is provided by indirect heat rather than the direct contact heat normally used.*

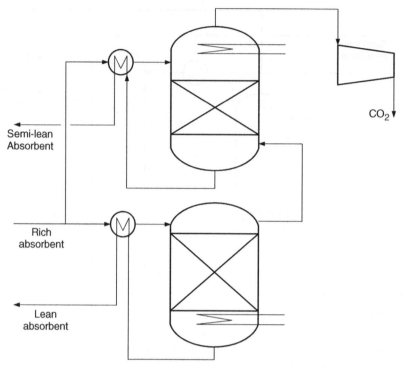

Figure 20.9   *The flashing feed stripper. It is known that for the standard process some CO_2 will flash off as soon as the rich absorbent enters the stripper. This process explores this road a little further.*

but it is an important feature of amine based absorption plants to keep the solution clean. A contaminated solution is more likely to foam and foaming would precipitate anti-foam treatment in the form of anti-foam agents.

Any amine solutions would also degrade over time. The degradation products will steadily build up and must be removed before they become a problem. The remedies include:

- Filtering
- Reclaiming
- Bleed (and make-up).

The presence of degradation products will, in general, further the formation of more degradation products. Keeping the absorbent clean is a question of process economics. It costs money to carry out but lost on-stream time if the process fails is even more expensive.

20.6   Flowsheet Variations to Save Desorption Energy

The advent of the great interest in capturing CO_2 from energy plant flue gases has provoked considerable thinking with respect to reducing the energy lost from capturing the said CO_2.

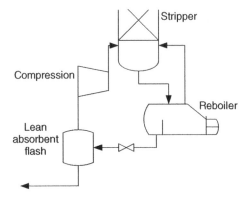

Figure 20.10 *Flash of lean absorbent from reboiler and re-compression of the ensuing water vapour with return to stripper where it would serve in the stripping process (Based on: Choi et al., 2005; Reddy, Johnson and Gilmartin, 2008).*

Some ideas are illustrated in the Figures 20.5–20.9. Their place relative to the desorption process has been touched upon in Chapter 17 but the actual flowsheets are summarised here. Further discussion of these may be found in the references (Aroonwilas, 2004; Oyenekam and Rochelle, 2007a,b). The essence of these process variations should be mainly self-explanatory to a chemical engineer.

In addition to the flowsheet variations shown in Figures 20.5–20.9, it is also possible to take the moist CO_2 stream off the top of the stripper and compress it instead of condensing the water vapour in a condenser where the associated energy is wasted to cooling water. Figure 20.10 shows how some stripping steam may be obtained by flashing the lean absorbent.

References

Aroonwilas, A. (2004) Evaluation of split-flow scheme for CO_2 absorption process using mechanistic mass-transfer and hydrodynamic model. Proceedings from GHGT-7, the 7th International Conference on Greenhouse Gas Control Technologies.

Choi, G.N., Chu, R., Degen, B., Wen, H., Richen, P.L. and Chinn, D. (2005) CO_2 removal from power plant flue gas – cost efficient design and integration study, in *Carbon Dioxide Capture for Storage in Deep Geological Formations – Results from the CO_2 Capture Project*, vol. **1** (ed D.C. Thomas), Elsevier, pp. 99–376.

Gas Processors Suppliers Association (GPSA) and Gas Processors Association (GPA) (2004) *Engineering Data Book*, 12th edn.

GPA Annual Convention. Proceedings are available from Engineering and Geosciences Programs University of Oklahoma Outreach, Norman, OK.

Kohl, A. and Nielsen, R. (1997) *Gas Purification*, 5th edn, Gulf Publishing (Laurance Reid Gas Conditioning Conference, Norman, OK. A collection of proceedings 1988-"to date" are updated annually and are available for purchase).

Maddox, R.N. (1977) *Gas and Liquid Sweetening*, John M. Campbell, Norman, OK..

Oyenekam, B.A. and Rochelle, G.T. (2007a) Energy performance of stripper configurations for CO_2 capture by aqueous amines. *Ind. Eng. Chem. Res.*, **45**, 2457–2464.

Oyenekam, B.A. and Rochelle, G.T. (2007b) Alternative stripper configurations for CO_2 capture by aqueous amines. *AIChE J.*, **53**, 3144–3154.

Reddy, S., D. Johnson, J. Gilmartin, (2008) Fluor's Econamine FG Plus[SM] technology for CO_2 capture at coal-fired power plants. Power Plant Air Pollutant Control "Mega" Symposium, Baltimore, MD, August 25–28.

Wagner, R., B. Judd, (2006) Fundamentals – gas sweetening. Laurance Reid Gas Conditioning Conference, Norman, OK.

21

Degradation

The object of this chapter is to give a primer with respect to the problem of degradation of the absorbent. A thorough review is beyond the scope of this text. It is a big field and major research is still needed to fully understand its chemistry.

21.1   Introduction to Degradation

It is observed that degradation products are formed in the various absorbent solutions used for gas treating. There is no question of 'if' they are formed, the questions are how much and what. This problem is old, degradation has been observed since the dawn of the application of alkanolamines for gas treating. However, the advent of flue gas treating means oxidative degradation has become the most important problem.

The accumulation of degradation products will lead to operational problems if left to their own devices. Solution management is needed and has been discussed in a previous chapter. We would obviously like to know how to best operate the plant to minimise degradation. Ideally all reaction paths and all the associated kinetics and/or equilibrium conditions should be known. Then it would be possible to indulge in an exact optimisation job. However, this is not the situation at the moment as there are merely a few pointers as to what are good measures to take to minimise the problem. These pointers will be discussed next.

There are many publications on treating degradation chemistry being published. With the advent of flue gas treating for greenhouse gas abatement this field is receiving accelerating attention. Although all the research work done is serious, we must also be allowed to say that it is somewhat fragmented in the sense that most of it seems to investigate the effect of isolated variations. The field would benefit from being subjected to the rigorous approach applied in organic chemistry. Most work being done today is done by

Gas Treating: Absorption Theory and Practice, First Edition. Dag Eimer.
© 2014 John Wiley & Sons, Ltd. Published 2014 by John Wiley & Sons, Ltd.

chemical engineers. A typical chemical engineering laboratory does not have all the analytical equipment and procedures that a good chemistry laboratory would have. However, until now it has been difficult to attract the interest of chemists to this problem. Research in the chemistry involved is something of the past in this field.

The aim of this chapter is not to review the field of degradation in depth. The literature is too fragmented to cover in one book chapter. There is an extensive discussion in the book by Kohl and Nielsen (1997). The reader's attention is also directed to an extensive report made for the Norwegian governmental body Gassnova (Fredriksen, Jens and Eimer, 2011; Fredriksen and Jens, 2013). Degradation is presently researched by many, but mainly by chemical engineers. A good starting point for reading up on this topic includes the papers coming out of University of Texas at Austin, University of Regina, NTNU in Trondheim and Telemark University College.

Degradation leads to waste products. Getting rid of them costs money. Since the waste is regarded as a 'problem waste', disposal can actually be quite expensive. This depends to a degree on the local situation, whether there are waste handling facilities nearby or not.

The waste is removed in the form of a bleed stream from the absorbent circulation system. It would be desirable if no fresh absorbent is lost with the bleed but that is not realistic. Hence the cost of disposing of waste also includes the replacement of perfectly good absorbent. Methods for minimising this fresh absorbent loss were discussed under solution management.

The whole process train may in the context of degradation be seen as one big and complex reactor. All liquid in the equipment represents a reaction volume, more volume means more reaction taking place. Reactions naturally go faster at higher the temperatures, and there is the question of which reactants are present where.

21.2   Carbamate Polymerisation

It has been known for a long time that the presence of CO_2 leads to degradation of the absorbent solution. This is now referred to as carbamate polymerisation. The reaction products include higher molecular weight compounds like oligomers/polymers and cyclic compounds. These reactions benefit from high levels of CO_2 and high temperatures.

Such conditions are particularly featured in the top of the desorption column before CO_2 is stripped from the solution. The temperature is a further 10 K or so higher in the reboiler region, but by then the CO_2 content has been significantly reduced although there is plenty left for a reaction.

Primary and secondary amines can form carbamates, but tertiary ones like methyldiethanolamine (MDEA) cannot. However, tertiary amine solutions, even if supposedly pure, are likely to have primary and/or secondary amines present because of their manufacturing process. Studies have also shown that transalkylation may take place and cause degradation anyway.

21.3   Thermal Degradation

Thermal degradation has often been listed as an important way to degradation. Actually, this form of degradation is of little importance at the temperatures normally used in the absorption–desorption process. It is reckoned that temperatures in excess of 200°C are needed to thermally degrade alkanolamines.

21.4 Oxidative Degradation

Oxidative degradation is particular to flue gas treating where very high oxygen levels are present in the gas. In the context of mass transfer, the oxidation reactions are 'slow', that is they do not cause enhancement of mass transfer rates as has been discussed for CO_2 and H_2S. So-called slow reactions may again be sub-divided into reaction limited and mass transfer limited.

A number of claims have been made with respect to oxidation products and reaction paths. The information is fragmented and partially contradictory. There is a need for a consistent study where all products and intermediates are identified and quantified in such a way that a proper claim to a reaction path, or paths, may be made.

21.5 Corrosion and Degradation

Corrosion and materials are discussed in another chapter. However, corrosion products are generally metal ions, Fe^{2+} and Fe^{3+} in particular. The literature on degradation discusses how the presence of these ions may accelerate the rate of other degradation reactions. The size of this effect remains to be quantified.

21.6 The Effect of Heat Stable Salts (HSSs)

Heat Stable Salt (HSS) will build up in the absorbent solution over time. There are many years of experience to back this up. With monoethanolamine (MEA) solutions these salts are removed in the reclaiming process since they are not volatile. One source for building these is the presence of organic acids in the solution. Natural gas may contain traces of such acids and in syngas or refinery gas this is even more common.

In flue gas treating HSSs may also be formed due to the presence of NO_x and SO_x in the gas. Since these become relatively strong acids in an aqueous solution, they will react irreversibly with the basic alkanolamines to form salts.

21.7 SOx and NOx in Feed Gas

SO_x and NO_x have not been a problem when treating syngas or natural gas. However, when treating flue gas these components would be expected to be present unless steps were taken to remove them upstream of the treatment. SO_x would be present if sulfur were present in the fuel gas. A typical natural gas specification is for less than 4 ppm of H_2S in the gas and SO_x levels would be lower due to the dilution with air for combustion. NO_x level in the flue gas is more likely to be caused by NO_x arising from the combustion process.

NO_x implies both NO_2 and NO. The latter is not easily absorbed and would tend to slip through the absorber unless it is oxidised on its way through.

When these compounds are present, their absorption rate will be accelerated by the alkaline absorbent. They also have high solubilities in water, which would help too. Once they are absorbed they would form nitrites and sulphites with the alkanolamines present. The formation of these salts represents a consumption of alkanolamine. Sodium or potassium carbonate solutions would also be affected by NO_2 and SO_2.

21.8   Nitrosamines

The problem of nitrosamines is very specific to the treatment of flue gas. There was no real discussion of nitrosamines until about 2010. At that point in time there was a worry arising in the development of the TCM (Test Centre Mongstad) in Norway over the possibility of releasing, directly or indirectly, cancerogeneous chemicals into the atmosphere. They would subsequently be washed out by rain and find its way into nature at large including the possibility of exposing humans to the risk of exposure.

A number of studies were launched. The general conclusion was that this was not a problem. It was stated that there was not a risk of such chemicals and it was equally stated that the wash process in the top of the absorber would take care of it anyway. Proving a negative is, however, hard. It is difficult to eradicate all ideas that this is a problem. Partially this may be due to communicating the results but there will also be elements of feeling that the conclusion is based more on judgement than on solid evidence. Anybody working in this area must expect to be confronted with this issue at some stage. This problem area is still under surveillance.

21.9   Concluding Remarks

A lot has been published on the degradation of alkanolamines. The report by Fredriksen *et al.* (2011) quotes in the order of 80 papers, and this review was probably not exhaustive. It seems that information in this field is still a bit fragmentary. Proper studies closing the chemical paths and mass balances are needed to settle this subject area.

In the meantime there is a significant quantity of research work going on that aims at quantifying the rate of degradation. Interesting as such work may be, and valuable for the documentation of the cases studied, it does not in general go into enough depth to enable full insight and the ability to extrapolate the knowledge into new operational domains and chemical systems that will undoubtedly see the light of day.

References

Fredriksen, S., K. Jens, D. Eimer, (2011) Theoretical Evaluation of the Probability to form and Emit Harmful Components from the Aqueous. Gassnova Project N&E TQP Amine, CCM Project, Frame Agreement 257430116, Contract 257430117. See web-link: www.gassnova.no/no/Documents/ProcessFormation_TELTEK.pdf (accessed 7 May 2014).

Fredriksen, S.B. and Jens, K.J. (2013) Oxidative degradation of aqueous amine solutions of MEA, AMP, MDEA, PZ. A review. *Energy Procedia*, **37**, 1770–1777.

Kohl, A. and Nielsen, R. (1997) *Gas Purification*, Gulf Publishing.

22

Materials, Corrosion, Inhibitors

22.1 Introduction

Amine based absorption processes were introduced in the 1930s. At that point in time chemical engineering and associated disciplines were much less theoretically based than today. In the 1950s a number of reports and publications were published on materials problems in amine based treaters. A lot of empirical observations were revealed and shared in the industry. Since then information has accumulated and an 'industry practice' has established itself. Early publications of this type include Paredes and Cronenberg (1954) and Hofmeyer, Scholten and Lloyd (1956). They presented observations made with the water-diethylene glycol (DEG) solvents with either monoethanolamine (MEA) or diethanolamine (DEA) used at the time. Some plants had severe corrosion problems within months of start-up while others operated for years without any trouble. Clearly there were effects in these plants not understood.

Kohl and Nielsen (1997) have reviewed this aspect of the process design. The object of this chapter is to make anybody studying gas treating or practising it, aware that there are materials challenges to be handled. More recent discussions have been published by a number of authors (Billingham *et al.*, 2011; Bosen, 2000; Kladkaew *et al.*, 2011; Kittel *et al.*, 2012; Rennie, 2006; Zhao, Yang and Fan, 2011). These references are by no means exhaustive, merely indicative. Corrosion rates (CRs) of carbon steel is said to increase by increase in temperature, concentration of amine and CO_2 loading in the solution (Zhao *et al.*, 2011). A few typical CRs are given by Cross *et al.* (1990) as 0.2–0.8 mm per year (8–32 mpy). The term mpy refers to milli-inches per year. Useful conversions between forms of reporting CRs may be found on a web site 'corrosion-doctors.org' (arranged by Roberge at the Royal Military College of Canada, Ontario).

It could be argued that the amine based absorption systems are relatively benign with respect to the need for exotic construction materials. By and large carbon steel is the default choice but there are areas in the process where more high grade materials are merited, or alternatively that the process conditions and equipment design are modified. It is common to specify stainless steel (304) for column internals and heat exchanger tubes. Reasonable

Gas Treating: Absorption Theory and Practice, First Edition. Dag Eimer.
© 2014 John Wiley & Sons, Ltd. Published 2014 by John Wiley & Sons, Ltd.

explanations of corrosion problems encountered may be found in the literature. It is in any case prudent to consult a corrosion expert if designing a full process. As pointed out by Kohl and Nielsen (1997), corrosion is the biggest operational problem experienced in gas treating in spite of the seemingly benign nature of the process. Most corrosion problems experienced in plants are localised problem rather than general problems (Dupart, Bacon and Edwards, 1993).

Flue gas treating represents a new field for CO_2 capture. It is low pressure, large flows and the presence of oxygen and the possibility of acidic trace components that make a new setting with respect to the choice of materials. Kladkaew *et al.* (2011) have provided an equation to predict the CRs for carbon steel in this case. The CR (mm per year) may be estimated from:

$$CR = 45.0 \times 10^7 \left\{ \exp\left(-\frac{5.955}{T}\right) \right\} \left\{ [SO_2]^{0.0011} [O_2]^{0.0006} [CO_2]^{0.9} [MEA]^{0.0001} \right\}$$
(22.1)

The first thing to define when considering potential corrosion problems is the composition of system. It may seem obvious but aspects of this are easily overlooked. Trace components can play a major role in corrosion mechanisms and these components are easily forgotten if the question is not raised. Hence, there is a need for awareness. In refineries, but not necessarily limited to, we may expect to find components like cyanide, ammonia, maybe oxygen and acids like formic, acetic and oxalic in the gas to be treated. In the newer field of flue gas treating there is likely to be traces of NOx, SOx and certainly oxygen included. The presence of H_2S and CO_2 are of course important, but less obvious is that a minimum presence of H_2S is beneficial because a protective sulfide layer may be established.

Four areas of the process are singled out for special attention with respect to the choice of material and/or special focus on process conditions.

• Vapour phase involving CO_2 and H_2S with water condensation.
• Under the bottom of the absorber packing/tray.
• Top of desorber and its reflux area.
• Reboiler.

There may be gas pockets appearing in unplanned places, for instance over the filters, and all of these will facilitate the occurrence of a sour condensate film or droplets.

22.2   Corrosion Basics

The most rudimentary basics of corrosion are described by the following reactions or half reactions. The reason why these reactions are proceeding is well described in books on corrosion (e.g. Stewart and Tulloch, 1968). The steel is being corroded by

$$Fe \rightarrow Fe^{2+} + 2e^-$$
(22.2)

This is accompanied by:

$$H^+ + e^- \rightarrow H$$
(22.3)

The last reaction forms atomic hydrogen. Most of this combines to molecular hydrogen but, given the opportunity, this atomic hydrogen may migrate into the steel where it will combine to molecular hydrogen in cavity-like faults in the material. This formation of molecular hydrogen may eventually lead to material cracking (hydrogen embrittlement).

These reactions are irreversible and their rates are controlled by temperature and the concentrations of the reactants.

When CO_2 dissolves in water, the liquid's pH is lowered, and the following reaction may occur leading to more corrosion by providing more H^+ to consume the e^- from reaction (22.2)

$$CO_2 + H_2O \rightarrow HCO_3^- + H^+ \qquad (22.4)$$

The carbonate can react with iron to form an iron carbonate layer but this layer is considered too porous to provide protection against further corrosion, although it can slow it down. In an oxidatively degraded MEA solvent, oxalate is expected to be present, which is suspected to dissolve any iron carbonate layer and hence accelerating the corrosion process (Rooney, Bacon and DuPart, 1996; 1997).

H_2S may react directly with the iron in the steel according to:

$$Fe + H_2S \rightarrow FeS + 2H \qquad (22.5)$$

The sulfide layer thus produced is compact enough to give protection against further corrosion. However, it is destroyed by the presence of HCN which reacts with the iron in FeS to form an iron cyanide complex.

An interesting observation is that corrosion is less in gas treatment plants that treat both CO_2 and H_2S than in plants just removing one or the other (Dupart, Bacon and Edwards, 1993). The same authors point out that corrosivity is increased by increasing the amine concentration, increasing the temperature and by increasing the gas loading. DEA is less corrosive than MEA, but not as little as methyldiethanolamine (MDEA) in comparison to either of them. Veawab, Tontiwachwuthikul and Bhole (1996) found in a study of carbon steel that 2-amino-2-methyl-1-propanol (AMP) with CO_2 was less corrosive than MEA, particularly at concentrations above 5 M. They also found that even small amounts of O_2 would lead to severe corrosion.

The presence of heat stable salts (HSSs) also enters these considerations. There are compounds formed that affect the pH of the solution and their nature also suggests that the solution's conductivity should be increased leading to the above reactions being able to go faster. Hence, the presence of HSSs increases CRs. They are always present to some extent in a practical system.

22.3 Gas Phase

Intuitively the immediate attention is directed towards the liquid phase and the gas phase is easily overlooked. However, both CO_2 and H_2S will make any aqueous condensate sour. When water condenses on the surfaces, it will absorb, for example, CO_2 and its pH is lowered into a region where there is potential for corrosion. This is referred to as wet CO_2 corrosion.

Areas affected by this in the process is underneath the lowest tray in the absorption column (given a tray column) and the overhead system of the desorber. Any other wash system in the process is also a section worth analysing with respect to such corrosion.

22.4   Protective Layers and What Makes Them Break Down (Chemistry)

When a material corrodes, there is often a protective layer formed that protects against further corrosion. However, the presence of other components in the system may lead to this protective layer being damaged whereby corrosion may progress after all. Such damage in a protective coat may lead to more severe damage locally than if the layer was not there in the first place. A protective layer may also be damaged by erosion. The sulfide layer caused by H_2S is liable to be damaged by high liquid velocities.

It is reckoned that a partial pressure of minimum 0.03 bar for H_2S (e.g. 600 ppm at 50 bar) will lead to the formation of a protective sulfide layer. This sulfide layer is dense enough to be protective. A similar layer formed by carbonate corrosion is porous and will not give effective protection.

In general anodic passivation layers can prevent corrosion but once a fault in the layer crops up the protective layer really does more harm than good. The fault area would be subject to increased corrosion and this would be more serious than an even corrosion attack.

22.5   Fluid Velocities and Corrosion

Fluid velocities must be watched. Overly high flow velocities may lead to erosion, not the least of a protective layer. Erosion becomes more severe when there is two phase flow involved. In an absorption–desorption process there are a number of areas in the process where gas and liquid appear together. These are the obvious process areas where gas escapes from liquid but there may also be erosion from liquid droplets entrained in the gas. Particles in the liquid may also appear, but that is a much smaller issue if the solution is properly maintained.

It is recommended to keep liquid velocities below 1 m/s (3 ft/s). There are also recommendations for twice this velocity. A maximum of $1.5\,\text{m·s}^{-1}$ is recommended by Rennie (2006). Dupart, Bacon and Edwards (1993) suggest less than 1 m/s when using carbon steel and 1.5–2.5 m/s if stainless steel is used. The picture is not clear, as may be expected. Seemingly conflicting information is readily available.

Two phase flow can lead to erosion corrosion. If particles precipitate, they will act as an abrasive and any protective film is particularly prone to damage. Hence, there is a need for filtering the absorbent solution. The evolution of gas bubbles may similarly cause erosion if the flow becomes too vigorous. Hence, the recommendation is to place the pressure letdown valve on the rich absorbent stream as near to the desorber as possible. There are also recommendations given for design of the reboiler. Gas blanketing around the tubes should be avoided. If the flow pattern leads to a vena contracta being formed, there would be a reduced pressure at this point and this might lead to gas being freed in the liquid.

Any obstruction such as an orifice plate may be the cause of this if not properly designed. Another possibility is liquid flow entering tubes or pipes.

Even minor features like a vortex breaker in the absorber bottom should be designed with care. It could cause enough pressure drop to create bubbles that would lead to two-phase flow in the piping and the lean-rich heat exchanger.

22.6   Stress Induced Corrosion

Hydrogen atoms diffusing into the material where hydrogen gas is formed and accumulated in material imperfections lead to stress cracking. This is caused by the hydrogen pressure building up in cavities in the material until it exceeds what the material can handle. It is a significant problem in industry. In ammonia plants where hydrogen is present at high partial pressures it is a concern that is taken very seriously. Hydrogen induced stress corrosion (HISC) as such is not a big problem in gas treating plants. H_2S can induce stress corrosion. Its presence enhances the migration of hydrogen in the steel. This may be a problem, but the circumstances will decide how big a problem it is. There is talk of Stress-Oriented HIC (SOHIC) and there is pure Sulfide Stress Corrosion cracking (SCC). All these cases involve migration of atomic hydrogen in the steel.

It is also possible for alkali to cause stress cracking, so-called alkaline stress-oriented cracking (ASCC). This is not seen as a big problem in gas treating plants.

22.7   Effect of Heat Stable Salts (HSS)

Some specific work has been done to investigate the effect of HSS on CRs in amine solutions. Thanthapanichakoon, Veawab and McGarvey (2006) studied the effect of six salts in 5 M MEA solution and concluded that they increased the CR of carbon steel with oxalate having the biggest effect followed by malonate and formate while there was only a small effect from acetate, glycolate and succinate salts. Neither of these salts had much effect on stainless steel (304).

22.8   Inhibitors

Corrosion problems in capture plants may be alleviated, or even controlled, by the use of inhibitors.

This field is likely to expand with the event of flue gas treating whenever this will take off. Inhibitors have been looked into by institutes heavily involved with CO_2 capture research as well as the typical corrosion research environments. A recent laboratory study by Kladkaew *et al.* (2011) included tests of six inhibitors as well as a general study of CRs. They have documented that there were inhibitors that could reduce CRs by 90–98%. However, the inhibitor identities were not revealed. The use of surfactants to prevent corrosion, but not specific to CO_2 capture, was reviewed by Malik *et al.* (2011). The protection of mild steel in electrolytic systems with CO_2 present has been discussed by Desimone *et al.* (2011). Inhibitors are also a part of anti-corrosion action in natural gas treating plants (Edwards and Rodriguez, 2000). An interesting review including historical trends is given in a paper by Zhao, Yang and Fan (2011b). These are but starting points for digging into this field.

22.9   Problem Areas, Observations and Mitigation Actions

Types of corrosion to look out for when reviewing design or trouble shooting include galvanic, crevice (where lack of oxygen or different concentrations exist), pitting and intergranular corrosion arising, for example, from poor welds. Erosion induced corrosion is another that has been discussed related to fluid flow. It is in this context good to have high radius elbows and avoid sudden changes in the direction of flow (Dupart, Bacon and Edwards, 1993).

In summary there are a number of actions that are prudently taken when designing a gas treating process. Such considerations needed for good equipment or hardware design are discussed by Bosen (2000) and Rennie (2006). It is clear that the devil lies in the detail. A short list of actions includes those summarised next but there is no *one* list that takes care of all problems to the extent that the engineer may be relieved of the need to use his insight and skills to foresee problems and deal with them. There will often, if not always, be special circumstances that could make the process more benign or aggravate problem areas.

Velocities of fluids should be kept at reasonable levels compared to past experience and recommendations. Sometimes recommendations may be challenged as they do tend to be on the conservative side. Flow impinging on surfaces may cause erosion, at least by destroying the protective film that may have been built up, and this again will increase local CRs.

The overall process design should always be considered. This includes the choice of amine and its concentration. Acid gas loadings are also a point to consider. Temperatures and pressures should be looked at.

The control strategy and the set points associated may be utilised to avoid process excursions into operating conditions that may lead to problems. If for instance loadings are to be kept within bounds, then the process conditions should be controlled to avoid harmful excursions.

The need for solution purification, bleed and make-up are all linked. Solution management is an important area also for corrosion prevention. At the design stage it is necessary to provide the operational staff the opportunity to attend to this in a proper manner as this is perhaps more of an operational problem than a design issue.

Finally, it is a good strategy to involve a corrosion expert with respect to the choice of materials. Proper interaction between disciplines must be strived for.

References

Billingham, M.A., Lee, C.-H., Smith, L., Haines, M., James, S.R., Goh, B.K., Dvorak, W.K., Robinson, L., Davis, C.J. and Peralta-Solorio, D. (2011) Corrosion and materials selection issues in carbon capture plants. *Energy Procedia*, **4**, 2020–2027.

Bosen, S. (2000) Causes of amine plant corrosion-design considerations. CORROSION 2000, March 26-31, Orlando, FL, Paper 00492 (abstract).

Cross, C., D. Edwards, J. Santos, E. Stewart, (1990) Gas treating through the accurate process modelling of specialty amine plants. Laurance Reid Gas Conditioning Conference, Norman, OK.

Desimone, M.P., Grundmeier, G., Gordillo, G. and Simison, S.N. (2011) Amphiphilic amido-amine as an effective corrosion inhibitor for mild steel exposed to CO_2 saturated solution: Polarization, EIS and PM-IRRAS studies. *Electrochim. Acta*, **56**, 2990–2998.

Dupart, M.S., T.R. Bacon, D.J. Edwards, (1993) Understanding corrosion in alkanolamine gas treating plants, parts 1 & 2. *Hydrocarbon Process.*, **93**, 75–80 and 89–94.

Edwards, M.A., E.F. Rodriguez, (2000) The use of an organic inhibitor to control corrosion in alkanolamine units processing gas containing CO_2 (abstract). CORROSION 2000, March 26–31, Orlando, FL, Paper 99260.

Hofmeyer, B.G., H.G. Scholten, W.G. Lloyd, (1956) Contamination and corrosion in monoethanolamine gas treating solutions. Symposium of Petroleum Division, American Chemical Society, Dallas, TX, April 8–13.

Kittel, J., Fleury, E., Vuillemin, B., Gonzalez, S., Ropital, F. and Oltra, R. (2012) Corrosion in alkanolamine used for acid gas removal: from natural gas processing to CO_2 capture. *Mater. Corros.*, **63** (3), 223–230.

Kladkaew, N., Idem, R., Tontiwachwuthikul, P. and Saiwan, C. (2011) Studies on corrosion for CO_2 absorption from power plant flue gases containing CO_2, O2 and SO_2. *Energy Procedia*, **4**, 1761–1768.

Kohl, A. and Nielsen, R. (1997) *Gas Purification*, 5th edn, Gulf Publishing.

Malik, M.A., Hashim, M.A., Nabi, F. and Al-Thabaiti, S.A. (2011) Anti-corrosion ability of surfactants: A review. *Int. J. Electrochem. Sci.*, **6**, 1927–1948.

Paredes, F. and Cronenberg, J.W. (1954) Combating heat-exchanger corrosion in glycol-amine gas-treating plants. *Oil Gas J.*, **53** (7), 159–161.

Rennie, S. (2006) Corrosion and materials selection for amine service. *Mater. Forum*, **30**, 126–130.

Rooney, P.C., Bacon, T.R. and DuPart, M.S. (1996) Effect of heat stable salts on MDEA solution corrosivity, Part 1. *Hydrocarbon Process.*, **75**, 95.

Rooney, P.C., Bacon, T.R. and DuPart, M.S. (1997) Effect of heat stable salts on MDEA solution corrosivity, Part 2. *Hydrocarbon Process.*, **76**, 65.

Stewart, D. and Tulloch, D.S. (1968) *Principles of Corrosion and Protection*, MacMillan, London.

Thanthapanichakoon, W., Veawab, A. and McGarvey, B. (2006) Electrochemical investigation on the effect of heat-stable salts on corrosion in CO_2 capture plants using aqueous solution of MEA. *Ind. Eng. Chem. Res.*, **45**, 2586–2593.

Veawab, A., Tontiwachwuthikul, P. and Bhole, A.D. (1996) Corrosivity in 2-amino-2-methyl-1-propanol (AMP)-CO_2 system. *Chem. Eng. Commun.*, **144**, 65–71.

Zhao, B., Sun, Y., Gao, J., Wang, S., Zhuo, Y. and Chen, C. (2011a) Study on corrosion in CO_2 chemical absorption process using amine solution. *Energy Procedia*, **4**, 93–100.

Zhao, X., Yang, J. and Fan, X. (2011b) Review on research on progress of corrosion inhibitors. *Appl. Mech. Mater.*, **44–47**, 4063–4066.

23

Technological Fronts

23.1 Historical Background

A bit of background is helpful for understanding how this field has developed and continues to develop. Gas treating, and most chemical engineering for that matter, is a conservative business with good reason. Plant failure is costly, especially lost production. However, the state-of-the-art has regularly been challenged and process improvements have been implemented.

The process for CO_2 capture as originally proposed by Bottoms (1930) made, interestingly enough, use of monoethanolamine (MEA) and diethanolamine (DEA). These absorbents are still in use today, which is remarkable in view of all development that has taken place over the last 80 years. The original practice of using glycol as part of the solvent has, however, gone.

In the period until the 1950s most publications dealt with operational problems and how to solve them. There was a lot of focus on corrosion issues. Even today, corrosion issues are probably the biggest source of operational costs.

Around 1950 and onwards, a lot of focus was put on the fundamentals of absorption with chemical reaction which is a key point in gas treating. Sixty years later there are still undetermined issues.

In the 1970s energy became much more expensive and reduction of energy expended became more of a focal point. Tertiary amines were revisited. These tertiary amines also received additional attention when an interest arose in adjusting the absorption process to make it more selective towards H_2S relative to CO_2.

From the 1990s research emphasis shifted towards flue gas treating where the object is cheap CO_2 capture related to the greenhouse gas issues.

Physical solvents have also been proposed from time to time. These have mainly been attractive for syngas treating as they have a tendency to co-absorb too much of the natural gas components.

Gas Treating: Absorption Theory and Practice, First Edition. Dag Eimer.
© 2014 John Wiley & Sons, Ltd. Published 2014 by John Wiley & Sons, Ltd.

Another parallel development is the use of the so-called hot carbonate solutions. They were introduced in the 1950s and are still around in parallel to the previously mentioned processes. The main development for these processes is the type of activator or catalyst used. Amines are now used also in these processes to speed up mass transfer. Their continued presence tells us there is not one all out winner for acid gas absorption processes.

23.2   Fundamental Understanding and Absorbent Trends

Rates of absorption are much higher with chemical reaction than without. Van Krevelen and Hoftijser (1948) were the first to account for and quantify the effect of chemical reaction on mass transfer. Like most processes described in chemical engineering their picture of the process is a model where simplifications are made, consciously or unconsciously. Danckwerts (1970) took this concept a lot further in the 1950s and 1960s while determining the chemical reaction rate constants in a variety of laboratory rigs. Brian, Hurley and Hasseltine (1961) looked further into the mathematics, and Astarita, Savage and Bisio (1983) joined in the theoretical work from the 1960s onwards writing his first of two books in 1967.

Later work done on kinetics at the University of California Santa Barbara (Al-Ghawas and Sandall, 1991), the University of British Columbia Vancouver (Dawodu and Meisen, 1992), the University of Regina that has roots from the Vancouver group (Tontiwachwuthikul and Idem, 2013) and the University of Texas at Austin (Bishnoi and Rochelle, 2000) have all worked along the same concepts. There is also the work of Sada and Kumazawa (1973) and Hikita and Asai (1964) in Japan where Hikita in particular has published a lot of mathematical modelling with respect to various cases of reaction kinetics and their influence on absorption rates. In Twente, van Swaaij and Versteeg and van Swaaij (1988) made substantial contributions to the understanding of reaction kinetics. Work has also been done by Svendsen and co-workers (Bruder *et al.*, 2012) in Trondheim, Norway (1990-present) and by Meng-Hui Li and co-workers (e.g. Sun, Yong, Li, 2005) in Taiwan (1990-present). There are more, but these may be argued to the major ones.

The reaction mechanism between CO_2 and primary and secondary amines is still under debate. The dominant idea is the zwitterion mechanism introduced to this field by Danckwerts (1979), but this has been challenged by Crooks and Donnellan (1989) who proposed a termolecular mechanism as explained in Chapter 4. More work is needed to settle this. This will need to be oriented towards identifying intermediates using various spectra and NMR rather than chemical kinetics, that is, it needs chemists rather chemical engineers to study the reaction.

For tertiary amines and CO_2 the reaction mechanism seems to have been accepted to follow the reaction paths outlined by Donaldson and Nguyen (1980) although Crooks and Donnellan (1989) have also discussed this without disagreeing.

Little work has been done to study the reactions as such from the viewpoint of fundamental chemistry. The efforts have been based on studies of reaction kinetics combined with mathematical modelling. It gives a good basis for using these mathematical models to estimate mass transfer processes but is weaker when it comes to basic chemistry. Hence the discussions going with respect to the reaction mechanisms. The field would benefit from the attention of pure chemists. A deeper insight would make the search for better absorbents easier.

The reaction between amines and H_2S is straightforward and there is no discussion going on here.

The reactions going on when using carbonate based absorbent systems are also understood with no fundamental discussions going on. The issue here would be with the use of so-called activators or catalysts to speed up the mass transfer.

The benefits of using mixed amines have received a lot of attention over the last 20 years or so. There is even older works where claims of 'activators' are made. No evidence of 'activation' as such has been tabled and the benefits are with the properties of the mixed amine solutions. The faster reacting amines of a mixture will be most active in the lean region of a column where the absorption driving force is lowest, and make their contribution where it matters most.

A recent trend is the study of ionic liquids and latterly also phase change liquids. There is also the Australian work on pH change discussed in the desorption chapter.

The methods for studying kinetics have only been subjected to minor modifications over last 30 years or so. The techniques are discussed in Chapter 15.

23.3 Natural Gas Treating

Natural gas has been treated along the lines discussed here for 80 years by now. Better design data have been in focus as exemplified by the Gas Processors Association (GPA) initiative for improved design data in the 1990s. Issues like higher contents of H_2S and BTEX (VOC) emissions have been worked on for 10–20 years or so. The challenges involved are mainly to do with process adjustments where better process models become important.

There is a continued search for process improvements. Incremental developments over time represent the expected main way ahead as the process and its application is mature. However, step changes may still take place. Equilibrium models that are key to providing absorption limitations could maybe benefit from being fitted to narrower data ranges reflecting design practice rather than embracing extremities in loadings. It is clearly an aim of academic research to achieve these all-embracing models and the good thing about them is that success means the underlying concepts are sound. If there are enough parameters, however, most behaviour patterns could be represented. We need regressed models where the parameters are physically significant.

23.4 Syngas Treating

There are no particular forums where acid gas removal as such is discussed through papers like there is within the field of natural gas treating. Most developments in this area are likely to come from the natural gas field. Introduction of physical solvents is more relevant to the syngas of course. Such research efforts have in the past come from industrial rather than academic research since the solvents have been proprietary. As an example the much published Selexol process has been around since the 1970s and came out of proprietary research.

23.5   Flue Gas Treating

Capture of CO_2 from flue gases is a new application still waiting to happen commercially. It is rooted in the greenhouse gas abatement challenge. Because of the enormity of the scale the cost issue as such is the major stumbling block. This is particularly so because a lack of any economic benefits arising for the operator. Any development that can reduce the total cost (capital expenditure: CAPEX + Operational Expenditure: OPEX) of CO_2 capture would be welcome.

The initial efforts from the 1990s were directed at lowering the capital cost. As was stated at the time, unless it becomes cheaper it is a non-starter. The progress made is limited and there is no reason to forget research to lower CAPEX. However, the main effort as per now is probably directed towards finding new absorbents that can lower the energy needs for regeneration as it is realised that this is a major cost element. Progress has also been made towards improved flowsheets that lower the energy consumption by better energy usage and/or recovery. International Energy Agency (IEA) work suggests that $3/4$ of the capture costs are with OPEX (IEAGHG, 2013).

Chemical reaction rates, possibly increased by 'activators', are also being researched as lack of reactivity means taller columns that will drive the cost up. There is a trade-off here as we expect to see bigger heat effects when there is a high reactivity and vice versa. However, research shows that there are possibilities, although a major breakthrough is in the future. Phase change solvents and ionic liquids have already been alluded to above.

Degradation of absorbent has been, and still is, a major issue. The cost of solvent replacement is significant. The chemistry of degradation is poorly understood although progress is being made and a number of papers have published mainly to quantify the effort. The next level, how to reduce degradation, has barely been touched. Recently the emission of degradation products to the environment has appeared as a major issue, particularly since there is concern that some of these may be cancerogeneous. Conclusions from these studies are that the problem is less than perceived and can be handled.

Because of the trace components and oxygen present in the flue gas, the corrosion issues are more prominent in treating flue gas than for the natural gas in particular. Current thinking is to use a lot of cost driving materials like stainless steels rather than carbon steel, which is a dominant feature of natural gas plants. Present research is along mapping corrosion rates and finding good inhibitors that will allow the use of cheaper materials.

23.6   Where Are We Heading?

Scientists being scientists there is no doubt that there will be fundamental research going on to enable more and more accurate predictions of absorption and desorption columns by the help of computer based models. There will also be laboratory work to accompany that to provide better and more complete physical and chemical data to enter into the models.

Somewhere along these lines there will also be attempts at making predictive models to estimate the data needed from properties of the chemical to be analysed. This will greatly reduce the need for experiments when looking for better absorbents.

Degradation research will progress. Hopefully this field will also draw the interest of chemists such that their rigorous chemical methods will help to clarify issues.

The look for new and better corrosion inhibitors will also continue.

Last, but not least, there is bound to be surprise results coming along. The clue may be to do interdisciplinary research. Serendipity is not dead.

References

Al-Ghawas, H.A. and Sandall, O.C. (1991) Simultaneous absorption of carbon dioxide, carbonyl cc and hydrogen sulphide in aqueous methyldiethanolamine. *Chem. Eng. Sci.*, **46**, 665–676.

Astarita, G., Savage, D.W. and Bisio, A. (1983) *Gas Treating with Chemical Solvents*, John Wiley & Sons, Inc., New York.

Bishnoi, S. and Rochelle, G.T. (2000) Physical and chemical solubility of carbon dioxide in aqueous methyldiethanolamine. *Fluid Phase Equilib.*, **168**, 241–258.

Bottoms, R.R. (1930) Process for separating acidic gases, US Patent 1,783,901.

Brian, P.L.T., Hurley, J.F. and Hasseltine, E.H. (1961) Penetration theory for gas absorption accompanied by a second order chemical reaction. *AIChE J.*, **7**, 226–231.

Bruder, P., Lauritsen, K.G., Mejdell, T. and Svendsen, H.F. (2012) CO_2 capture into aqueous solutions of 3-methylaminopropylamine activated dimethyl-monoethanolamine. *Chem. Eng. Sci.*, **75**, 28–37.

Crooks, J.E. and Donnellan, J.P. (1989) Kinetics and mechanism of the reaction between CO_2 and amines in aqueous solution. *J. Chem. Soc., Perkin Trans. 2*, 331–333.

Danckwerts, P.V. (1970) *Gas-Liquid Reactions*, McGraw-Hill, New York.

Danckwerts, P.V. (1979) The reactions of CO_2 with ethanolamines. *Chem. Eng. Sci.*, **34**, 443–446.

Dawodu, O.F. and Meisen, A. (1992) The effects of operating conditions on COS-induced degradation of aqueous diethanolamine solutions. *Gas. Sep. Purif.*, **6**, 115–124.

Donaldson, T.L. and Nguyen, Y.N. (1980) CO_2 reaction kinetics and transport in aqueous amine membranes. *Ind. Eng. Chem. Fundam.*, **19**, 260–266.

Hikita, H. and Asai, S. (1964) Gas absorption with (m,n)-th order irreversible chemical reaction. *Int. Chem. Eng.*, **4**, 332–340.

IEAGHG (2013) Incorporating Future Technological Improvements in Existing CO_2 Post Combustion Capture Plants Technical Review, 2013/TRS, May 2013.

Sada, E. and Kumazawa, H. (1973) Gas absorption accompanied by a complex chemical reaction: variation of the enhancement factors with the orders of the reaction. *Chem. Eng. Sci.*, **28**, 1903–1905.

Sun, W.-S., Yong, C.-B. and Li, M.-H. (2005) Kinetics of the absorption of carbon dioxide into mixed aqueous solutions of 2-amino-2-methyl-1-propanol and piperazine. *Chem. Eng. Sci.*, **60**, 503–516.

Van Krevelen, D.W. and Hoftijzer, P.J. (1948) Kinetics of gas-liquid reactions. *Part I. General theory. Rec. Trav. Chim.*, **67**, 563–599.

Tontiwachwuthikul, P. and Idem, R. (eds) (2013) *Recent Progress and New Developments in Post-Combustion Carbon-Capture Technology with Reactive Solvents*, Future Science Ltd.

Versteeg, G.F. and van Swaaij, W.P.M. (1988) On the kinetics between CO_2 and alkanolamines in both aqueous and non-aqueous solutions I. Primary and secondary amines. *Chem. Eng. Sci.*, **43**, 587–591.

24

Flue Gas Treating

24.1 Introduction

There have been a number of cases in the past, particularly small applications, where it has been considered worthwhile to recover CO_2 from flue gas. The reason has most often been the need to find a source of CO_2 for industrial gas usage but there are better sources if they are available, for example the CO_2 from ammonia trains where it is available ready recovered because of process needs. Often this CO_2 is used to make urea. There have also been a few cases when CO_2 has been wanted for enhanced oil recovery (EOR) purposes.

Flue gas is at atmospheric pressure and it is usually highly undesirable to allow any pressure build-up to deal with any pressure drop needed for the removal of CO_2. If the gas were from a combined cycle gas turbine (CCGT) power plant, it would mean increasing the outlet pressure from the gas turbine and this would reduce its efficiency with a power loss as a consequence. There is also the integrity of the flue gas channel.

A flue gas will normally also contain components that are detrimental to the absorbent used for catching CO_2. The alkaline absorbent needed would naturally also pick up acid gas components like NO_2 and SO_2 that are more acidic than CO_2 itself and would react irreversibly with the presently available absorbents. Warm gas and particle content are also bad news. Some form of pretreatment is usually necessary.

The CO_2 capture challenge has already been discussed. In this text only post-combustion capture has really been dealt with although the CO_2 capture in pre-combustion decarbonisation approach has implicitly been dealt with through the discussion of syngas treating. The so-called oxy-fuel case is about burning fuel with oxygen and the CO_2 separation issue is merely the condensation of water vapour from the flue gas leaving a reasonably pure CO_2 behind.

Gas Treating: Absorption Theory and Practice, First Edition. Dag Eimer.

24.2   Pressure Drop and Size Issues

Absorption columns are the workhorse for CO_2 recovery from flue gas. Other techniques have been discussed since the dawn of interest in greenhouse gas abatement but so far none have given serious competition. Packed columns are used for absorption in this application since they have a much lower pressure drop than tray columns. Even so, pressure drops of 4 mm W.C. or more per metre of packed height is needed. There is a balance between column diameter and specific pressure drop. For a 400 MW CCGT power plant a single train column would typically have a diameter around 16 m to put it into perspective. Such large diameter columns are known from other environmental applications, but are new in the context of alkanolamines. Packing heights of 10–15 m have been normal in designs discussed up until now.

The consensus is that it will be necessary to install a fan or blower, usually as the flue gas enters the treatment plant, and upstream of even the pre-treatment. The more the gas must be increased in pressure, the higher the power consumption will be for the blower. Attempts to manage without a blower have so far been unsuccessful, but there is promising work going on (Åsen *et al.*, 2011).

24.3   Absorbent Degradation

The absorbent solutions proposed so far have been aqueous solutions of amines. The so-called base case process uses an absorbent consisting of 30% (wt) monoethanolamine (MEA) in water, but there are a number of amine mixtures that have been proposed. The search for better absorbents are mainly for reducing the energy needed to regenerate the solution but there are also issues with degradation of amines. There is a significant consumption of MEA expected due to this. It is significant in terms of money but there are also related environmental issues.

Based on a typical absorber design combined with estimating mass transfer coefficients with the model due to Onda, Takeuchi and Okumoto (1968), it is estimated that the rich absorbent solution is roughly 1/3 saturated with oxygen. Equation 7.7k is used for this estimate assuming that the packed height is 12 m and 2″ Stainless Steel (SS) Pall rings are used. This oxygen will gradually be desorbed from the solution when the liquid is heated up and eventually stripped in the desorber. In the meantime this oxygen can contribute to oxidative degradation of the absorbent, a quite slow reaction in this context.

Components like NO_2 and SO_2, if present, form stronger acids than CO_2 and will be irreversibly absorbed. Their absorption mass transfer coefficients are much higher than for oxygen because they react very fast with the amine in the solution. However, both NO_2 and SO_2 are Lewis acids and will not react instantaneously like H_2S.

24.4   Treated Gas as Effluent

The treated gas coming out of the top of the absorber will be vented to the atmosphere. This is different from the treatment of natural gas or syngas where the treated gas is the actual product of the plant. Since the gas is vented, it must not carry with it components that may be environmentally harmful. The absorbent itself is recovered also for economic reasons, and, for example MEA as such is not particularly harmful to the environment. However, there are degradation products from all of the amines proposed used, a number of possible degradation products are either harmful in themselves, or they could give rise to harmful products once released. That is the current fear. Recent studies have been done addressing these concerns, and the conclusions drawn do not support the notion of harmful emissions. This is work not necessarily generally published and there may still be discussions over these conclusions. There are still unresolved issues in this area.

It is also claimed that the internal treatment section in the top of the absorber is capable of preventing harmful emissions.

To prove a negative is difficult, however.

24.5 CO_2 Export Specification

The only reason to recover CO_2 from flue gas on the scale envisaged is to ensure that the CO_2 is not released to the atmosphere. Hence, it is the CO_2 that is the real product from these capture plants. It is foreseen that CO_2 will be gathered and piped to some sink, probably a disused gas field or an aquifer, for storage. This is already being done in a few places around the world, usually in situations where CO_2 has to be separated from the gas anyway and often because it is part of the permit to operate. There are also EOR applications.

The export CO_2 must be compressed to 150–250 bar depending on the situation. Since CO_2 may form hydrates, water needs to be removed to avoid condensation in the export system. There are also CO_2 purity issues, but those are of little consequence for a post-combustion absorption plant.

Some nitrogen will be co-absorbed in the absorber, and this nitrogen will find its way along with any surviving oxygen into the export CO_2.

24.6   Energy Implications

Any energy consumption arising from the operation of a CO_2 post-combustion capture plant is a parasitic consumption on the power plant itself. The main energy consumption is thermal energy used in the form of steam to drive the desorption column. Roughly 30% of this could have been electricity output via a steam turbine. The 30% reflects the level of efficiency expected for the steam turbine for the steam quality in question. Then there is the electricity needed to drive the blower discussed earlier, pumps to circulate liquids in the plant and also to provide cooling water.

The base case absorbent plant reduces the power efficiency of a CCGT by roughly 10 percentage points from 58%. That means that the electrical energy output is reduced by around 17%. The so-called base case is absorption in 30% (wt) aqueous MEA solution and a conventional plant design. So far various proposals for reduction of this parasitic energy consumption have been to reduce the energy use by order of 30% (corresponding to 7% points reduced efficiency for the power plant remaining). This situation seems to have remained unchanged for the last 10 years, possibly because no one seems to have tried to combine proposals. Since there are proposals that cover both process configuration and solvent, there ought to be some additive effect to be found.

24.7 Cost Issues

When evaluating technical proposals for CO_2 capture, it is paramount that all items needed to operate the capture plant are included in the equipment list. This is often not true when people are promoting their technologies. The net present value of operational costs must be calculated and added to the investment costs. A typical distribution of costs is shown in Figure 24.1. Beware that this distribution will be influenced by choice of battery limit for estimate.

A pitfall often encountered is the transfer of costs from investment to operational. A lower investment in energy recovery would naturally lead to more energy being used. The return on capital may be differently handled for various commodities. There is also the situation that the CO_2 emissions associated with the production of say steam. If this is produced by burning oil, it is not irrelevant if this CO_2 is allowed to be emitted without abatement. Likewise oxygen needed in an oxy-fuel process will have a climate gas impact unless the electricity used to produce the oxygen is so-called 'green'. Any comparison of technologies or plants should be made on the basis of 'CO_2 emission avoided'. Sounds easy, but it is easy to go astray, and the conclusion can quickly become wrong. This was discussed extensively by Melien (2005).

It is not possible to reduce capture costs to the extent desired by addressing one item or feature in the process alone. Figures 24.1 and 24.2 show that the costs are widely distributed

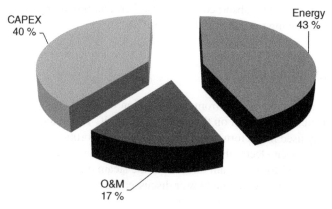

Figure 24.1 *Distribution of costs based on net present value (NPV) between capital, energy and other operational costs. This is for a post-combustion capture plant for a CCGT based on a very conservative case, the so-called base case as defined by the CCP around 2000. Steam for desorption reboil is counted as lost electricity production with 25% efficiency. (Reproduced with permission from Eldrup (2014). © N.H. Eldrup.)*

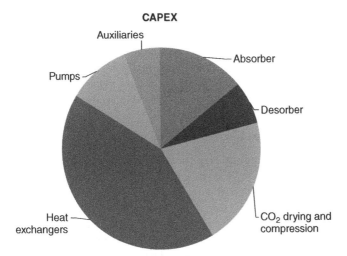

Figure 24.2 *Relative costs between equipment segments in the absorption–desorption process as defined for Figure 24.1. Costs are installed costs (Eldrup, 2014). (Reproduced with permission from Eldrup (2014). © N. H. Eldrup.)*

and a broad approach is needed if the costs are to be reduced by say 75%, which must clearly be a target.

The direct equipment costs of a process represents perhaps only about a fourth of the investment cost. The other $3/4$ of the costs are mainly due to piping, instruments and structures. If the cost of an item itself is cut in half, it will probably not have much impact on the cost of the $3/4$. This aspect does normally not receive much attention. Clearly if an item was eliminated from the flowsheet, then the rest of the cost associated with this item would also be expected to disappear.

In general claims to cost improvements by technology stakeholders must be looked at with healthy scepticism. It is imperative that the costs of processes being compared are estimated on the same basis. This means that the same cost estimator must be used to estimate both processes. Different approaches to estimates and what is included would clearly distort comparisons. There are some studies available on such unified basis as mentioned next.

The International Energy Agency (IEA) was formed as a Western counter-weight to Organization of the Petroleum Exporting Countries (OPEC). IEA's role is to support efforts to secure energy supplies. In 1991 a subsidiary was formed called International Energy Agency Greenhouse Gas program (IEAGHG). Their role is to provide technological information to enable the world at large to make educated choices in the field of CO_2 capture and storage (CCS). They commission a number of studies to assess technologies, and endeavour to make sure technologies are compared on a like-for-like basis.

The CCP (The CO_2 Capture Project; www.CO2captureproject.org) is another organisation that has worked extensively with cost comparisons of various processes on all three fronts of CO_2 capture. This organisation is a co-operation of a few international oil and energy companies and has been operative for more than 10 years.

24.8   The Greenhouse Gas Problem

The intention with this section is to provide some overall picture of what this problem is all about, who is who and to a degree, what is actually going on. It is a complex and very political field and there are few, if any, definite answers to the questions raised.

Some websites of interest are listed in the References section.

24.8.1   Global Warming and Increased Level of CO_2

There is CO_2 in the global atmosphere. The present level is about 390 ppm and rising by approximately 2 ppm each year. The presence of CO_2 has a greenhouse effect and contributes to global warming. There is no serious opposition to this. The issue is really to what extent this situation is man-made or not.

There is an international group of experts organised under the auspices of the United Nations known as the Intergovernmental Panel Climate Change (IPCC, Figure 24.3). The IPCC's mission is to provide the world with a scientifically based judgement of the progress of climate change. This is a large and complex body that involves renowned experts, bringing them together and enabling them to make judgements on available data related to climate change. It is this body that issues at roughly 6 year intervals a series of reports on global warming and was awarded the Nobel Peace Prize in 2007: 'Contribution of Working Group III to the Fourth Assessment Report of the Intergovernmental Panel on Climate Change' (Metz *et al.*, 2007). The present consensus view of the IPCC is that the world is in for serious warming unless steps are taken to reduce CO_2 emissions. A new 2013 report does not change this picture, but this report is not yet finally approved per June 2014.

Another good source of information is the CICERO (Center for International Climate and Environmental Research – Oslo) unit at the University of Oslo. It is an academic institution and they are concerned with facts rather than promoting solutions.

There is no shortage of sceptics. Some feel they have the ability to make a better judgement than the panel referred to, some feel the IPCC is putting too much weight on the

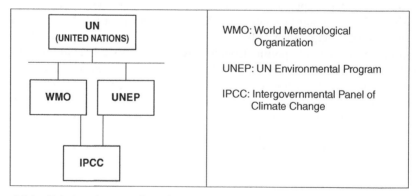

Figure 24.3   *Relations between key players judging climate change under the United Nations umbrella. WMO, World Meteorological Organization; UNEP, UN Environmental Program; IPCC, Intergovernmental Panel of Climate Change.*

wrong data and some have a vested interest in unabated use of fossil fuels. Joining this debate is beyond the scope of this text.

24.8.2 Geological Storage

Given that CO_2 is captured, it would naturally make sense not to release it. A number of options for storage have been discussed including empty gas reservoirs and aquifers. Dissolving it at great ocean depths has also been discussed but seems an unlikely solution as of now because of worries regarding marine life and also how long the ocean would actually keep the CO_2 dissolved.

At this point in time there is CO_2 being injected into both aquifers and rock formations to do away with the captured CO_2. The longest running such project is at the Sleipner gas field where CO_2 is injected into an aquifer (Chadwick *et al.*, 2008). There have also been a good number of EOR projects where CO_2 is used to produce more oil from the field.

There are now serious follow-up programmes established to monitor the process. The most widely publicised case is the Norwegian Sleipner field and the injection of CO_2 into the underlying Utsira aquifer formation.

Legal issues are still being sorted out. There are a number of international conventions that were established in the past to deal with other environmental issues that make it difficult to dispose of CO_2 this way.

24.8.3 Transport of CO_2

Transport of CO_2 from the point of capture to the point of injection is also a challenge. It could be done both by pipeline and by ship, but either way it would cost a significant amount of money. Pathways for CO_2 will need to be established in a manner reminiscent of the distribution of natural gas, but in the reverse order. Small emitters, and most are, could probably not finance the establishment of a CO_2 pathway on their own and they would lack the technical skills to do so.

There are legal issues involved also with transport. At the moment, international pipelines seem to be accepted but transport by ship is contested.

24.8.4 Political Challenges

At this moment any motivation for CO_2 capture may be traced back to the so-called Kyoto convention that dates back to the 1990s. Several attempts have been made to develop this convention further, but have failed. Big CO_2 emitting countries like the USA have so far refused to ratify the convention, and they are thus not bound by it. The political arena where these discussions are held is known as COP (Conference Of Parties). COP-17 was held in Durban, South Africa in 2011 while the famous Kyoto conference was COP-3 (1997). Progress in not fast.

Some legislation has been introduced in a number of countries but this has not made much impact on the situation regarding CCS.

There is also political disagreement on how to attack the problem. Some feel that CCS is no good, and the only worthwhile action is to do away with the use of fossil fuels. This is of course a limited resource but the size of this industry means it would take a long time to phase it out even if there were cost effective solutions to enable this.

To make life a little harder for politicians, there are so-called lobby groups and NGOs (Non-Governmental Organisations) campaigning for their particular views. The electorates are influenced by this and politicians' freedom to operate is thus limited. If re-election is to be achieved, they can't move faster than what the electorate is prepared to accept.

References

Åsen, K.I., T. Fiveland, D.A. Eimer, N.H. Eldrup, (2011) Method and apparatus for CO_2 capture. WO Patent WO/2011/005116.

Chadwick, A., Arts, R., Bernstone, C., May, F., Thibeau, S. and Zweigel, P. (eds) (2008) *Best Practice for the Storage of CO2 in Saline Aquifers. (European Commission, and IEA Greenhouse Gas R&D Programme)*, British Geological Survey, Keyworth, Nottingham.

Eldrup, N.H. (2014) Personal Communication.

Melien, T. (2005) Economic and cost analysis for CO_2 capture costs in the CO_2 capture project, in *Carbon Dioxide Capture for Storage in Deep Geological Formations – Results from the CO_2 Capture Project* (ed D.W. Thomas), Elsevier, pp. 47–90.

Metz, B., Davidson, O.R., Bosch, P.R., Dave, R. and Meyer, L.A. (eds) (2007) *Contribution of Working Group III to the Fourth Assessment Report of the Intergovernmental Panel on Climate Change*, Cambridge University Press, Cambridge and New York.

Onda, K., Takeuchi, H. and Okumoto, Y. (1968) Mass transfer coefficients between gas and liquid phases in a packed column. *J. Chem. Eng. Jpn.*, **1**, 56–62.

Web Sites

http://ieaghg.org/
http://www.ipcc.ch/index.htm
http://www.wmo.int/pages/index_en.html
http://unep.org/
http://www.un.org/en/
http://www.cicero.uio.no/home/index_e.aspx

25

Natural Gas Treating (and Syngas)

The object of this chapter is to give a primer for understanding the intricacies of natural gas treating. This includes some background, problems arising and limitations or challenges posed by the interacting plants and pipelines.

A few extra points related to trains for processing synthesis gas will also be made with the same target in mind.

25.1 Introduction

Natural gas as it comes out of the wellhead is not fit for transport to gas customers and most probably not for immediate processing. It has to be treated first to render it suitable for the next step in the value chain.

The content of natural gas varies widely. Typical values of major components may be

- Methane: 80–95% (mol).
- Ethane: 1–10%.
- Propane: 1–5%.
- Hydrocarbons heavier than propane: 0–2%.
- Carbon dioxide: 0.5–50%.
- Hydrogen sulfide: 0–40%.
- Nitrogen: 0.1–5%.
- Water: low to saturated.

Its pressure may vary widely. Wellhead pressure could be 100 bar or more but often the gas is associated with oil and/or condensate that most likely will be targeted for recovery. This will involve a pressure let-down to bring the pressure below the mixture's pseudo

Gas Treating: Absorption Theory and Practice, First Edition. Dag Eimer.
© 2014 John Wiley & Sons, Ltd. Published 2014 by John Wiley & Sons, Ltd.

critical point to enable phase separation. The gas would thereafter need to be recompressed before export and the treatment pressure may be chosen in this perspective.

The temperature would depend on the upstream processing.

The closer the treatment plant is to the wellhead in the treatment train, the more likely it is to be exposed to salts coming from the reservoir brines.

It should also be realised that natural gas treating plants vary widely in size from small plants at the wellhead to big plants dealing with several wells. The latter could be represented by plants preparing natural gas for liquefied natural gas (LNG) trains and these have capacities of several million tonnes of gas per year.

25.2   Gas Export Specification

Specifications of export gas will vary between locations. It is normally dictated by the customer or the pipeline operator. If the gas is to be piped away, it will be necessary to ensure that the gas will not precipitate any liquids that may impair the operation of the pipeline. This is often referred to as *flow assurance*. The pipeline may be subjected to very low temperatures and it is the lowest temperature foreseen that will dictate the need for low dew points. It is undesirable to have hydrocarbons condensing in the pipeline even if the immediate effect is small, as the condensate may accumulate and lead to operational problems. Free water that is condensed could easily lead to the formation of solid hydrates that would grow and completely block the pipeline. Hydrates can be formed by both methane and CO_2.

The presence of CO_2 and H_2S has an impact on the choice of materials in the pipeline. Since the pipeline often exists, the gas must be conditioned to pipeline needs. Furthermore, the customers down the line also have some expectations for gas quality. H_2S is very poisonous and it is normally required to be present in concentrations less than 3–4 ppm (<0.25 grains per $100 S ft^3$ often used in the USA). Whatever H_2S that is present in the gas being burnt will end up as SO_2, another potentially corrosive gas. The specification for CO_2 in sales gas may vary a bit more but is usually between 1 and 3.5%. If there was a big customer like a power plant next door to the well, the situation may obviously be different if this power plant was custom built to the gas being produced. This is rarely the case.

The content of hydrocarbons and inert gas may need to be adjusted if there are defined values to be met with respect to heating value and Wobbe index.

25.3   Natural Gas Contaminants and Foaming

The gas industry defines amine based absorbents as 'foaming' meaning that these liquids have a significant tendency to foam. This foaming may be caused by either high fluid shear forces and/or by contamination. The shear forces are dealt with by proper design of the plant. The contamination is an operational issue that must be handled by proper solution management (see Chapter 19). However, some gas sources are more likely to cause foaming problems than others.

Traces of 'heavy' hydrocarbons in the gas *will* to a degree be absorbed in the absorbent over time and accumulate as they will not be boiled off in the stripper. The gas may also be a so-called retrograde gas where hydrocarbons will condense and thus precipitate when the pressure is reduced. An absorption column has a pressure drop associated with it and that is enough to cause such condensation. Foaming will result. In the latter case the gas should be pre-heated, but that is expensive and is something that should be avoided if at all possible. This situation is especially associated with condensate fields.

25.4 Hydrogen Sulfide

Books could be written about H_2S and natural gas, and there is certainly an abundance of papers on the subject. It must be removed from the gas and properly handled. What proper handling is will vary and sometimes this is a compromise.

A key factor when deciding how to deal with H_2S is 'how much sulfur?' Small quantities, in the parts per million range, may be dealt with using scavengers for removal. These are chemicals bought and injected into the process stream at a convenient point. The resultant product will often go out with other streams. For higher quantities, still counting in parts per million, dedicated scavenger units may be set up. Here, the end product stays in the unit and must be shipped out when the scavenger is spent.

The next step up from using scavengers, maybe up to 5–20 tonnes of sulfur per day, could be dealt with by using liquid redox processes. The 'original' process of this type is the Stretford process but today there are more modern varieties like LO-CAT and SulFerox available. In these processes the H_2S reacts directly with a reagent that is selective with respect to H_2S, and this reagent is regenerated while elemental sulfur is precipitated.

For really big loads of sulfur, the absorption process dominates with the captured H_2S being piped to a Claus plant for conversion to elemental sulfur.

25.5 Regeneration by Flash

Hydrogen sulfide must be desorbed in a proper desorption column since the absorbent regeneration must be near complete to enable meeting the very low specification for H_2S in the treated gas. However, CO_2 may be present with a high partial pressure and a benign treated gas specification, and this may make it possible to regenerate merely by using a flash vessel in a pressure swing process. Hybrids are of course possible. This is the case with the so-called activated methyldiethanolamine (MDEA) process. This process has been popular in ammonia plants as it lowers the energy use and still enables CO_2 to be removed to levels of 150 ppm and less.

25.6 Choice of Absorbents

Thousands of absorption–desorption plants have been built to treat natural gas. The plants based on aqueous amine solutions seem to be the subject of most papers on gas treating but there are also many plants out there based on hot carbonate solutions.

Physical solvents have also been used but they have been less popular due to significant co-absorption of gas. This is a problem in natural gas treating as this co-absorption represents a loss of gas. In trains treating synthesis gas this is not such a big problem since hydrogen in particular is less soluble.

Further Reading

Campbell, J.M. (1976) *Gas Conditioning and Processing*, Campbell Petroleum Series, vol. **1 & 2**, Campbell Petroleum Corporation, Norman, OK.

Engineering and Geosciences Programs Proceedings Annual Laurance Reid Gas Conditioning Conference (1951–1987, 1988–present), University of Oklahoma Outreach, Norman, OK. See https://www.ou.edu/content/outreach/engr/lrgcc_home/paper_index/2013.html.

Kohl, A. and Nielsen, R. (1997) *Gas Purification*, Gulf Publishing.

26

Treating in Various Situations

The previous chapters have discussed the really big and publicised challenges related to gas treating. The object of this chapter is to pinpoint a few further problem areas that are perhaps less spectacular but definitely offer challenges for a chemical engineer involved with these problems. In some of these areas there is also to a lesser extent a tradition for involving chemical engineering expertise.

26.1 Introduction and Environmental Perspective

When environmental concerns came to the fore in the 1960s, as exemplified by the launch of the ACS's journal *Environmental Science & Technology*, there was a legacy in the form of operational plants where substantial sums of money had been invested. These plants had many waste gases exiting to the environment. It would be wrong to say that had been no consideration for the environment prior to this but it is clear that emission standards have been improved over time. Most environmental agencies quickly introduced a demand for BAT (Best Available Technology) to be used by those having emissions rather than giving permissions for certain amounts of effluents. This was followed up and permissions were steadily tightened. In some cases, mainly the more obnoxious species, the emission limits followed closely the state of the art in sampling and analysing.

26.2 End of Pipe Solutions

Early work on cutting emissions had to involve so-called end-of-pipe solutions. They are referred to as such since they treated the gas as it left the plant before emission. A lot of good chemistry has been made use of in numerous such processes, more often than not in an absorber to prevent targeted chemicals from being emitted to the atmosphere. The absorbent would have to be routed to a suitable point in the process unless there was chemical reaction taking place to render the targeted chemical harmless. A case in point would be the capture of HCl by a NaOH solution. At the seaside the end product of this would hardly be noticed as emitted to the ocean.

Very quickly it was realised that processes could be tweaked to reduce the total flow of the waste streams and in the process increase the concentration of the species to be caught. This enabled much smaller absorption columns, and with the larger mass transfer driving force they could also be shorter. Waste streams are seldom available at pressure, which means that pressure drop must be kept to a minimum. (One case of waste gas at pressure is the bleed stream from an ammonia synthesis loop where ammonia must be removed before the gas can be let out).

Most of these waste streams would be easy enough to deal with and estimates could show very short columns could be used. However, adding a metre or two of packing to a column is a marginal cost since column ends, packing support and liquid distributors would be unchanged. Hence, any design for less than 3 m of packing height would easily be rounded off to 3 m to add a bit of safety margin or even better than specified performance. After all, design data were usually uncertain and maybe in short supply. Sometimes, however, the treatment had to be done in a confined space and that could be a bit more challenging.

Present day demand for CO_2 capture represents such an end of pipe solution. Like before there has been an effort to see if the volume flow of gas could be reduced and CO_2 concentration increased as discussed previously.

26.3   Sulfur Dioxide

The emission of SO_2 at various plants around has been coupled to so-called acid rain. It could lead to lowered pH of fresh water lakes and death of fish in them. There was also a debate of its effect on forestry. Some argued that sulfur was a growth factor, others that there were detrimental effects in the way acid rain was applied through acidity and quantity.

There are today a myriad of processes being offered for capturing SO_2 from gases.

26.4   Nitrogen Oxides

Nitrogen oxides as a group is another effluent that is worth a few comments (nitrogen oxides comprise N_2O, NO and NO_2). All combustion processes using air for oxidation is liable to produce NOx. Modern burner design addresses this as far as possible, which is referred to as low NOx burners. Even so, modern demands for reducing NOx emissions mean that an end of pipe solution must also be used. The presently dominating solution is a catalytic reactor, but absorption systems can also be used. In car exhaust there is a catalyst based solution.

NO_2 absorbs readily in water while NO does not. Given an oxidative system, like air for instance, NO will oxidise to NO_2 but there is an equilibrium limitation. Laughter gas, N_2O, can be dealt with catalytically.

Nitric acid plants have historically been very big emitters of NOx. Here, the upstream change to higher process pressure in the nitric acid absorption column has alleviated this situation. A modern plant does not need to have big emissions.

26.5   Dusts and Aerosols

Dusts will mostly be dealt with by filters, and they can be very big, but there is no real technical problem with large dust particles although there are always challenges. The big problem arises when the dust particles become very small with sizes similar to aerosols.

A similar problem is the advent of aerosols. These arise from a couple angles. One is the instant creation of a mist when hot and cold streams are contacted and small liquid particles are formed. The other is when liquid splashes against surfaces with force or when sprays are used.

Aerosols are notoriously difficult to remove from a gas. Very special filtering techniques have been the name of the game. Normal demisters will not work. The best approach is to prevent their formation as far as possible.

26.6   New Challenges

Human civilisation moves on and the demand for a good life and good health is forever increasing. The importance of the environment in relation to this has long since been understood although some of the detailed interactions are not fully appreciated. Biodiversity is certainly a focal point. For this reason there will continue to be increased focus on reduction, not say elimination, of harmful emissions. The zero emission vision has already been introduced in some sectors.

International transport has been a field that has been difficult to target when it comes to emissions. Jurisdiction has been an issue, and policing it another. This is not to say that there are no rules. There are regional rules afoot for curbing SO_2 emissions. Ships sailing in the Baltic Sea will be required to reduce their SO_2 emission very significantly. This could be done by the use of low sulfur fuel but that is very expensive and may even be impossible to obtain in the quantity needed. There is a trend towards installing SO_2 scrubbers on board. Only time will tell as to where this will lead.

Any other sectors? Your guess is as good as mine. However, whatever will happen and whatever opportunity that may arise, we have an obligation as engineers to find good and cost-effective solutions.

Index

Printed and bound by CPI Group (UK) Ltd, Croydon, CR0 4YY

16/04/2025

14658398-0001